Cultural Encounters with the Environment

Cultural Encounters with the Environment

Enduring and Evolving Geographic Themes

EDITED BY ALEXANDER B. MURPHY AND
DOUGLAS L. JOHNSON

WITH THE ASSISTANCE OF VIOLA HAARMANN

ROWMAN & LITTLEFIELD PUBLISHERS, INC.
Lanham • Boulder • New York • Oxford

ROWMAN & LITTLEFIELD PUBLISHERS, INC.

Published in the United States of America
by Rowman & Littlefield Publishers, Inc.
4720 Boston Way, Lanham, Maryland 20706
http://www.rowmanlittlefield.com

12 Hid's Copse Road
Cumnor Hill, Oxford OX2 9JJ, England

British Library Cataloguing in Publication Information Available

Library of Congress Cataloging-in-Publication Data

Cultural encounters with the environment : enduring and evolving geographic themes / edited by Alexander B. Murphy and Douglas L. Johnson with the assistance of Viola Haarmann.

p. cm.
Includes bibliographical references and index.
ISBN 0-7425-0105-1 (cloth : alk. paper)—ISBN 0-7425-0106-X (paper : alk. paper)
1. Human geography. 2. Geographic perception. 3. Landscape assessment.
I. Murphy, Alexander B., 1954– II. Johnson, Douglas L. III. Haarmann, Viola.
GF41.C85 2000
304.2—dc21 99-087405

Printed in the United States of America

™ The paper used in this publication meets the minimum requirements of American National Standard for Information Sciences—Permanence of Paper for Printed Library Materials, ANSI/NISO Z39.48-1992.

This book is dedicated to the
extraordinary life and career of
Marvin W. Mikesell—
mentor, colleague, and friend
of the contributors to this volume.

Contents

List of Figures and Tables

FIGURES

TABLES

Preface

Most projects have complex, multifaceted origins. This volume is no exception. In the case of *Cultural Encounters with the Environment,* at least four sources of inspiration deserve mention. At the Association of American Geographers' meeting in Chicago in the spring of 1995, Marvin Mikesell received honors for his scholarly contributions and years of meritorious service to the field of geography. On that occasion, a number of his students and colleagues had occasion to reflect on the role that Mr. Mikesell played in their growth and development. From such contemplation there developed an as yet vague, but growing conviction that there was more to be said about the significance of his example as both scholar and adviser.

Second, Michael Conzen, in his exquisitely gentle and gentlemanly way, encouraged the editors to translate these vague notions of recognition into more concrete reality. It is due to Michael's encouragement, enthusiasm, and persistent admonition that no mere festschrift would suffice that the current shape of this volume emerged. For he repeatedly urged that only a thoughtful set of scholarly papers would be a fitting reflection upon the career and contributions of Marvin Mikesell.

The third driving force behind the emergence of the present volume was the editors' involvement in the Geography Advanced Placement project of the College Board. There is every reason to assume that *Cultural Encounters with the Environment* would have either been stillborn or suffered through a much longer period of conception and gestation had not the AP Geography committee of which the editors were members met on a regular basis beginning in 1995. In addition to the stimulation provided by our geography colleagues and the Educational Testing Service (ETS) staff involved in the AP Geography project, the editors were given regular opportunities to meet and develop the rationale and content of this volume. To Sheila Ager of the ETS, whose enthusiasm for and commitment to the enhancement of geographic literacy has been boundless, and who served as a somewhat unsuspecting midwife for this volume, we owe a great debt.

Susan McEachern, our editor at Roman & Littlefield, provided a final pillar of support for the project. Her positive response to the concept when it was first broached to her, and the consistent pattern of support and encouragement that followed, helped immeasurably in moving this volume from bright idea to tangible product.

Along the way, many others contributed. Mary Hartman, Irene Walch, and Ed McDermott, reference librarians extraordinaire at Clark University's Goddard Library, solved numerous seemingly intractable bibliographic questions. Much helpful advice and cartographic skill were made available by Anne Gibson of the Clark Labs for Cartographic Technology & Geographic Analysis. Dimitri Varlyguin and Laura Schneider helped with Russian and Spanish citations, respectively. Sarah Shafer provided invaluable assistance in formatting tables. Madelein Grinkis, Heidi Mundell, Denise Robertson, and Matt White served as computer format gurus, casting light upon hidden codes. To all of these essential contributors we say a heartfelt thank you.

In the end, Viola Haarmann, the third member of the editorial triumvirate that worked on this volume, was absolutely essential in bringing the project to completion in a timely fashion and scholarly condition. Her efforts extended far beyond copyediting, although the meticulous attention to detail and uncompromising commitment to consistency that characterize her work have improved the quality of the volume immeasurably, as have her critical judgment and substantive contributions.

Finally, the authors of the following chapters deserve our resounding thanks, for without their efforts and patience no book would have been forthcoming. The editors, who as Marvin Mikesell's students in the late 1960s and late 1980s, respectively, span the arc of much of their mentor's teaching career, feel honored to have been entrusted with this rewarding project and pleased that it could be brought to successful fruition.

Introduction

Encounters with Environment and Place

Alexander B. Murphy and Douglas L. Johnson

By almost any reckoning, cultural geography is one of the most dynamic fields within geography today. Indeed, culture has become such a pervasive theme in human geography that the boundaries around the topical area are increasingly indistinct. Even among those who self-identify as cultural geographers, there is a great heterogeneity of work being produced—surely a sign of strength in a subdiscipline that was regarded by many as marginal as recently as the 1980s. The last few years alone have seen the publication of an extraordinary array of new work in cultural geography (summarized in part in Mathewson 1998, 1999), the founding of several new journals focused on cultural geographic work (*Ecumene*, *Géographie et Cultures*, and *Gender, Place and Culture*), and the virtual assimilation of "social geography" within the cultural geographic project (Gregson 1995).

CULTURE AND GEOGRAPHY IN CULTURAL GEOGRAPHY

Not surprisingly, the recent blossoming of cultural geography has been accompanied by efforts to make sense of where the field has come from and where it is going (Hugill and Foote 1994; Mathewson 1996). In charting the emergence of cultural geography in the United States, attention is invariably directed to the role of Carl Sauer and his colleagues and students at the University of California, Berkeley, in the early distillation of the field in the decade before World War II, and then to the seminal role of Philip Wagner and Marvin Mikesell's (1962) *Readings in Cultural Geography* in giving cultural geography shape and form through the 1960s and beyond. Accounts of the recent history of the subdiscipline are more complicated and varied. As cultural geography's traditional orientations came to coexist with a variety of other complementary and competing perspectives, the coherence of cultural geography became much more difficult to map. Indeed, there was a sense of fragmentation within the subfield, as some of its practitioners sought to chart a course for cultural geography that was distinct from its earlier roots.

There were several initial catalysts for the different directions that cultural geography took. One was the behavioral and humanistic work of the late 1960s and early 1970s, which led to calls for a cultural geography focused on the role of human understanding and meaning in the creation of places, landscapes, and cultural practices (Ley and Samuels 1978). Another was the rise of structuralist theories, particularly Marxism, which prompted work aimed at elucidating the ways in which structures of power and struggles between classes shaped the emergence of cultural landscapes (Cosgrove 1984). Yet another came from work in ecological anthropology, producing a stream of work focused on human ecological adaptation (Nietschmann 1973).

In the 1980s and early 1990s, the growing salience, first of newly reworked political economy perspectives and then of poststructuralist ideas further complicated and enriched the scene, producing a literature that sometimes seemed far removed from the cultural geography of the earlier generation. Some began to speak of a "new cultural geography" (Johnston 1986), and explicit attempts were made to articulate the differences between the traditional and the new. In the process, attention came to focus on the sorts of things that were indeed new about the cultural geography of the 1980s and early 1990s.

To understand what was new about the cultural geography of the more recent generation, it is useful to consider Mikesell's summary of the orientations of cultural geographers as seen from the perspective of the late seventies (Mikesell 1978). He suggested that the cultural geography of the preceding generation had been dominated by: "(1) a historical orientation; (2) a stress on man's role as an agent of environmental modification; (3) a preoccupation with material culture; (4) a bias in favor of rural areas in this country and non-Western or preindustrial societies abroad; (5) a tendency to seek support in anthropology; (6) a commitment to substantive research and a consequent attitude of extreme individualism; and (7) a preference for field work rather than 'armchair geography' " (Mikesell 1978: 4). Although most of the cultural geographers who came on the scene after 1978 were sympathetic to the historical orientation of their predecessors—and shared with them a substantive interest in the evolution of landscapes—many of the newer generation challenged the other pre-1978 preoccupations identified by Mikesell. They argued that cultural geography needed to embrace a dialectical approach to human–environment relations; that it should address nonmaterial, as well as material, culture; that it was important to focus on the urban, industrial world; that practitioners should look beyond anthropology to work in social theory and political economy for insights; that they should pursue theoretical, as well as empirical, research; and that they needed to break out of a mold in which "muddy boots" research was the mark of legitimacy (Jackson 1989; Crang 1998).

The tension between these new orientations and those of traditional cultural geography initially created an atmosphere in which concerns about dichotomization and misstatement were voiced (see Price and Lewis 1993, and the subsequent Commentary 1993). Before long the debate over old versus new was overshadowed by the very heterogeneity of modern cultural geography itself and the sense that the diversity of work being produced by cultural geographers was broad enough to defy easy categorization (Duncan 1995). New projects were mounted that sought to build bridges between different strands of cultural geographic research—most notably the publication of *Re-Reading Cultural Geography*, designed as a successor to the Wagner and Mikesell classic (Foote et al. 1994).

At its best, the new intellectual climate in cultural geography came to be characterized by efforts to reach across orientations and traditions to develop innovative, insightful approaches to research questions. It is in this spirit that the present volume is conceived. It is inspired by the promise of studies that draw on some of the traditional themes in the cultural geography literature, while bringing new insights to them through the application of recent ideas and approaches. Representative examples of this merger of new and old approaches include Donald Mitchell's (1996) work on landscape change in California and Karl Zimmerer's (1996) examination of cultural and environmental change in the Peruvian Andes. Mitchell's and Zimmerer's goal is to further the cultural geographic enterprise by bringing together work that engages persistent themes, but looks at those themes in new ways.

In pursuit of this goal, an initial challenge was to consider which enduring themes might merit consideration. The themes around which the original *Readings in Cultural Geography* (Wagner and Mikesell 1962) was organized—culture, culture area, the cultural landscape, culture history, and cultural ecology—were not appropriate to the task, as some had continued to be the subject of discussion and work (the cultural landscape, cultural ecology), whereas others were so broad as to defy efforts at synthesis (culture, culture history). Hence, we concluded that the most promising way forward was to consider which traditional areas of inquiry had continuing promise as research themes, and yet had been pushed somewhat to the side in the wake of more recent ideas and orientations. At least four areas of inquiry seemed to fit these criteria: (1) the interplay between the evolution of particular biophysical niches and the activities of the culture groups that inhabit them; (2) the diffusion of cultural traits; (3) the establishment and definition of culture areas; and (4) the distinctive mix of geographical characteristics that gives places their special character in relation to one another. There are overlaps between these areas of inquiry, and each is represented in some way in the contemporary cultural geographic literature. Yet there are ways in which each has been relegated to narrow corners of the field, and as a result cultural geographers have tended either to overlook their potential significance or to ignore how they might relate to, and be enriched by, recent ideas and approaches.

During their careers, the contributors to this volume have all produced work that touches on one or more of the four themes identified above. The chapters that they have written for this book seek to move our understanding of these themes forward by considering them with reference to the perspectives and ideas of contemporary cultural geography. In particular, the authors direct attention to such contemporary foci of inquiry as cultural values and ideologies, structures and institutions of power, and regimes of conflict. Moreover, they do so in both urban and rural settings in the industrialized world, and in rural or more traditional settings elsewhere.

A principal inspiration for the book—and its particular mix of contributors—is the life and career of Marvin Mikesell. As a leading student of Carl Sauer, coauthor of the influential *Readings in Cultural Geography* (Wagner and Mikesell 1962), and author of dozens of influential books, monographs, and articles, Mikesell is closely identified with the development of cultural geography in the United States during the second half of the twentieth century. At the same time, his work evolved substantially through his career, shifting from an early emphasis on the impress of human agency on the landscape (Mikesell 1969a) to a later focus on cultural conflict and the construction of nationalist territorial ideologies (Mikesell 1983; 1985). Mikesell is an inspiration for the volume because of his persistent effort to promote a cultural geography that is intellectually rigorous yet accessible, catholic in its scope and diversity, and forward looking even as it is cognizant of its intellectual origins.

FROM LANDSCAPE TO CULTURE CONFLICT: THE SPECIAL CONTRIBUTIONS OF MARVIN MIKESELL

As this volume is being published, Marvin Mikesell is closing out a career of more than forty years' duration as a geography professor at the University of Chicago. He came to the University of Chicago in 1958 after completing bachelor's and master's degrees in geography at the University of California, Los Angeles, and while finishing up a Ph.D. in geography at the University of California, Berkeley. His doctoral dissertation, written under the direction of Carl Sauer, was a landmark study of rural settlement and landscape change in northern Morocco. Based on extensive fieldwork and an in-depth knowledge of the literature, the published version of the dissertation (Mikesell 1961) has long been viewed as a superb example of cultural geographic work in the human–environment tradition.

Mikesell's first two decades at the University of Chicago were ones in which he carved out a distinctive niche in the geography literature, both as a researcher in cultural geography/cultural ecology and as a commentator on geography. Mikesell has sometimes said, of his own work, that he is particularly fond of an article on the dispersal of the dromedary that he published while still a graduate student (Mikesell 1955). The article was emblematic of much of what made Mikesell's work in cultural geography so influential. It was a thoughtful, carefully constructed analysis that drew on contributions written in several different languages and that cut across the intellectual divide between physical and human geography. It was sophisticated yet accessible, and it was thoroughly geographic. It presaged a scholarly career that, in 1969, would lead the American Geographical Society to award Mikesell an honorary fellowship for his "sensitive articulation of both science and humanism in cultural geography," and in 1995 would prompt the Association of American Geographers to bestow on him its highest honors for contributions that provided "clear and compelling insights into the cultural geography of our planet and the nature of our discipline."

Mikesell went on from his graduate work to produce a collection of studies that are still shaping our understanding of the relationship between people and place in the Middle East and North Africa. Among other topics he wrote on traditional markets in Morocco and Spain (Mikesell 1958; 1960a) and deforestation in Morocco and Lebanon (Mikesell 1960b; 1969a). The clarity of Mikesell's observations, his ability to draw on works written in several different languages, and his coauthorship (with Philip Wagner) of the influential editors' essays in *Readings in Cultural Geography* led to opportunities to comment generally on the development of geography as a discipline and more specifically on trends in cultural geography and cultural ecology. He penned the much-cited article on "landscape" for the *International Encyclopedia of the Social Sciences* (Mikesell 1968), he wrote about geography's relationship to anthropology and other disciplines (Mikesell 1967; 1969b), and he defined the terrain of cultural ecology from the perspective of 1970 (Mikesell 1970).

A good deal of Mikesell's empirical and synthetic writings dealt in some way with human–environment relations. Thus, when public attention came to be focused on human despoliation of the environment in the late 1960s and early 1970s, Mikesell was the person to whom the Association of American Geographers turned to put together a disciplinary response. It took the form of a collection of essays on geographic perspectives on the environment, and included an introductory essay reviewing the evolving role of environmental themes in the geographical literature (Mikesell 1974). The lack of much disciplinary follow-up from that endeavor was both a frustration for Mikesell and a reflection of an unanswered challenge that is only recently beginning to be addressed. Mikesell saw other challenges as well, most particularly the risk of parochialism at a time when language training was on the decline and British and American geographers were turning their attentions away from the wider world in the (often laudable) pursuit of understanding their more immediate milieus. He addressed this risk in another edited compilation (Mikesell 1973), which focused on the "problems and prospects of field research in foreign areas."

Not surprisingly, Mikesell's emergence as a major contributor to the geographical literature led to prestigious appointments, both at the University of Chicago and in the discipline of geography. Just ten years after receiving his Ph.D., Mikesell was appointed chair of Chicago's prestigious Department of Geography, a position he occupied for five years and then assumed again in the mid-1980s. By the early 1970s he was also elected as a national councilor of the Association of American Geographers, and then moved on to become its vice-president and then president in the middle of the decade. Mikesell's presidential address was another landmark statement on cultural geography (Mikesell 1978) that once again demonstrated his extraordinary capacity for synthesis and understanding.

Mikesell's position as one of the leaders of a cultural geography with roots in the Berkeley of Carl Sauer created for him a conundrum of sorts. On the one hand, he clearly felt an intellectual debt to his mentor and graduate school. On the other hand, his work evolved substantially from its early underpinnings, and by the latter part of his career was far indeed from some of the (often inaccurate) characterizations of Berkeley cultural geography. *Readings in Cultural Geography* itself was never seen as a codification of Sauerian or Berkeley School cultural geography, and Mikesell himself was just as interested in looking forward as he was in looking back (Mikesell 1992). Indeed, he had a certain impatience with a literature that seemed to be devoted primarily to lionizing Sauer (see his comments on "Sauerology" in Mikesell 1987) preferring instead to look to the substance of what Sauer accomplished and consider where that substance should go. Through similar logic, Mikesell would likely view even this brief tribute to his career with a bit of impatience, anxious to move on to the subject matter of the volume.

Perhaps the clearest evidence of Mikesell's unwillingness to be defined by the past was his own growing interest during the late 1970s and early 1980s in what he termed cultural conflict. Although Mikesell has continued to devote some time and energy to research themes from the early part of his career, over the past two decades his attention has turned increasingly to the comparative study of cultural frontiers and conflicts among groups with differing ethno-national ambitions and senses of territory. A career observing the nature and character of culture and cultural differences apparently convinced Mikesell that territorialized conflicts over culture were among the most fundamental dynamics shaping the peoples and places of the contemporary world. As such, Mikesell turned his attention to matters that bridged cultural and political geography. His pioneering statements during the first half of the 1980s (Mikesell 1983; 1985) called for a cultural geography that was explicitly focused on the often discordant relationship between cultural patterns and the political organization of territory.

Mikesell was certainly not the only one looking at cultural conflict at the time, but he approached the subject in his characteristically thorough and thoughtful fashion. He read widely in the literatures of several different languages and he sought to "cut through" the cloud of contentions being made about culture and nationality in a set of writings that brought insight and clarity to the subject. His early general statements led to a 1991 piece, written in collaboration with one of us (Mikesell and Murphy 1991), which set forth an approach to the study of cultural–political conflict that reflected a concern not just with traditional distributional and locational issues, but with the political contexts within which cultural relations unfold, the symbolic attributes of culture, and the discourses of exclusion and inclusion. Mikesell had clearly come some way from the cultural geography of his graduate student days, and he was picking up on some themes that were to define the work of a new generation of cultural geographers. Yet he has not followed any bandwagon. Instead, he has balanced enduring and evolving themes in his typically provocative, insightful fashion. This is why his career is such a fitting inspiration for the volume at hand.

Mikesell's career accomplishments are not the entire story, however. Throughout his forty years at the University of Chicago, Marvin Mikesell touched countless students and colleagues who do not see him solely as an accomplished researcher and writer, but as a humane individual who was sympathetic yet challenging, supportive yet demanding. The premiums he placed on directness, clarity of expression, and careful scholarship were inspirations to those around him. Over the years he supervised thirty-nine master's theses and eighteen doctoral dissertations, and he was a wonderful colleague and friend to many professional geographers who spent time in the Department of Geography at the University of Chicago. The contributors to this book are among those whose lives and careers were most influenced by Marvin Mikesell in his complementary roles as teacher, adviser, and colleague. They joined together for this project out of a sense of gratitude and a desire to honor Marvin Mikesell with a volume dedicated to the search for a better understanding of nature and society through geographical scholarship. Since the goal of the volume is not simply to revisit traditional interests but to do so in new ways, its organizational structure reflects more recent preoccupations of cultural geographers: how humans are constructing cultural spaces, remaking their environments, and claiming places.

INTERPRETING ENVIRONMENTAL ENCOUNTERS

How to make sense of the human experience of environment, both historically and in the contemporary world, is a profoundly important issue. As an institutionalized endeavor, cultural geography's initial efforts to confront this issue were shaped by ruminations on history and culture, and were little concerned with matters of relevance and policy. The diffusion of the subject from its early Berkeley base to other venues, and its encounters with new and different intellectual currents, produced great changes. Untroubled, or unconvinced, by the proposition that a road's location might be influenced by witchcraft, second- and third-generation cultural geographers looked to political, economic, and social arrangements for explanation. Moreover, at places such as the University of Chicago, where policy-oriented pragmatism commanded widespread respect and adherence, few could (or wanted to) remain aloof from matters of pragmatic relevance. The students and associates of Marvin Mikesell were no exception—yet they also were constantly reminded that the political, the economic, the social, and the pragmatic were all conditioned in some way by culture. Thus, it is not surprising that the essays in this volume reflect an interest in themes and influences that extend beyond the traditional domain of cultural geography, yet the meditating role of culture is never far from view.

Three themes, each comprising a major section of this collection, dominate the papers in the volume. A diverse array of specific case studies, each deeply rooted in and sensitive to historical processes, comprise the first group of papers. Here the concern is with how cultural spaces are constructed, often through extended periods of struggle and trial and error experimentation, by different actors on the local scene. In this instance the locale is hardly local, but instead draws on a number of examples from the North American cultural experience. The role of humankind in habitat modification and the human endeavor to control, remake, and understand its place in the natural world dominate the second group of essays. Mixing a concern for cultural historical processes and for the cultural prism through which nature is viewed, these essays also are closely attuned to the implications of human action—both for contemporary environmental health and for the well-being of future generations. In the concluding section, the authors explore how places are claimed in different ways in a number of cultural settings. Almost invariably considerable conflict is present in this effort to impress claims upon the landscape, and seemingly innocuous symbols can contain implicit significance far beyond their actual importance. Largely situated in cultural settings distant from North America, these essays provide a coherent statement on the universality of culture as well as its unique local manifestations.

Constructing Cultural Spaces

Our first theme addresses the construction of cultural spaces by asking how contemporary landscapes became what they are today. All of the examples offered here come from the North American culture realm, and all bring a distinctly historical perspective to the issue. In Mexico, Karl and Elisabeth Butzer consider the role that material culture artifacts in the landscape play as evidence of cultural development. How much of the contemporary Mexican cultural landscape is Spanish in origin and how much is the product of indigenous cultural development to which the Spanish ultimately adapted? Were processes of independent cultural invention predominant, or did diffusion from distant centers of innovation govern what we see today? In raising this explanatory controversy within cultural geography, the Butzers present a cautionary challenge to conventional wisdom based upon a careful examination of evidence contained in historical documents and ordinary landscapes.

Peter Goheen and Michael Conzen explore two radically different stories of the construction of culture landscapes. In the rural Hill Country of Texas, six different cultural visions contested for the dominant role in shaping the region's landscape. While several proved transitory and left little long-term impact, others, particularly practitioners of small-scale farming of Anglo-Saxon and German heritage, made substantial contributions and ultimately created the cultural environment from which Lyndon Johnson ascended to the presidency. In nineteenth-century Toronto, the waterfront became the focus of struggle between economic and political forces. Led by the Canadian Pacific Railroad, business interests viewed the waterfront in exclusively utilitarian terms. Initially most Canadians shared that view, failing to attach any significant value to public water-edge uses beyond those that created jobs. Ultimately less utilitarian values promoted the emergence of alternatives to industrial use of the waterfront. Today the issue of how to value and redevelop the ordinary landscape of the lakefront has reemerged, demonstrating that few culture landscapes and issues are ever frozen in time or settled forever.

If Conzen and Goheen deal with conflict over access to resources in specific cultural settings, in the chapter that concludes this section Carville Earle tackles a powerful myth in American culture history: the seminal role of the frontier in the American experience. As deconstructed by Earle, the frontier becomes a highly variable experience both temporally and spatially. Disaggregated by region and by period, the frontier proves to be a much more volatile phenomenon than that extolled by Fredrick Jackson Turner in his classic formulation.

Remaking the Environment

The effort to construct cultural spaces, and to make sense out of that endeavor, invariably involves the use and modification of the earth to meet human needs. In an age when increasing demands are made on the earth's resources, and when an impending water crisis places constraints on even the best-endowed areas, it is perhaps not an accident that three of our environment-oriented essays are concerned with issues of wetness—or its absence. Efforts to overcome the consequences of aridity have led to the development of elaborate civilizations constructed around and dependent upon irrigation technology. Jim Wescoat examines the importance of irrigation as a tool for modifying habitat in two novel ways. The first is to counter the conventional wisdom about the diffusion of twentieth-century irrigation technology. Today it is taken as axiomatic that Third World countries depend on industrialized states (particularly the United States) for the technology that makes economic growth and development possible. This time-limited perspective ignores the degree to which a century ago American irrigation technology found its inspiration and models in the elaborate and sophisticated irrigation practices of South Asia, and could well do so again. The second thread in Wescoat's discussion is a larger issue in nature/society relations: what impact does a production system such as irrigation agriculture have on the structure and spirit of a society that is dependent on it for survival? Karl Wittfogel, the former Marxist theoretician whose concept of "oriental despotism" enlivened political and cultural debates in the 1950s and 1960s, believed that the political control needed to run an irrigation system could, once developed, diffuse independently from its natal setting, contaminating the cultures with which it came in contact.

A more contemporary perspective is offered by Jim Schmid on the role that naturally, as opposed to artificially, wet places play in American culture. Having left academics for the life of an ecological consultant, Schmid has made an applied career out of North American culture's dealings, or failures to deal, with muddy morasses. With more than half of the wetlands in the lower forty-eight states filled, drained, or otherwise modified, there are few ecological systems that have felt the impact of humankind's modifying touch more severely. And there are few habitats where, at the local scale, messages about values and goals are more confused and the conflicts over resource use are more difficult to resolve. John Kirchner examines conflicts over riverine navigability and the role that the historical geographer can play in their resolution. As an exercise in applied geography and as a demonstration of the utility of archival skills, Kirchner mirrors Schmid's translation of academic skills into the arena of business, law, and political conflict resolution. That proof of former navigability can also mean funding for future development transforms an esoteric concept into an intensely practical and controversial issue.

David Lowenthal's examination of the history of environmentalism in the English-speaking world places these issues of the conquest, subjugation, management, and valuation of nature in a broad cultural context. Ironically, today little if any real nature survives, since even wilderness is largely a human artifact. Today nature is managed (or at least we attempt its management), Lowenthal points out, in order to achieve a perceived harmonic, human-construed balance that is generally ephemeral and is most often absent in a world of endemic disequilibrium. In a laudable effort to avoid unrelieved pessimism, Lowenthal offers principles of future conduct that are rooted in cultural values and in lessons drawn from the last three centuries of environmental experience: value community over unbridled individualism; demand intergenerational equity; promote stewardship of nature; favor generalization over narrow specialization in resource management.

Claiming Places

From a broad view of environment and the meaning of humankind's impact upon nature, our third theme turns to consideration of the human propensity to develop attachment to place. Claims to place are an essential part of individual and collective assertions of existence. Many of these claims to place are mirrored in and shaped by artists and writers, a theme explored by Anne Buttimer. Her analysis of Ernest Hemingway's use of place as both rooted to specific locales and as a metaphor for spontaneous events that, if not timeless, transcend space specificity offers insight into how space is imaged. By exploring the connections between an author's lifeworld and the evocation of place, we gain a better understanding of how, often less articulately, our experience is connected to the reality of place.

Medicine and culture often are much more place specific than literature, since there are fewer possibilities for rearranging elements into composite images that may project truths about places but may not actually exist in the real world. Thus Charles Good describes the relationship between healthy landscapes and the people who act in those places, which may be differently interpreted (the practitioner of female circumcision and the critic of the practice are unlikely to find common ground on either the practice's appropriateness or meaning). Through examples drawn from colonial and post-colonial Africa, Good explores how context, continuity, and change pervade efforts to assert claims to place. That these claims are often based on ethnicity is an insight developed by Chauncy Harris and affirmed by a host of tragic conflicts that have erupted in many parts of the globe in the last decade of the second millennium. In the former Soviet Union, efforts to acknowledge ethnic communities through varying levels of internal autonomy promoted centrifugal tendencies when the power of the central government weakened. As Harris points out, the resulting political rearrangements have left millions of Russians as minorities in independent republics while still leaving large numbers of minorities embedded within the Russian Federation, thus providing a fertile breeding ground for future conflicts.

No place is more replete with conflicting claims than Israel/Palestine, sacred to three religions and contested by two ethnic communities. Both Chad Emmett and Shaul Cohen analyze the complexities of reconciling conflicting claims to space by shared, although not necessarily conflict free, arrangements or by more exclusionist approaches to overlapping and apparently uncompromisable assertions of priority. In either instance, landscape symbols, often as trivial as who has the right to wash what wall, serve as metaphors for claims to privilege and preference.

Finally, place counts in profound ways for all practitioners of cultural geography, regardless of persuasion or approach. The importance of place in this cultural encounter with the environment is one of the enduring themes that informs our conclusion and the epilogue offered by Phil Wagner. Wagner revisits the notion of place, how it is constructed and how it operates at various scales, as well as the seminal role that his Berkeley experience played in launching his personal exploration of place as heuristic device and metonymic model. We hope that the epilogue's last word sparks not closure but contemplation about cultural geography's future direction. If this collection of papers opens new doors to innovative future work that constructs creative bridges between enduring themes and contemporary concerns, then we will have assembled a fitting tribute to the efforts and career of Marvin W. Mikesell.

REFERENCES

Commentary on "The Reinvention of Cultural Geography" by Price and Lewis (contributions from Denis Cosgrove, James S. Duncan, Peter Jackson, Marie Price, and Martin Lewis). 1993. *Annals of the Association of American Geographers* 83 (3): 515–22.

Cosgrove, Denis E. 1984. *Social Formation and Symbolic Landscape*. London: Croom Helm.

Crang, Mike. 1998. *Cultural Geography*. London: Routledge.

Duncan, James S. 1995. "Landscape Geography, 1993–94." *Progress in Human Geography* 19 (3): 414–22.

Foote, Kenneth E., Peter J. Hugill, Kent Mathewson, and Jonathan M. Smith, eds. 1994. *Re-reading Cultural Geography*. Austin: University of Texas Press.

Gregson, Nicky. 1995. "And Now It's All Consumption?" *Progress in Human Geography* 19 (1): 135–41.

Hugill, Peter J., and Kenneth E. Foote. 1994. "Re-reading Cultural Geography." In *Re-reading Cultural Geography*, ed. Kenneth E. Foote, Peter J. Hugill, Kent Mathewson, and Jonathan M. Smith, 9–23. Austin: University of Texas Press.

Jackson, Peter. 1989. *Maps of Meaning: An Introduction to Cultural Geography*. London: Routledge.

Johnston, R. J., ed. 1986. *The Dictionary of Human Geography*, 2nd ed. London: Basil Blackwell.

Ley, David, and Marwyn S. Samuels, eds. 1978. *Humanistic Geography*. Chicago: Maaroufa Press.

Mathewson, Kent. 1996. "High/Low, Back/Center: Culture's Stages in Human Geography." In *Concepts in Human Geography*, ed. Carville Earle, Kent Mathewson, and Martin S. Kenzer, 97–125. Lanham, MD: Rowman & Littlefield.

———. 1998. "Cultural Landscapes and Ecology, 1995–96: Of Oecumenics and Nature(s)." *Progress in Human Geography* 21 (1): 115–28.

———. 1999. "Cultural Landscape and Ecology II: Regions, Retrospects, Revivals," *Progress in Human Geography* 23 (2): 267–281.

Mikesell, Marvin W. 1955. "Notes on the Dispersal of the Dromedary." *Southwestern Journal of Anthropology* 11 (3): 213–45.

———. 1958. "The Role of Tribal Markets in Morocco." *Geographical Review* 48 (4): 494–511.

———. 1960a. "Market Centers of Northeastern Spain." *Geographical Review* 50 (2): 247–51.

———. 1960b. "Deforestation in Northern Morocco." *Science* 132: 441–48.

———. 1961. *Northern Morocco: A Cultural Geography*. Berkeley: University of California Press.

———. 1967. "Geographic Perspectives in Anthropology." *Annals of the Association of American Geographers* 57 (3): 617–34.

———. 1968. "Landscape." In *International Encyclopedia of the Social Sciences*, vol. 8, 575–80. New York: Macmillan and Free Press.

———. 1969a. "The Deforestation of Mount Lebanon." *Geographical Review* 59 (1): 1–28.

———. 1969b. "The Borderlands of Geography as a Social Science." In *Interdisciplinary Relationships in the Social Sciences*, ed. Muzafer Sherif and Carolyn W. Sherif, 227–48. Chicago: Aldine.

———. 1970. "Cultural Ecology." In *Focus on Geography: Key Concepts and Teaching Strategies*, ed. Philip Bacon, 39–61. Washington, DC: National Council for the Social Studies.

———, ed. 1973. *Geographers Abroad: Essays on the Problems and Prospects of Research in Foreign Areas*. Chicago: University of Chicago, Department of Geography Research Paper No. 152.

———, ed. 1974. "Geography as the Study of Environment: An Assessment of Some Old and New Commitments." In *Perspectives on Environment*, ed. Ian R. Manners and Marvin W. Mikesell, 1–23. Washington, DC: Association of American Geographers.

———. 1978. "Tradition and Innovation in Cultural Geography." *Annals of the Association of American Geographers* 68 (1): 1–16.

———. 1983. "The Myth of the Nation State." *Journal of Geography* 82 (6): 257–60.

———. 1985. "Culture and Nationality." In *Geography in Internationalizing the Undergraduate Curriculum*, ed. S. J. Natoli and A. R. Bond, 67–90. Washington, DC: Association of American Geographers.

———. 1987. "Sauer and 'Sauerology': A Student's Perspective." In *Carl O. Sauer: A Tribute*, ed. Martin S. Kenzer, 144–50. Corvallis: Oregon State University Press.

———. 1992. "Reflections on a Shared Venture: Readings in Cultural Geography." In *Person, Place, and Thing: Interpretive and Empirical Essays in Cultural Geography*, ed. S. T. Won, 31–45. Baton Rouge: Department of Geography and Anthropology, Louisiana State University.

Mikesell, Marvin W., and Alexander B. Murphy. 1991. "A Framework for Comparative Study of Minority-Group Aspirations." *Annals of the Association of American Geographers* 81 (4): 581–604.

Mitchell, Donald. 1996. *The Lie of the Land: Migrant Workers and the California Landscape*. Minneapolis: University of Minnesota Press.

Nietschmann, Bernard Q. 1973. *Between Land and Water: The Subsistence Ecology of the Miskito Indians, Eastern Nicaragua*. New York: Academic Press.

Price, Marie, and Martin Lewis. 1993. "The Reinvention of Cultural Geography." *Annals of the Association of American Geographers* 83 (1): 1–17.

Wagner, Philip L., and Marvin W. Mikesell, eds. 1962. *Readings in Cultural Geography*. Chicago: University of Chicago Press.

Zimmerer, Karl S. 1996. *Changing Fortunes: Biodiversity and Peasant Livelihood in the Peruvian Andes*. Berkeley: University of California Press.

I

CONSTRUCTING CULTURAL SPACES

1

Domestic Architecture in Early Colonial Mexico: Material Culture as (Sub)Text

Karl W. Butzer and Elisabeth K. Butzer

CHANGING VIEWS OF CULTURAL GEOGRAPHY

More than a century ago, geographers turned to material culture for analytical tools to characterize features and patterns of the cultural landscape. This strategy was borrowed from geomorphology, which sought to identify forms linked with specific processes, in order to categorize landscapes. The resulting settlement geography initially favored ethnic explanations for differences in the spatial arrangements visible within villages or farmsteads and in the construction of dwellings and outbuildings. But it became apparent that many such differences were temporal rather than ethnic. Reflecting changing defensive and social needs, land tenures, technological skills, and stylistic tastes, places occupied for two millennia could be expected to differ from others with a continuity of only 500 years. It is now widely recognized that differences in traditional rural landscapes across Europe from France to Russia reflect regional settlement histories rather than ethnic imprints.

An alternative approach developed within cultural geography focused on dynamic phenomena. Different economic pursuits should be reflected in activity patterning, spatial expressions of resource utilization, and related living or processing structures. In this more comprehensive vision, differences between or within regions would be embedded within a variable environment, as a function of topography, soil and forest distribution, availability of building materials, and the productivity of competing economic modes and mobility patterns. Satisfactory ecological studies of this tradition of *géographie humaine* required intensive local investigations, such as Marvin Mikesell's study of northern Morocco

(1961), but tended to be superficial when generalized to larger regions, let alone global distributions.

A very different direction came with the quantitative revolution, in the guise of a spatial geography emphasizing distance-decay and the economic rationales for location and connectivity. This served to explain the clustering of population nodes, and provided practical tools to study patterns of economic integration. But it tended to ignore environmental variability and cultural proclivities, so as to become the antithesis of settlement geography and *géographie humaine*.

As a consequence, alternative directions emerged within geography to focus once more on traditional landscapes. One borrowed from anthropology to examine the environmental parameters of resource utilization, land-use change, community structures, and the sociopolitical constraints imposed upon the decision making of small or rural populations. Exemplified by cultural ecology, this process-oriented approach is complemented by a "new" cultural geography that, inspired by the arts and humanities, underscores the nonrational elements of human behavior and experience, by highlighting taste, emotion, gender, attitudes, identity, and worldviews. By including iconographic landmarks, it emphasizes that communities build their houses and places of worship or authority as an expression of their own values, aesthetics, and ways of life.

This linear but overlapping progression of thesis and antithesis in geography is complicated by the latent difficulties of conceptualizing culture as a material repertoire, a behavioral sphere, an ideological realm, or a combination of all three. Problems become apparent in attempting to deal with culture change, which reflects stimuli that may be internal (innovation, adaptation, evolution, "revolution") or external (innovation diffusion, migration, conquest). These are complex and interrelated processes that can only be studied as component parts and within stated parameters. But the temptation is to gloss over the complexities and champion simplistic conclusions.

Some of the most egregious shortcomings relate to the diffusion of ideas, a phenomenon discussed by social scientists well before it was studied by geographers. The problem in historical/cultural geography is that diffusion is all too often used as a "black box" explanation—without examining the internal structure of a cultural system, directing attention to results which are assumed to derive from particular inputs. For example, new people are believed to introduce new ideas, whereas attention should instead focus on how such information is selected, tested, adapted, or in fact rejected. Black box diffusionism, whereby a "rabbit in the hat" explanation is disguised as social science, is by no means limited to geography. A classic prototype is the well-received book by historian Andrew Watson (1983), who portrays the Medieval Islamic Conquest as the propagation of a beneficial "Green Revolution" that greatly expanded and improved irrigation technology and food crops. Quite apart from the many erroneous facts or generalizations of his empirical database (Adams 1981; Butzer et al. 1985), Watson limits his theoretical frame to deductive economic principles that favored the acceptance of new ideas, while ignoring the processes of cultural resistance to change.

Conquest societies have enduring interest as test cases for cultural change (Foster 1960; Mikesell 1960), and it is appealing to posit that a dominant minority, particularly one with economic and technological superiority, will impose its methods and ideas upon the conquered. That is either the claim or the subtext of many writings on the "ecological imperialism" of the Columbian Encounter, even though the Spaniards faced major problems in transplanting their agrosystem to the New World (Butzer 1992) and, in most regions, their food plants were unable to compete, in economic or cultural terms, with indigenous counterparts (Butzer 1995).

Domestic architecture remains a prime arena for advocates of diffusion, by migration or conquest, in which construction technologies or structural types inexorably move from the Old to the New World (for example, Jordan and Kaups 1989; Jordan, Kilpinen, and Gritzner 1996). In the case of Colonial Mexico, similar assumptions underlie the writings of Robert West (1948, 1975), West and Augelli (1989: 252), and John Winberry (1974, 1975). It is not our intent to single out such works as straw men, because they represent measured and thoughtful studies. The problem is the pervasive assumption that items such as adobe construction or flat-roof adobe houses in the American Southwest or Mexico, for example, were introduced from the Near East via Spain. That is almost a truism among educated Southwesterners, and it comes up all too often in big-picture undergraduate courses in cultural geography. At a recent Gran Quivira Conference in Nuevo Laredo (10 October 1998), the distinguished Mexican landscape architect Francisco López Morales made similar claims, as purportedly documented by a series of slides from Egypt and elsewhere.

This is then not a dead horse. It is grounded in the premise of European cultural dominance, flying in the face of a growing appreciation of indigenous resistance to Colonial appropriation and domination (Andrien and Adorno 1991). It is also devoid of historical context in that it ignores the roles of agency and innovation by all the social participants in the complex processes of transculturation (Adorno 1994).

This chapter examines the substantial body of Spanish and indigenous materials on domestic architecture from the first century of Spanish Colonial dominion in Mexico, as a source of information on construction methods and domestic architecture, and within a much larger, sociohistorical context. These data demonstrate that indigenous resistance to assimilation was paramount, and that it was primarily the conquerors who changed their ways. Even that icon of the Colonial urban imprint, the grid-plan town, raises reasonable questions about the essential dichotomy between Spanish and indigenous built environments. We conclude that the subject of domestic architecture is embedded within a number of different discourses—between different indigenous classes, between different indigenous groups, and between the indigenous people and the Spaniards.

The empirical core of this chapter represents the distillation of a monographic study that will form part of a larger book, currently in preparation. It is here that the detailed arguments and documentation will eventually be found. In the present medium we can only identify some of the basic sources.[1]

DOMESTIC ARCHITECTURE IN EARLY COLONIAL MEXICO

This section summarizes the database within a broad cultural and historical context and examines the claims and assumptions of the diffusionist school.

Wattle-and-Daub Construction

The most common technique in sixteenth-century New Spain to build the walls of domestic structures was to fix screens of sticks, wickerwork or matting on a frame of poles. Everything was tied together by fiber cords, and the screen-panels were commonly, but not necessarily, sealed with adobe mud and then possibly painted over with murals. Opportunistic use of different materials in different environments required different frame rigidities and dimensions, as well as variable binding methods and knots.

That implies divergent stylistic development, which is verified not only by modern ethnographic data but also by the unusual *pinturas* of the *Relación de Michoacan* (Alcalá 1980), which represent both common housing and ritual or elite buildings made of wattle and daub. These display both a refined aesthetic as well as a highly ritualized use of construction forms, materials, color, and space. There are hints that this may also have been true to some degree within the Otomí culture domain. While thatched-roof houses used adobe walls in some areas, in others they were probably combined with wattle-and-daub construction; we must assume that the latter was commonplace everywhere except in the Nahua heartland of Central Mexico. This technique served as a very flexible medium for functional and ritualized expression that was by no means concentrated in the *tierra caliente* and that was carried to high levels of craftsmanship and aesthetic distinction. What survive today in mountainous backwaters or marginalized communities of Mexico are comparatively impoverished traditions that do little credit to what was found at the time of the Conquest.

With the exception of one claim that round (corbeled) huts on the Guerrero coast were introduced by African slaves, diffusionists have generally not disputed that wattle-and-daub construction was truly indigenous, presumably because it was perceived as too humble for Europeans to bother about. Nonetheless, there are some similar applications of this technique in the Old World, such as in the former Yugoslavia, which illustrate the existence of evolutionary parallels conditioned by the convenience and specific tractability of available raw materials.

Thatched Roofs

The dominant indigenous roof type of sixteenth-century New Spain was thatched. Sophisticated frames were built with a steep pitch and adjusted for different types and lengths of covering thatch. Some were two-shed, gabled roofs, but most were probably of the four-shed, hipped category. Circular or oval thatched roofs, today still visible in the Yucatan, the Huasteca, and in Oaxaca and Guerrero, were known but attributed to "primitive" peoples or to the prehistoric past. As in the case of wattle-and-daub walls, cords and knotting were critical for both the roof frame and the thatch itself, varying according to the raw material available. To avoid leakage at the summit of the roof where the gables met, several kinds of tie-beams were used, and in some regions the tie-beam flared out to project over the shorter sides of a hipped roof.

The *Relación de Michoacan* (Alcalá 1980) illustrates stylistic variation, the importance of decoration, and presumed ritual or ethnic details. Thatched roofs prevailed everywhere except among the Nahua (Aztecs) of Central Mexico, who preferred flat-roof structures. The Spaniards fairly consistently labeled thatched roofs as *techos* (or *casas*) *de paja* (straw roofs or straw houses), regardless of the materials, but sometimes used *jacal* to describe a thatched roof, as opposed to a thatched house, as became common later. Nahuatl usage appears to have been ambiguous, with *xacalli* describing either a roof or a house unless clarified by further adjectives.

Diffusionists accept thatched roofs as indigenous, even though reed thatching was traditionally used in Europe, near the coasts of southwestern Spain (parapet gabled houses near the marshes below Sevilla), in England, the Frisian estuaries of the North Sea, and even in Austria. Clay pots are now sometimes found at the end of a tie-beam in southern Mexico and Morelos, and presumably have some ritual significance. A similar custom is evident in parts of West Africa, but such features are already suggested by sixteenth-century representations in New Spain.

Wood Shingles

Shingle or "shake" roofs had become fashionable by 1580 in some towns of New Spain, primarily in mining centers. Such shingles were exclusively made in central Michoacan, from fir wood (*oyamel*) found there in montane forests. Purépecha (Tarascan) craftsmen not only fashioned and bundled these wood shingles (just as today), but apparently also transported and applied them to make roofs in distant places.

Although the accepted but uncommon term for such wood shingles in northern Spain was *ripia*, in Mexico Spaniards used *taxamanilli* (or *tejamanil*), borrowed from the Nahuatl *tlaxmanilli*, as early as 1531 to describe wood shingles used to cover a Spanish residence built in Mexico City during the 1520s. Only a few informants of the 1580s were aware of *ripia* as a Spanish equivalent, and most of these note similarity rather than equivalence. In fact, the Mexican *tejamaniles* were (and remain) two or three times as long as any European shingle, wood or slate, and they apparently were applied by the same techniques used for thatching. The diffusionist interpretation is that wood shingles were introduced from Spain by Basque miners, primarily because these were familiar in northern Europe and were used in Spanish towns of Mexico during the 1580s. Since such 80 to 120cm-long slats were basically produced by wedging rather than sawing, the demonstrated introduction of a facilitating Old World iron technology is not material to the argument.

In our view, the persistence of wood shingle manufacture and use as a specifically Purépecha craft is as telling as the Spanish commissioning of a shingle roof within a decade of the Conquest. We suggest that *tejamaniles* evolved in Central Mexico from the basic thatched roof technology, prior to the Conquest, but that its increasing popularity was a result of Spanish demand and facilitated by Spanish tools and transportation.

Indigenous skills in woodworking were already abundantly noted in 1519, and iron tools facilitated the diversification of such craftsmanship to the new ritual and practical needs of the Colonial caste. Although there were some plank houses by 1580, they did not use corner notching, and corner-timbered wood houses are not verified until the eighteenth century.

Flat Roofs

The indigenous, flat-roofed house (Nahuatl, *tlapancalli*) was a familiar feature in parts of New Spain from as early as 1519. A sturdy, horizontal ceiling of wood beams and planks was covered by a bed of soil, drained by a conduit, and surrounded by a low parapet enclosing a terrace, which served as an extension of the living space.

The antiquity of the flat-roofed house is evident from glyphs or pictographs for such houses, symbolized as such, with or without parapets, just as there were glyphs for steep, thatch-roofed houses. But their distribution was circumscribed, found as a dominant type in the Valley of Mexico and as a more incidental one in Puebla, Morelos, and Central Oaxaca, or in the Bajío district of Guanajuato. Initially an elite house form, the flat-roofed house was apparently first disseminated among still independent or tributary elites around the perimeter of the expanding Aztec imperium, and later perhaps implanted in a few Aztec garrison towns. After the Conquest, Indian veterans from Tlaxcala, assisting the Spaniards in reducing more distant provinces, established colonies in ravaged towns, while Otomí colonists founded new towns in the Bajío, where flat-roofed houses were also noted in the 1580s.

While a solid link between indigenous colonists from Central Mexico and flat-roofed houses in outlying areas is apparent, the Spanish elite in Mexico City adopted flat-roofed houses even though these were uncommon in Spain. These were houses of status that presumably would enhance the prestige of Spanish settlers in Mexico City, quite apart from the practical implications of the fact that the construction workers were indigenous.

The diffusionist perspective, despite this broad evidence, holds that although there were indigenous flat-roofed houses, the Spaniards probably introduced a different type, ultimately derived from Berber North Africa. This was then propagated by Spanish stockherders to the frontier and eventually up to northwestern Mexico. In Spain, a flat roof commonly has a parapet and is called an *azotea* or *terrado*. That also is what the indigenous type was called by Spaniards in Mexico. However, in that one part of Andalucía where flat roofs are common or characteristic—the Alpujarras district of Granada and Almeria—these terms are only used for midlevel porches, and none of the flat roofs has a parapet or is used as a living space. Yet these houses are demonstrably the ones reoccupied by Christian settlers following expulsion of the Muslims in 1572, and identical types are also found in Morocco. There is no evidence of New World colonists coming from the Alpujarras. Finally, the modern flat-roofed house in northern Mexico with its walls of adobe brick bears no resemblance to those of Spain or the Alpujarras, which are built of mortared rock.

The New World flat-roofed house was a cultural icon not only of indigenous technology but also of social interaction, integral to urban living, and mirroring distinctions of wealth and class. Disseminated by identity-conscious indigenous settlers to the frontier, it eventually changed in function, status, and appearance within a very different social context.

Adobe Bricks

Adobe is the Spanish term for a sun-dried brick, made from a mix of mud and vegetal temper. Already in 1519, Cortés reported that where stone was uncommon, indigenous houses were built of adobes, a fact confirmed by an indigenous report of 1582 that the adobes of two century-old buildings in Texcoco were in sufficiently good condition to reuse. It is not surprising therefore that about 1580 the most common building material in New Spain was the adobe brick. Since adobes in direct contact with the subsoil are prone to absorbing moisture and meltdown in the rainsplash zone, such walls were commonly raised on stone foundations or platforms in many areas, with elaborate stone platforms a visible marker of status. Adobe walls also seem to have been plastered over and painted with murals, and their durability will have been further enhanced by overhanging roofs.

It is widely believed that prior to the Spanish arrival, Mesoamerican builders only knew hand-formed adobes that were loaf-shaped. In Europe a variety of multiple form boxes were used to produce a set of bricks until fairly recently. Given the assumption that such boxes can only be efficiently made with iron tools, it is inferred that the Spaniards introduced them, so as to spawn an explosion of adobe-brick construction. That assumption is faulty because, in the Old World, mold-box adobes were being produced in Mesopotamia and Egypt long before the invention of iron saws and nails. The form boxes were presumably made of split wood boards, held together with resin glues.

A body of sixteenth-century sources raises serious questions as to whether indigenous builders had not already invented some sort of form box. The original Nahuatl for mud brick was *xamitl*, with complex cognate forms used for "to make adobes," "adobe mold," or "molded into adobes"—all verified for the period 1555–1576. Spanish sources of 1519, 1533, and 1540 all refer to adobes in the plural, that is, bricks, and none mentions that they were handmade or loaf-shaped. But they were much larger than adobes familiar from the Old World, although dimensions are cited in only one case: 42 by 84cm, but made in "square molds." Modern adobes in Morelos typically measure 25 by 50cm, those in New Mexico 25 by 35cm (but up to 30 by 40cm in size), compared with 13 by 25cm in Spain or Egypt. These New World examples have three to six times the mass of their Old World counterparts. Although today made in Mexico with Mediterranean-style, multiple-form boxes, an indigenous illustration of the 1570s shows a single-unit box that may indeed be of pre-Hispanic origin.

Statements in the *relaciones* to the effect that people in a town "now" live in adobe houses, whereas previously they had lived in wattle-and-daub huts, are only four in number and are framed with reference to the benefits of evangelization. That linkage is not entirely propagandistic, because in one area it was noted that a son would not move into the "hut" of a deceased father; until recently among the Totanacs and Huastecs of the eastern Sierra Madre a deceased person's hut would be burnt rather than reoccupied, so as to avoid the spirit of the dead. The one area where wattle-and-daub construction was demonstrably replaced with adobe brick by about 1580 was Michoacan, where the *pinturas* of around 1540 show that wattle and daub had traditionally been used by the native ruler and in cult buildings in a ritually significant fashion. The notion that Spanish form boxes facilitated a revolutionary change in domestic construction can be seriously questioned.

An integral part of the diffusionist argument is that adobe-brick construction carried a North African, Berber technique via Spain to Mexico. That is untenable. Muslim houses in Spain were built of mortared rock or rammed earth (*tapial*), as evident from medieval archaeology throughout eastern Spain or reused Muslim houses of the sixteenth century in the Alpujarras. The same is true of traditional houses in much of Morocco.

Rammed Earth or *Tapial*

Unlike adobe bricks, rammed earth was a mainstream construction method in western Europe until the 1800s. It was predicated on a removable form box, into which a concrete-like mix was incrementally filled, to produce a block that typically measured 45cm wide, 75cm high, and 2.5m long and was called a *tapia*. When used for house walls or fortifications, the mixture consisted of lime, sand, and gravel that could support multistory structures, lasting for many centuries. Alternatively, less durable walls for garden enclosures or outbuildings could be made by ramming a simple adobe mix into a similar slip box; in this case the load-bearing capacity was limited to three courses (2.1–2.4m). Both kinds of wall, however different, tend to be called *tapia* in Spanish or rammed earth in English.

During the 1520s, lime-cemented *tapias* were used to build the first palace of Cortés in Mexico City, but forty years later usage seems to have been limited to simple rammed mud, with reference to enclosing walls and single-story houses on the mining frontier. In Mexico, rammed earth construction was eventually displaced by mud brick, but it survives in New Mexico and Andean South America, continuing along with the characteristic form boxes introduced from Spain. But in the borderlands a number of pre-Hispanic monuments and towns had already been built with regular but tapered courses of "puddled mud," apparently hand-shaped and smoothened with adzes.

Tapias represent a good case for a major Iberian construction technique that was rejected in New Spain, where there apparently was no pre-Hispanic analogue and adobe bricks were more economical.

Kiln Bricks, Roof Tiles and Cut Stone

Although kiln-fired bricks and roof tiles were already made by Cortés' men on the coast of Mexico in 1519, high-temperature kilns were unfamiliar to the indigenous people, and these building materials remained quite uncommon during the first century after the Conquest. It was the conquerors who abandoned the familiar, red-tiled roofs of southern Spain and adapted indigenous-style, flat-roof construction.

Cut or trimmed stone was widely used on temple platforms and other centers of power a millennium before the Spaniards introduced iron tools, which should remind us that the Egyptian pyramids also did not benefit from an "advanced" technology to quarry and cut rock. Most extensively used in royal palaces, cut stone was nonetheless increasingly adapted to the doorways, windows and corners of elite houses during the eighty years before the Conquest. Trimmed rock, two stories, and enclosed courtyards became hallmarks of wealth and power, and were largely limited to the Valley of Mexico and western Puebla, representing the Aztec heartland and its persisting rivals in Tlaxcala and Cholula.

The conquering Spaniards also saw cut stone as a symbol of status and, with the rebuilding of a European city within the core of Mexico City (Tenochtitlan) in 1523, they began to elaborate their own self-myth on an unprecendented scale. Once humble captains of the Conquest aspired to stone palaces of their own, making use of their newfound power to mobilize the same great crews of construction workers and specialized craftsmen, who had once built and maintained temples. Prior to the catastrophic pandemic of 1545, indigenous labor was a "cheap" commodity. After 1538 the center of Mexico City was rebuilt a second time on an even grander scale, according to the new Renaissance ideals. In the process, the "comrades-in-arms" of the Conquest were themselves segregated into a steep socioeconomic hierarchy, creating a small Colonial oligarchy with privilege, money, and power. It was spatially circumscribed, concentrated near the ceremonial plazas of Mexico City and Puebla.

The great bulk of the former soldiers and later immigrants did not share in this redistribution of wealth, and lived in much more modest houses, probably comparable to the lower scale of indigenous elite structures. Early in the 1600s, well-to-do provincial centers, such as Guadalajara and Querétaro, consisted only of adobe houses. In the mining towns, Spanish entrepreneurs lived in houses of rammed mud and shingle roofs, and further afield Spanish stock raisers and bureaucrats had to make do with thatched roofs and sometimes even wattle-and-daub walls. Many immigrants of limited means and their mestizo descendants could not even afford wheat bread and had to switch to indigenous staples. All too many were probably worse off than they or their "illustrious" ancestors had been in Spain.

The "Planned" Colonial Town

While it is commonplace to emphasize the geometrically-structured grid town—the *traza*—as the normative Colonial imposition on the New World, such grid plans had more structural and functional similarities with their Mesoamerican counterparts than differences. Indigenous urban centers were arranged according to astronomical bearings, dictated by cosmological criteria. They were focused on great squares that served ceremonial and commercial needs and functions, in direct proximity to prominent temples and palaces, to project a particular social order and proclaim dynastic power. As the visible markers of wealth and status dissipated with increasing distance from such centers, more crowded residential quarters of the commoners were organized about more modest sacred places. Beyond the urban perimeter, the landscape dissolved into less structured villages and hamlets, surrounded by market gardens.

Missionaries and the Colonial government sought to transform such indigenous built environments into something culturally familiar. The ideological markers of the old religion were built over by Christian churches and shrines, and the former symbols of secular authority were replaced by a Spanish government palace or *casa real*. Some towns, originally located in defensive settings, were rebuilt on the plain, and late in the sixteenth century populations were congregated into larger towns or new places as the demographic collapse set in train by recurrent epidemics made countless settlements unviable. The built environment and the wider cultural landscape were indeed modified, but with the exception of Mexico City and Puebla, Renaissance ideals did not permeate the indigenous built environment.

Closer examination of the metamorphosis of the city of Tlaxcala (1537–1562) shows that while the indigenous elite took pride in the splendid cityscape that emerged, there was a protracted debate over the desirability of rebuilding, with the indigenous government strongly resisting further congregation after 1560. The central grid at its margins dissolved into an indigenous landscape that recreated the past, so much so as to frustrate the Colonial authorities. But in the end, the segregation and acculturation of the indigenous elite undermined the cohesion and authority of the community. That is the basic, historical text, which can be interpreted as a narrative of resistance. But there is an equally important implication that may not be obvious, namely that the Tlaxcalans could and did work within an imposed Colonial framework that seemed reasonably familiar to them.

At the institutional level, the Spanish model of elected executive officers and town council had sufficient analogues with traditional, participatory government that this deliberative institution continued to function effectively, even under a different guise. The built environment of the new city functioned much like the old, despite some structural changes, to channel everyday activities, commercial interchange, and ritual behavior. The *casa real*, as the symbol of secular authority, was built in Spanish style and represented Spain, but it was occupied by the indigenous governor and his family. The new churches that replaced the old temples were built by Tlaxcalans as a community effort, and thus were indigenous, even though the officiating clergy was Spanish. The *hermitas* that replaced the rural shrines of the old religion were frequented and decorated by indigenous devotees, and sometimes housed a native holy person. In effect, the Tlaxcalans were able to adapt to their new circumstances because they indigenized the structural or functional components and, behind the screen of their own language, they interpreted their meaning and role in their own terms, which often diverged substantially from what the Spaniards had intended.

For an imposed, foreign power, the built environment could indeed be most effectively used as an instrument of acculturation when the structural and functional differences between indigenous and alien urbanism were so small that the natives could make ready mental transpositions between old and new component parts or dynamics. That is why the Colonial grid town ultimately did prove to be of importance, as both an institution and an environment. Since its role and function were readily intelligible, its images and their underlying ideology would become increasingly familiar with time, thus contributing in a cumulative manner to sociocultural transformation.

DIFFUSION OR RESISTANCE TO INNOVATION?

Given the rich database of the texts and *pinturas* it has been possible to focus on analytical categories, based on construction technologies, rather than descriptive ones such as house "types." Nine such categories plus one composite are examined and presented in Table 1.1, which is organized from the perspective of information exchange. The implications are anything but simple, with a great deal more similarity between available Old and New World technologies than is generally assumed, and with "innovations" as likely to be rejected as accepted.

The indigenous peoples did not adopt any new construction methods from the Old World, but their production of wood shingles, adobe bricks and cut stone was eventually facilitated by Spanish iron tools and by increasing demand in the Spanish sector of the Colonial economy. Important Spanish construction methods such as rammed earth, kiln bricks and roof tiles were also rejected by indigenous craftsmen in their role as workmen for the Spaniards. Rammed earth failed, and high-temperature kiln products only gained some importance during the seventeenth century, as the brick and tile operations of the Puebla area became increasingly indigenized.

The Spaniards, fundamentally dependent on indigenous labor, not only gave up on some of their mainstream technologies but also soon adopted flat-roof construction in most urban contexts, as well as adobe brick in working-class barrios or provincial towns. Even thatched roofs and wattle and daub were adopted in smaller towns or rural contexts of the outback.

Table 1.1 Information Exchange on Construction Technologies in New Spain, 1520–1620

Technology	*Familiar to I?*	*Familiar to S?*	*Changes Documented*
Wattle-and-daub	Yes, common	Yes, but local and marginal	Decline among I
Thatched roofs	Yes, common	Yes, but local or marginal	Little change among I
Wood shingles	Yes, but local	Yes, but uncommon	Increasing popularity (I, S)
Flat roofs	Yes, common	Yes, but not primary	Widely adopted by S
Form-box adobes	Probably yes	Yes, in some regions only	Partly adopted by S, expanded use by I
Rammed earth	No (New Spain)	Yes, common	Rejected by I and failed
Kiln bricks	No	Yes, common	Dispersal delayed (I, S)
Roof tiles	No	Yes, common	Dispersal delayed (I, S)
Cut stone	Yes, elite	Yes, elite	Expanded use by S
Geometric towns	Yes	Yes	Expansion with imposition of S model

I = indigenous peoples; S = Spanish population.

In other words, there was no straightforward, unidirectional movement of innovations in house construction from the Old World to the New. The Spanish settlers were far more open to new ideas than were the indigenous people. The latter were willing to adapt new tools that facilitated standard operations, which over time began to modify basic construction methods and forms. Indigenous stonecutters rapidly learned to imitate Renaissance stonework and ornamentation, even while imbuing it with stylistic elements and symbols of their own cultural heritage in the facades of churches and secular buildings. Finally, it is apparent that domestic architecture in Mexico continued to evolve across the centuries, until new forms emerged that were rooted in both cultures.

A narrow focus on a highly selected number of questions about historical innovation diffusion can lead to considerable misunderstanding. For one, by limiting the discussion to a few traits, the distinct impression is given that ideas were simply transferred from the Old to the New World. That may not be an author's intent, but it is what the uncritical reader may well infer. Second, the premises of the diffusionist hypotheses offered, and the structure of the supporting arguments used, all imply that advanced technologies were European, and that New World peoples were incapable of mastering technical problems that had been resolved in the ancient Near East before the invention of iron tools. That is a blatant ethnocentrism, however unintended. Third, while innovation diffusion is an important variable in cultural change, it is probably true that most new ideas in the cross-cultural domain were greatly modified or even rejected. But diffusionists tend to sidestep or ignore such outcomes, although they are arguably far more challenging.

While ideas do travel through time and space, they are culturally and historically anchored. The notion that the Iberian Peninsula was a conduit through which adobe bricks and flat roofs moved on to the New World does a great disservice to the settlement archaeology, history, and geography of Iberia. Its settlement patterns have a time depth of five millennia and are grounded in complex ecological adaptations, ethnic continuity and discontinuity, and historical experience. Some diffusionists see innovations as unique and never consider ecological convergence, a basic biological principle in considering geographical versus adaptive radiation. A key aspect of house construction is the convenience and specific tractability of available raw materials. There are only a finite number of ways that soil, rock and plant materials can be incorporated into making a house, whether round or rectangular, and they have all been tried on every continent at some time or other, regardless of stylistic and ideological details. Those ecologically based, evolutionary parallels deserve to be explicitly considered, as a sine qua non for any cultural interpretation. Diffusion studies must be based on properly informed historical perspectives, drawn from multidisciplinary sources, and studied within a sophisticated and flexible theoretical framework.

BEYOND DIFFUSION: CAN MATERIAL CULTURE SERVE AS A METANARRATIVE?

In the world of postmodernity, material culture has become unfashionable as a subject of study. Stripped of its historical and cultural context, material culture may indeed be reduced to a nonproductive alpha-taxonomy. But it can also serve as a medium in which to examine underlying discursive relations among individuals, groups, and larger institutions that form the basis for the channeling and transference of power. In this concluding section we show that domestic architecture can be used as a metanarrative situating knowledge in a productive way to approach the study of changing social relations and the transfer of power in early, post-Conquest Mexico.

The subject of domestic architecture in the texts and *pinturas* is embedded within a number of different discourses—between different indigenous classes, between different indigenous ethnic groups, and between the indigenous peoples and the Spaniards. These discourses were set in train by the Conquest, which shattered the pre-Hispanic political system, threatened the normative social order, and marginalized traditional ethnic differences. It also created a powerful, new polarization between the Mesoamerican and the European that played itself out in every arena of human interaction from the institutional to the cosmological. These several discourses marked the dismantling of the old sociopolitical order and the incremental transfer of power to a new Conquest society.

Indigenous Class Distinction

The text and illustrations of the Florentine Codex (Sahagún 1963) use quality and type of house as a direct correlate of social status. Just as the palace proclaims the residence of the ruler, various kinds of houses define the socioeconomic rank of their elite residents and distinguish them from commoners, who could only use single-story dwellings. But a number of the "captions" are either switched or incorrect, for example, identifying elite houses as hovels or small pigsties. The younger artist-scribes, who added the illustrations at some point after the text had been prepared by older, elite informants, either failed to understand this traditional method of class differentiation or, less likely, deliberately subverted it. There was a generational disparity that implies the old, hierarchical social structure was breaking down by the 1570s, through either the ignorance or disapproval of artist-scribes, who had grown up in a Colonial world and had never experienced the traditional order.

The *relaciones* (Acuña 1984–88) do little to illuminate this particular discourse, since descriptions of housing were almost always supplied by Spanish respondents. But many of these do note the striking difference between elite and commoner houses, suggesting that, as outsiders, Spaniards were to some degree aware of the class cleavage among the indigenous people. Several *relaciones* claim that the quality of indigenous housing had improved after conversion, but the only two indigenous reports instead extoll the quality and splendor of pre-Hispanic architecture in Texcoco and Tlaxcala.

Indigenous Social Evolution and Ethnic Superiority

Both the *pinturas* and various texts identify a model of social evolution that is directly tied to house types. Primitive foragers once lived in caves or small grass huts, but early farmers lived in wattle-and-daub houses, and civilized craftsmen resided in urban adobe structures while the elite lived in palaces (Mota y Escobar 1940: 33). In the Aztec oral histories, first tapped during the 1530s by Motolinía (1989) and recorded during the 1560s and 1570s by Sahagún (1963) and Durán (1967), only the Aztecs achieved this higher level of sociocultural evolution. Such "earlier" stages are explicitly linked to other ethnographic groups, namely Chichimecs living in huts and Otomí or Purépecha in modest rural houses. Among the Mixtecs and Purépecha there was a similar but less explicit tradition, whereby the primeval population is shown as living in round huts. The indigenous historian of Tlaxcala, Muñoz Camargo (Acuña 1984, vol. 4), offers a biased ethnography of the Otomí and Purépecha, saying that they used thatched "huts," but contradicts himself by elsewhere claiming that Otomí houses were very large and occupied by extended families. Both here and in a Spanish report from western Mexico, the image of a group of men and women sleeping together on the floor of a common room carries a subtle allusion to promiscuity in pre-Hispanic times.

Domestic architecture was a central part of the indigenous discourse about class, ethnicity, and social evolution. Even more surprising is that class distinctions and internal ethnic stereotyping remained a key part of the dialectical tensions within the indigenous society two generations after the Conquest, even under the burden of an increasingly disruptive Colonial hegemony. That supports the contention of Lockhart (1992), based on an extensive study of other archival sources, that rivalries between indigenous communities or other microethnicities during the first century of subjugation overshadowed the dialectic between Mesoamerican and Spaniard.

Spanish Stereotyping of the Indigenous World

There are more than a few examples in the *relaciones* of negative Spanish stereotyping of indigenous housing as small, low, primitive, flimsy, smoke-filled, unusual (*curioso*), or imitative. The most prejudicial, however, is the imaginary conversation between three Spaniards walking through Mexico City in 1554, as constructed by Cervantes de Salazar (1964: 61). Upon leaving the Spanish quarters, one explains:

> "Here we come to the hovels (*casuchas*) of the Indians that are so humble they scarcely rise above the ground. . . ."

Comment: "They are arranged without order."
Reply: "That is an ancient custom of theirs."

Indigenous Tenochtitlan was neither a large slum nor "disorderly." But Cervantes de Salazar was a confirmed elitist, enamored with the rigid enforcement of identical house elevation or Renaissance ornamentation within the Spanish *traza*. Like many other Spaniards, he probably was unable to grasp the explicit orderliness of another culture. All this gives the impression that, in the Spanish discourse, indigenous housing served as a prominent marker for ethnic circumscription, both in segregating the cultural landscape and in shaping attitudes to the other.

Unlike the wonderment of Cortés (1985) and Diaz del Castillo (1984) in 1519 at the splendid indigenous towns and structures, the second generation of secular Spaniards was somehow blind to indigenous talents and accomplishments. They did not recognize that they themselves were actively adopting indigenous technologies and house types. Cervantes de Salazar (1964) expresses the irony that Spaniards, who in Spain lived in houses with tile roofs, in Mexico City had houses with flat roofs. But he did not recognize that this was the indigenous way of living, claiming instead that the Greeks had invented flat-roof construction, which was currently gaining favor in southern Italy. Similarly one searches in vain for a secular, as opposed to a missionary, source acknowledging that the imposing monastery churches or the fine architecture of Mexico City were the product of indigenous labor, skill, and talent. The educated tour guide of Cervantes de Salazar (1964: 34, 51–52, 71–72) remarks on exquisite woodwork or cut stone, but the single reference to craftsmen implies a Spanish craft guild. The Spanish discourse on this aspect of material culture focused on what it deemed to be negative, and simply omitted what might have cast indigenous accomplishments in a positive light.

Projecting the Policy of "Civilized Behavior"

The most explicit Spanish discourse centered around *policía* or civilized behavior. Part of this is expressed in various decrees about the founding of new towns, eventually codified in the comprehensive ordinances for settlement. Such towns were to be modules for a Mediterranean-style social ambience in which Spanish settlers could enjoy all the opportunities and amenities of a familiar urban environment, but within an alien countryside.

The other part of the equation is to be found in the substantial but scattered literature advocating resettlement or *congregación* of the indigenous population. The new indigenous towns were to be equivalent structures, but with a different didactic twist. Here the imposed, built environment and its social institutions were to play the role of "civilizing" and acculturating the natives, so that they might share and profit from the same orderly temporal and spiritual living as their Spanish counterparts (Licate 1981). Whether pragmatic or utopianist, many voices pleaded, argued, or decreed congregation as indispensable for religious and secular administration. We interpret this multivocal Spanish "preaching" of *policía* as another distinctive discourse, incorporating not only urban form and dynamics, but also the imposition of Spanish or Italian architectural forms.

Domestic architecture and the built environment were an integral, rather than peripheral, part of the Spanish discourse in regard to the indigenous people. That discourse was marked by a dialectical tension between the mendicant orders, the Colonial government, and the settlers, with significant underlying differences as to goals and methods. Those tensions led to periodic shifts of both policy and praxis, but the basic discourse was about affecting change and imposing a new social order. It was a paternalistic discourse, directed at subject peoples widely deemed to be inferior, and did not seek to engage in an exchange of views.

An Indigenous Discourse on Colonization

In domestic architecture there is little written material to illuminate the tension between the native population and the conquerors, as expressed by indigenous voices. For this we must turn to those *pinturas* that show settled landscapes in which natives do battle with Spaniards, where Spanish cattle and *vaqueros* fill the voids between towns, or where a symbolic foreign magistrate sits like an observer in the corner of a map. More directly still, there are many examples of indigenous towns symbolically represented, in European style, by a church, suggesting a passive acceptance of cooptation.

But a much larger perspective emerges clearly from Table 1.1. The indigenous people selectively adapted facilitating technologies, but did not accept new construction forms. They continued to live in their traditional dwellings. It was the Spaniards who abandoned or shelved their mainstream construction techniques and increasingly used modified forms of indigenous houses, built by indigenous workers with mainly traditional technologies. This is a classic example of "resistance"; that is, a subconscious unwillingness to change or be changed. Nonetheless, over time the cultural distance between Conqueror and Conquered was reduced, and each became more like the other. With this transculturation came changes in the architectural sphere that took shape only during the eighteenth century.

There was then an indigenous discourse on the Conquest and its aftermath, manifest in the medium of domestic architecture. Its voices remain largely mute, but it had a significant impact on the Colonial transformation. The Spaniards refused to hear those once audible voices. Little did they anticipate how powerful that discourse of resistance would be.

NOTES

1. The raw materials for the empirical structure include the great body of comparative information recorded by Spanish observers in the *relaciones geográficas* (geographic rela-

tions) of New Spain, compiled in 1579–85 (Acuña 1984–88: vols. 2–9). The corpus of pictorial maps of the period before 1620, which present the indigenous perspective, is diverse; these include those of the *relaciones* (Acuña 1984–88; Mundy 1996), some 450 indigenous litigation maps (Catálogo 1979; Gruzinski 1987; Leibsohn 1996; Butzer 1998), and other indigenous maps known as *lienzos* or *códices* (Glass 1964; Glass and Robertson 1975; Alcalá 1980). The text and illustrations of Sahagún's Florentine Codex, compiled from indigenous informants during the 1570s (Sahagún 1963; Klor de Alva 1988; Peterson 1988), present both a foil and a source. The early chronicles of the Conquest (Cortés 1985; Díaz del Castillo 1984) provide an essential datum, while the indigenous historian of Tlaxcala, Diego Muñoz Camargo (Acuña 1984: vol. 4) illuminates many subtleties. Beyond the Eurocentric architectural history of Kubler (1948), modern vernacular architecture is treated in volumes such as Moya Rubio (1982) and McHenry (1989), pre-Hispanic archaeological evidence by Evans (1991) and Smith (1992). Islamic architecture in Spain and Morocco has received specific attention (Mikesell 1961; Butzer, Butzer, and Mateu 1986; Delaigue 1988, 1990). Forced indigenous resettlement continues to be examined (Gerhard 1977; Licate 1981; Cordova and Parsons 1997), as does the grid-plan town (Morse 1987; Tóvar 1987), with a fresh appreciation emerging for the geometry of pre-Hispanic urban plans (Tichy 1974, 1991; Broda, Carrasco, and Moctecuma 1987; Tyrakowski 1989). The rebuilding and subsequent remodeling of Tlaxcala has a growing literature (Meade 1986; Lockhart, Berdan and Anderson 1986). Original Spanish texts were invariably used, in conjunction with Molina's (1944) Spanish–Nahuatl dictionary of 1571, and Lockhart's (1992) analysis of Nahuatl linguistic accommodation to Spanish.

REFERENCES

Acuña, René, ed. 1984–88. *Relaciones geográficas del siglo XVI, 9* vols. Mexico City: Universidad Nacional Autónoma de México. [complied 1579–85]

Adams, Robert McC. 1981. *Heartland of Cities: Surveys of Ancient Settlement and Land Use on the Central Floodplains of the Euphrates.* Chicago: University of Chicago Press.

Adorno, Rolena. 1994. "The Indigenous Ethnographer: The 'indio ladino' as Historian and Cultural Mediation." In *Implicit Understandings*, ed. Stuart B. Schwartz, 378–402. Cambridge: Cambridge University Press.

Alcalá, Jerónimo de. 1980/(circa 1541). *La relación de Michoacan*, ed. Francisco Miranda. Morelia: Fimax.

Andrien, Kenneth J., and Rolena Adorno, eds. 1991. *Transatlantic Encounters: Europeans and Andeans in the Sixteenth Century.* Berkeley: University of California Press.

Broda, Johanna, David Carrasco, and Eduardo Matos Moctecuma. 1987. *The Great Temple of Tenochtitlan: Center and Periphery in the Aztec World.* Berkeley: University of California Press.

Butzer, Karl W. 1992. "Spanish Conquest Society in the New World: Ecological Readaptation and Cultural Transformation." In *Person, Place and Thing: Interpretative and Empirical Essays in Cultural Geography*, ed. Shue Tuck Wong, 211–44. Baton Rouge: Louisiana State University.

———. 1995. "Biological Transfer, Agricultural Change, and Environmental Implications of 1492." In *International Germplasm Transfer—Past and Present*, ed. R. R. Duncan, 3–29. Madison, WI: Crop Science Society of America.

———. 1998. "Cartographies of Colonial Mexico: A Dialogue Between Spanish and Indigenous Map-Makers." Alexander von Humboldt Lecture, University of California at Los Angeles.

Butzer, Karl W., Elisabeth K. Butzer, and Juan F. Mateu. 1986. "Medieval Muslim Communities of the Sierra de Espadán, Kingdom of Valencia." *Viator: Journal of Medieval and Renaissance Studies* 17: 339–413.

Butzer, Karl W., Juan F. Mateu, Elisabeth K. Butzer, and Pavel Kraus. 1985. "Irrigation Agrosystems in Eastern Spain: Roman or Islamic Origins?" *Annals of the Association of American Geographers* 75: 479–509.

Catálogo. 1979. *Catálogo de Ilustraciones*. Mexico City: Universidad Nacional Autónoma de Mexico.

Cervantes de Salazar, Francisco. 1964. *México en 1554,* 3rd ed., trans. Julio Jiménez Ruido. Mexico City: Universidad Nacional Autónoma de Mexico.

Cordova, Carlos E., and Jeffrey R. Parsons. 1997. "Geoarchaeology of an Aztec Dispersed Village on the Texcoco Piedmont of Central Mexico." *Geoarchaeology* 12: 177–210.

Cortés, Hernán. 1985. *Cartas de relación*, 2nd ed. Mexico City: Editores Mexicanos Unidos.

Delaigue, Marie-Christine. 1988. *Capileira, Village Andalous: Un habitat montagnard à toits plats.* Oxford: British Archaeological Reports, International Series 466.

———. 1990. "Deux exemples d'habitat rural en Andalousie Orientale: Apprôche ethnoarchéologique." In *La casa hispano-musulmana: Aportaciones de la arqueología*, 21–45. Granada: Publicaciones del Patronato de la Alhambra y Generalife, Coloquio de la Casa de Velázquez.

Díaz del Castillo, Bernal. 1984/1564. *Historia verdadera de la conquista de la Nueva España*, 2 vols., ed. Miguel León-Portilla. Madrid: Historia 16.

Durán, Diego. 1967/1579–81. *Historia de las Indias de Nueva España de tierra firma,* 2 vols., ed. Angel María Garibay. Mexico City: Porrúna Hermanos.

Evans, Susan T. 1991. "Architecture and Authority in an Aztec Village: Form and Function in Tecpan." In *Land and Politics in the Valley of Mexico*, ed. Herbert E. Harvey, 63–97. Albuquerque: University of New Mexico Press.

Foster, George. 1960. *Culture and Conquest: America's Spanish Heritage.* Chicago: University of Chicago Press.

Gerhard, Peter. 1977. "Congregaciones de indios en la Nueva España antes de 1570." *Historia Mexicana* 26: 347–95.

Glass, John B. 1964. *Catálogo de la colección de Códices*. Mexico City: Museo Nacional de Antropología, Instituto Nacional de Antropología e Historia.

Glass, John B., and Donald Robertson. 1975. "A Census of Native Middle American Pictorial Manuscripts." In *Handbook of Middle American Indians*, ed. Robert Wauchope, vol. 14, 81–252 plus 103 plates. Austin: University of Texas Press.

Gruzinski, Serge. 1987. "Colonial Indian Maps in Sixteenth-Century Mexico: An Essay in Mixed Cartography." *Res: Anthropology and Aesthetics* 13: 47–62.

Jordan, Terry G., and Matti Kaups. 1989. *The American Backwoods Frontier: An Ethnic and Ecological Interpretation*. Baltimore: Johns Hopkins University Press.

Jordan, Terry G., Jon T. Kilpinen, and Charles F. Gritzner. 1996. *The Mountain West: Interpreting the Folk Landscape*. Baltimore: Johns Hopkins University Press.

Klor de Alva, J. Jorge. 1988. "Sahagún and the Birth of Modern Ethnography: Representing, Confessing, and Inscribing the Native Other." In *The Work of Bernardino de Sahagún: Pioneer Ethnographer of Sixteenth Century Aztec Mexico*, ed. J. J. Klor de Alva, H. B. Nicholson, and E. Quiñones Keber, 31–52. Austin: University of Texas Press.

Kubler, George. 1948. *Mexican Architecture of the Sixteenth Century*, 2 vols. New Haven: Yale University Press.

Leibsohn, Dana. 1996. "Mapping Metaphors: Figuring the Ground in Sixteenth-Century New Spain." *Journal of Medieval and Early Modern Studies* 26: 499–523.

Licate, Jack A. 1981. *Creation of a Mexican Landscape: Territorial Organization and Settlement in the Eastern Puebla Basin, 1520–1605*. Chicago: University of Chicago, Department of Geography, Research Paper No. 201.

Lockhart, James. 1992. *The Nahuas After the Conquest: A Social and Cultural History of the Indians of Central Mexico, Sixteenth through Eighteenth Centuries*. Stanford: Stanford University Press.

Lockhart, James, Frances Berdan, and Arthur J. O. Anderson, eds. 1986. *The Tlaxcalan Actos: A Compendium of the Records of the Cabildo of Tlaxcala (1545–1627).* Salt Lake City: University of Utah Press.

McHenry, Paul Graham, Jr., 1989. *Adobe and Rammed Earth Buildings: Design and Construction*. Tucson: University of Arizona Press.

Meade de Angulo, Mercedes. 1986. "Fundación de la ciudad de Tlaxcala." In *Historia y Sociedad en Tlaxcala*, 42–47. Tlaxcala: Gobierno del Estado de Tlaxcala, Memorias del Ier. Simposio Internacional de Investigaciones socio-históricas sobre Tlaxcala.

Mikesell, Marvin W. 1960. "Comparative Studies in Frontier History." *Annals of the Association of American Geographers* 50: 62–74.

———. 1961. *Northern Morocco: A Cultural Geography*. Berkeley: University of California Press, Publications in Geography 14.

Molina, Fray Alonso de. 1944/1571. *Vocabulario en lengua castellana y mexicana*, facsimile edition. Madrid: Ediciones Cultura Hispánica, Colección de Incunables Americanos 4.

Morse, Richard M. 1987. "Urban development." In *Colonial Spanish America*, ed. Leslie Bethell, 165–202. Cambridge: Cambridge University Press.

Mota y Escobar, Alonso de la. 1940/circa 1605. *Descripción geográfica de los reinos de Nueva Galicia, Nueva Vizcaya y Nuevo León*, 2nd ed., with an introduction by J. Ramírez Cabañas. Mexico City: Editorial Pedro Robredo.

Motolinía, Fray Toribio de. 1989/circa 1541. *El Libro perdido: Ensayo de reconstrucción de la obra histórica extravieda de Fray Toribio*, ed. Edmundo O'Gorman. Mexico City: Consejo Nacional para la Cultura y las Artes.

Moya Rubio, Victor J. 1982. *La Vivienda indígena de México y del mundo*. Mexico City: Universidad Nacional Autónoma de México.

Mundy, Barbara E. 1996. *The Mapping of New Spain: Indigenous Cartography and the Maps of the Relaciones Geográficas*. Chicago: University of Chicago Press.

Peterson, Jeanette Favrot. 1988. "The *Florentine Codex* Imagery and the Colonial *tlacuilo*." In *The Work of Bernardino de Sahagún*, ed. J. J. Klor de Alva, H. B. Nicholson, and E. Quiñones Keber, 273–93. Austin: University of Texas Press.

Sahagún, Fray Bernadino de. 1963. *General History of the Things of New Spain*, trans. Arthur J. O. Anderson and Charles E. Dibble. Salt Lake City: University of Utah Press. [compiled in the 1570s]

Smith, Michael E. 1992. *Archaeological Research at Aztec-Period Rural Sites in Morelos, Mexico*. Pittsburgh: University of Pittsburgh Memoirs in Latin American Archaeology 4, vol. 1.

Tichy, Franz. 1974. "Deuting von Orts- und Flurnetzen im Hochland von Mexiko als kulturreligiöse Reliktformen altindianischer Besiedlung." *Erdkunde* 28: 194–207.

———. 1991. *Die geordnete Welt indianischer Völker*. Wiesbaden: F. Steiner, Das Mexiko-Projekt der deutschen Forschungsgemeinschaft.

Tóvar de Teresa, Guillermo. 1987. *La Ciudad de México y la utopia en el siglo XVI*. Mexico City: Seguros de México.

Tyrakowski, Konrad. 1989. "Autochthone regelmässige Netze vorspanischer Siedlungen im Mexikanischen Hochland." *Geographische Zeitschrift* 77: 107–23.

Watson, Andrew M. 1983. *Agricultural Innovation in the Early Islamic World: The Diffusion of Crops and Farming Techniques, 700–1100*. Cambridge: Cambridge University Press.

West, Robert C. 1948. *Cultural Geography of the Modern Tarascan Area*. Washington, DC: Smithsonian Institution, Institute of Social Anthropology Publication 7.

———. 1975. "The Flat-Roofed Folk Dwelling in Rural Mexico." In *Man and Cultural Heritage*, ed. H. J. Walker and W. G. Haag, 111–32. Baton Rouge: Louisiana State University.

West, Robert C., and John P. Augelli. 1989. *Middle America: Its Lands and Peoples*, 3rd ed. Englewood Cliffs, NJ: Prentice Hall.

Winberry, John J. 1974. "The Log House in Mexico." *Annals of the Association of American Geographers* 64: 54–59.

———. 1975. "Tejamanil: The Origin of the Shake Roof in Mexico." *Proceedings, Association of American Geographers* 7: 288–93.

2

The Clash of Utopias: Sisterdale and the Six-Sided Struggle for the Texas Hill Country

Michael P. Conzen

> Those were . . . very good times for the energetic Germans in Texas, whether they held slaves or not; for the hunter far up in the hills, where no cow ventured, for the German freeholder on the frontier, for the well-to-do German citizens of San Antonio and New Braunfels, for the Hanseatic trading barons and shopkeepers on the coast; "that was a happy time for Texas under old Uncle Sam, under slavery."—Edgar von Westphalen (1887: 210)[1]

Thus did Karl Marx's brother-in-law sum up the would-be utopian quality of life in west-central Texas before the Civil War for the ethnic group to which he belonged—as filtered through his own experiences centered on Sisterdale, a little farming community tucked away in the Texas Hill Country. His nostalgia had some basis, although he was more bystander than contributor to the social well-being of the community in which he resided, being, as one Marx family chronicler has noted, irresolute and impressionable.[2] Nevertheless, Edgar von Westphalen captured at least in part the German optimism for a better life which had once held sway in west-central Texas during the fateful decade of the 1850s.

Sisterdale and the surrounding Hill Country—viewed widely today as the very epitome of German Texas—were the loci of multiple embryonic utopias, and few locales exist anywhere in the United States where more settlement ideologies competed for expression on the ground at the same time. These ideologies were not rooted simply in ethnic difference, but attracted followers from multiple groups. Within the span of a mere decade, the Hill Country was fought over for possession by six distinct social ideologies, or worldviews, as represented by pre-European indigenes, small-scale communists, humanistic "Latin Farmers," schismatic Mormons, cotton-based slaveholders, and capitalist yeoman farmers.[3] No one in the early 1850s could safely have predicted which ideology would prevail and which would fail. Hence, the region offers a tempting case study in the historical and geographical contingencies that shaped the territorial contours of such a contest and its ultimate outcome. This chapter illuminates the complex geographical circumstances underlying this unusual, multicultural clash of utopias. Some emphasis will be given the Latin farmers, because theirs was in some ways the most subtle and the least materialistic vision, and it played itself out more successfully here than anywhere else in Texas. The interaction between the competing groups and their ideas of livelihood, and its progressive outcomes, fall into three phases distinct in their shifting dynamics.

APACHERÍA AND COMANCHERÍA PRIOR TO 1845

For more than two centuries prior to the 1840s the Hill Country provided migratory range for several Indian peoples. Wichita and Waco Indians moved over territory north of the Colorado and east of the San Saba rivers. Over the southern Hill Country Lipan Apache bands held sway, except when Penateka ("Honey-Eater") Comanches pushed south from the "Comanchería" plains of north Texas, which they and other Comanche bands controlled. All these nomadic peoples followed the buffalo, living in tent lodges which they moved periodically to keep close to the migrating herds. Expert horsemanship had given the Comanches general superiority over their neighbors, long permitting them to penetrate Apache regions almost at will. Since the advent of the Spanish, these different peoples had maintained variable relations with each other and Europeans, alternately at war or peace with some neighbor or other. Successive European authorities, Spanish, Mexican, and Republic of Texas, sought peaceful coexistence with the Indians, and, while white settlements remained few, minuscule, and far between, this worked, albeit fitfully (Kavanagh 1996: 279–91; Schilz 1987: 46–49; Richardson 1933: 106–07, 262–63). But the mounting intrusion of white colonists into Apachería and Comanchería from the 1830s on rendered such policy untenable, and peace treaties became fewer and shorter-lived. San Antonio had long provided a focus for both Indian raids and peace councils. The Hill Country to the north, of no special interest to Europeans before the American annexation of Texas in 1845, reposed in virtual oblivion. The territory served simply as the adjunct to hunting grounds of the major tribes who circulated widely around and through it. As a distinct ecosystem between the high plains of the Llano Estacado and the coastal plains of Texas, it provided dietary and medicinal supplements in the form of roots and small game (Bollaert 1956). It was part of the Indians' de facto estate. They sought no more than to keep it that way.

EARLY EUROPEAN COLONIES: SWEAT AND FROLIC, 1845–1854

The Yeoman Vision

Since 1824 Mexico had welcomed immigration, and Texas soon received an infusion of settlers from the United States, a population shift that would by 1836 produce an independent, English-speaking state. By the 1830s, some Germans were finding their way to the coastal plain, but for the Hill Country it was not until the organized colonization efforts of the "Verein zum Schutze Deutscher Einwanderer in Texas,"[4] beginning in 1844, that European settlers penetrated deep into the interior (Jordan 1966: 41–47). The Verein, led by German noblemen keen to do social good by easing Germany's population problem, purchased the Fisher-Miller Grant, situated north and west of the Llano River, and planned to settle thousands of German colonists there. The grant turned out to be an arid grassland expanse remote from market connections. The group bought additional land for town sites at intermediate points between the coast and the grant to ease access and create way stations for commerce. Thus New Braunfels (1845) was founded at the base of the Balcones Escarpment which defines the Hill Country's southern and eastern margin, and also Fredericksburg (1846), in the Pedernales Valley. A 146-mile wagon road led from the little port of Indianola on the coast to New Braunfels, whence a trail continued another 78 miles to Fredericksburg and then to the edge of the grant at the Llano River. Much of this northern link used the old Pinta Trail northward from San Antonio into the high country. One of the rivers it forded is the Guadalupe, a spot where the rural community of Sisterdale would shortly develop (Figure 2.1).

The Verein's grandiose colonization effort fell quick prey to financial and logistical problems, having shipped too many immigrants with too little support. Heroic exertions by a replacement commissioner, John O. Meusebach, turned a certain fiasco into moderate triumph, and eventually New Braunfels and Fredericksburg survived and prospered. In 1847, Meusebach concluded a crucial treaty with the Comanches which secured relative peace for the Hill Country Germans during the critical early years of colonization. But the ultimate price paid was the realization that the interior grant was not colonizable then, so the colonists settled in the valleys between the grant and the coastal plain.

The impulse for this organized colonization was aristocratic, but its premise was egalitarian and individualistic. Immigrant families were to be helped with the short-lived migration and settling process, but the society they were to create in Texas was to be capitalist and family-based. The migrants included landless peasants and small farmers, artisans, small businessmen and some professionals (such as apothecaries and doctors). The lure of Texas was a new start in life for individuals and families and expanded material possibilities. This movement sought to establish a new yeomanry in the uplands of Texas, where cheap land, healthy climate, and German industry would combine to produce a model society based on personal responsibility and ambition. After a calamitous start, the effort took root. Families gained a foothold in the region, and through dogged effort expanded their domains, eventually settling their children on additional farms (Figure 2.2). New Braunfels and Fredericksburg also advanced, adding services as more tradesmen and artisans accumulated (Biesele 1930: 161–77).

Figure 2.1 Texas Hill Country, 1845–1850: Colonies Engage the Indigenes.

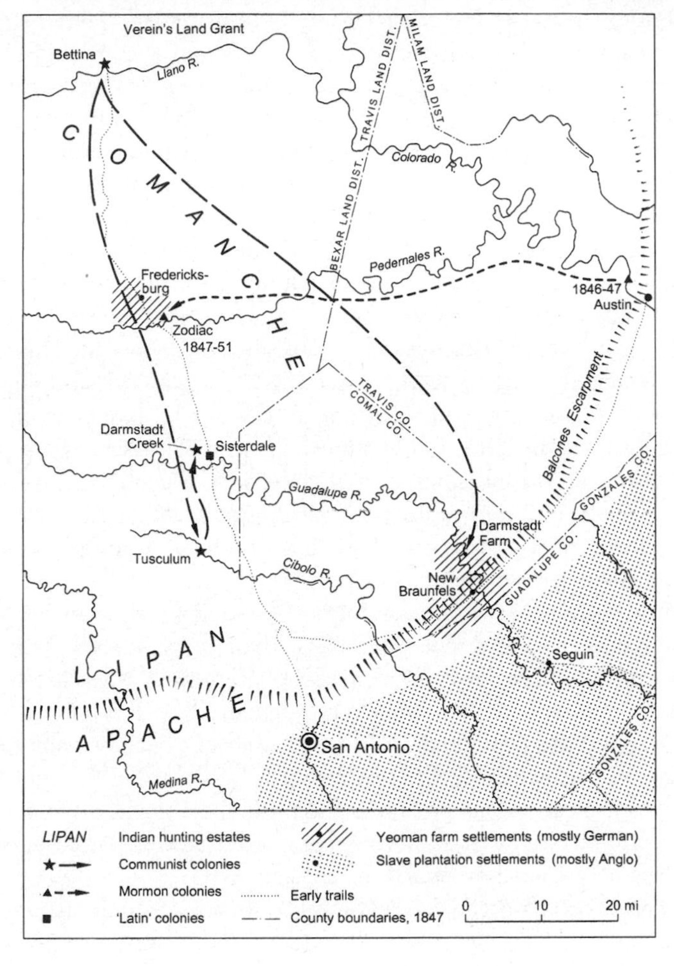

The Communal Vision

While the Verein's activities framed colonization of the Hill Country in the mid-1840s, settlement also proceeded under other auspices. Although the Verein's settlers occupied the land on the way to the grant rather than within the grant itself, one group did indeed reach just inside it, occupying land on the northern bank of the Llano River. The "Vierziger" (the Forty), actually a group of thirty-six young men who gathered in Darmstadt, Germany, settled on the Verein's grant to build a communist paradise (Heinemann 1994). They were idealistic, seeking to create a model community out on the Texas frontier for the world to admire and emulate. They had some practical leadership and wrested excellent

Figure 2.2 Texas Hill Country, 1852–1858: Heyday of the Latin Farmers.

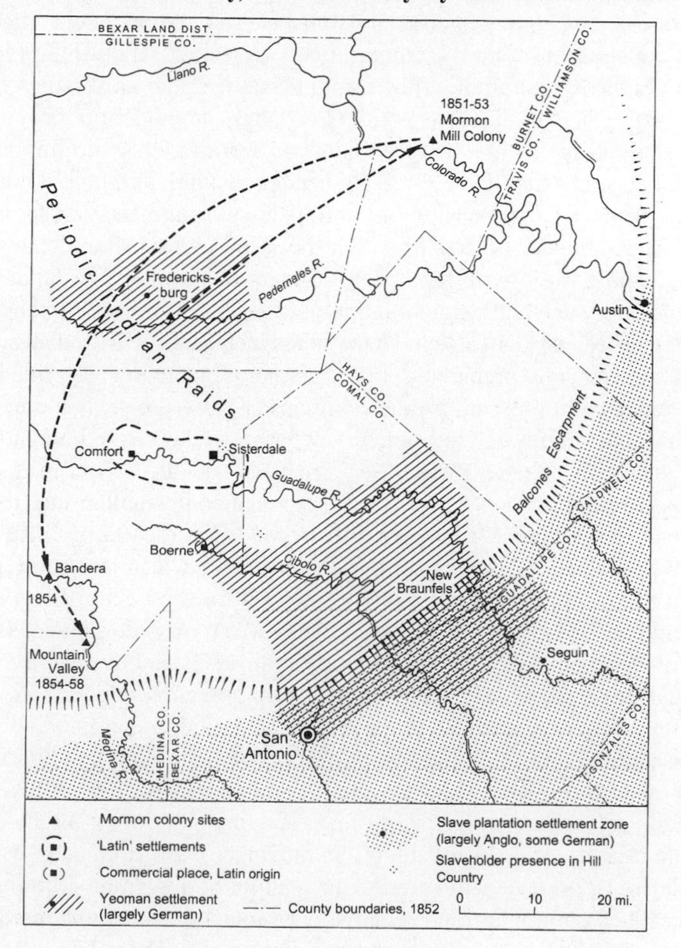

conditions from the Verein as inducements to settle within the grant, mainly food supplies until the first harvest and farming equipment. The group espoused communal principles of living, based on brotherly love, adding only one woman to the settlement, who seems to have done the cooking. This was hardly the sort of socialist experiment Karl Marx envisioned, though his brother-in-law later claimed to have been briefly a participant. The location was the most isolated of any European or American settlement in Texas at that time, precociously situated within the southern margins of Comanchería.

The group took up residence on a farm it named Bettina (Figure 2.1) in honor of Bettina von Armin, Goethe's friend, on October 1, 1847, and proceeded to practice its principles, which included avoidance of any decisions not reached communally and voluntarily. By the following August, this voluntarism, combined with creeping sloth and class divisions latent in the members' social backgrounds, conspired to rend the community's cohesion. About half came from middle- or higher-class families, having studied at German universities, with theoretical training in such fields as political economy, architecture, law, medicine, and forestry management. The other half had working-class origins and knew about cabinetry, blacksmithing, brewing, saddle making, and butchering. As the first group displayed a propensity for hunting, composing poetry under trees, and formulating community objectives which the second group should carry out, especially anything to do with the actual business of farming, the latter failed to see the justice of such a division of labor, and dissension grew (Reinhardt 1899). The colony lasted less than a year, as members increasingly drifted away, disillusioned and presumably prepared to melt back into regular society. Internal combustion, not external pressure from the Indians or anyone else, had crushed it.

If the Bettina Colony had presumed too heavily on positive human nature to function effectively, the chimera of a perfectable socialist world in this part of Texas was not extinguished. Fifteen diehards retreated in small groups to two less isolated environs (Figure 2.1). A handful regrouped at the Darmstadt Farm outside New Braunfels, a Verein facility developed as an insurance policy against failure on the grant, though not initially intended for the Vierziger. Six or eight others resettled in 1849 on the upper Cibolo River on a communal farm they named Tusculum, aiming to learn from recent mistakes and determined to persevere (Ostermeyer 1850: 100–01, 106–07; Heinemann 1994: 337; Haas 1968: 250). Tusculum lasted perhaps two years, but again individual interests drew the members apart. Three Tusculumites moved to the Sisterdale neighborhood and started a joint farm in Darmstadter Creek (Esselborn 1937: 129–30). Eventually these efforts petered out, too small to make any dramatic mark on the course of settlement at large. As Prince Paul Wilhelm, Duke of Württemberg, a visitor to Sisterdale in 1855, noted in relation to Bettina and similar colonies across America: "Such communist and socialist experiments, which give material form to European fantasies, normally founder in the New World and sink through the requirements of colonization" (Prinz Paul Wilhelm 1986: 285).

The Aesthetic Vision

Another strand woven into the Hill Country settlement fabric of this period was the extraordinary flowering of an intellectual community centered at Sisterdale (Figure 2.1). Its origins were thoroughly hybrid, but its *Weltanschauung* became well defined, comparatively durable, and briefly but, given its size, stunningly prominent in Texas politics. It was begun by Baron Ottmar von Behr, a cultivated German with private means and son of a onetime prime minister of Anhalt-Köthen. Moving to America where utopian dreams seemed more feasible, Behr toured the country helping found socialist organizations from Philadelphia to

Missouri (*Der Volks-Tribun* Jan. 24, 1846: 4; Petermann 1988). He visited Texas perhaps as early as 1846, and in February 1848 bought a large tract of land at Sisterdale on which he established a sort of communist colony, although, as one participant recalled wryly, while the labor was communal, the ownership was certainly not (*Der Volks-Tribun*, April 25, 1846: 3; Anonymous 1848). Behr's farm became less known for its communism than for Behr's "hospitality," and while he continued to support socialist causes in Germany and America, he must have been influenced in the management of his own affairs by the quick failure of communist colonies he had known at first hand, such as New Helvetia in Missouri and the Bettina Colony just fifty miles north of him. Several refugees from Bettina "decompressed" for a while as his guests (Douai 1876; Fischer 1980: 74, 82).

For nearly a decade, Behr's farm became, until his untimely death in late 1855, the nucleus of a growing community of similarly educated political theorists and aesthetes, would-be farmers almost to a man without practical experience. He worked hard to attract "quality" settlers to become his neighbors, some from his home region in Germany, others he met arriving in New Braunfels (Dresel 1943: 904). Many were "forty-eighters," liberal-minded, often idealistic refugees from Germany's failed social revolutions of 1848. This strategy was so successful that in five years, by the early 1850s, Sisterdale boasted six barons (including Edgar von Westphalen), nineteen settlers with university educations (including three Ph.D.s), a medical doctor, a professor, an architect, a vintner, a mining engineer, an ex-customs official, a lawyer, a merchant, a squire–politician, a foundry manager, a family of industrial chemists, and a former military officer.

Geographically, Sisterdale (Figure 2.2) consisted of about twenty to thirty farms belonging to these men and their families, distributed within a radius of about six miles around the ford that carried the Pinta Trail over the unnavigable Guadalupe River. Economically, Sisterdale developed as a community raising wheat, corn, cotton, tobacco, grapes, cattle, and sheep, much of the time on a subsistence basis with occasional surpluses to export. There was no commercial store, but Behr officially distributed mail and dispensed local justice. Socially, the majority of the families had no prior experience with agriculture or ranching, but plenty with music, literature, philosophy, politics, and learned discourse (Siemering 1878). Visitors on occasion witnessed Cicero being read out loud to the birds in a field from a book held in one hand, while the other guided the plow behind the ox team, giving name to such agriculturalists as "Latin farmers." One new neighbor wrote to relatives in Germany in early 1850: "Of books and educational aids we have many; if I show interest in cattle and corn and fence rails and such, I also do not forget the spirit that lies in material things, as cultural matériel. In short, I dedicate myself with body and soul to my new profession, and, as one needs to work on the whole at most three months of the year, enough spare time will remain to work on the intellectual development of my children" (Kapp 1936: 37). Remarkably, Behr's recruiting success, combined perhaps with the glowing reputation of the district for its health and beauty (reminiscent of the Rhineland for some), resulted in the largest concentration in Texas of well-educated, well-financed, and public-spirited settlers outside any city. Even San Antonio in this period was hard-pressed to match Sisterdale's cultured assembly.

If Sisterdale's Latin farmers had a unitary social vision, its focus centered explicitly on the brotherhood of man, the cultivation of learning for self-improvement, love of the arts, political engagement, and vigilance against denials of personal freedom, which encompassed both slavery and temperance. What further distinguished the community from others in the region was the time and energy given to these interests. "A library of the ancient and modern classics was to be found in almost every house and the latest products of literature were eagerly discussed at the weekly meetings of these gentlemen farmers at the school house. It sometimes occurred at these meetings that Comanches stood listening gravely at the open door, while one of the Latin farmers was lecturing on the socialist theories of St. Simon or Fourier" (Tiling 1913: 123).

The years 1850 to 1854 represented the "blossom time" of Sisterdale as a cultural community. Newcomers of like persuasion gathered in the district, which consequently expanded westward where land was still available. A singing society was formed, which participated in statewide choral festivals once or twice a year. Behr recorded daily meteorological observations, a practice carried on by his neighbor, Professor Ernst Kapp, who was later recruited to the Smithsonian Institution's first national network of weather observers. Kapp established a water-cure spa at his farm, which was advertised all over west-central Texas (Siemering 1878: 59; Smithsonian Institution Archives 1849–1875; *San Antonio Zeitung*, Aug. 20, 1853). In 1853, Sisterdalers organized the "Freie Verein" (Free Society), a political club with the goal of debating the social issues of the day and working for the full expression of German opinion in state politics. This club was to ignite a political firestorm across Texas, as slavery became the defining public issue of the times. Ten miles up the river, a new town, Comfort, was established in 1854, which as a stronghold of freethinkers and abolitionists—not normal for Texas then or now—quickly took on a character as Sisterdale's daughter colony (Figure 2.2; Ransleben 1974).

The philosophical rationalism and political radicalism of the upper Guadalupe Valley found varying echoes in the San Antonio, New Braunfels, and Fredericksburg triangle. San Antonio, the old Spanish capital of Texas, was a thoroughly Mexican place in the 1850s and looked it. But it was steadily acquiring Anglo and German residents, mostly merchants, artisans, and professional people. While the Americans commandeered the political power, the Germans set about building institutions serving their interests, such as the Casino Club. Sisterdale residents frequented the city and the club, and many travelers to San Antonio with scientific and sociological interests were directed to visit Sisterdale.[5] There can be little doubt that the opinions of Sisterdalers were closely followed in the German salons and taverns of San Antonio before the Civil War. Similarly, leaders in

Fredericksburg and New Braunfels had frequent contacts with men from Sisterdale, if for no other reason than that the small number of educated people in all these neighboring places formed a loose network for social and political purposes, lubricated by the towns' newspapers. The *San Antonio Zeitung* and the *Neu Braunfelser Zeitung* circulated throughout the Hill Country, and Sisterdalers contributed to both.

The Theocratic Vision

Lyman Wight, a former member of the Council of Twelve Apostles of the Mormon Church, led a band of 150 followers into the easternmost Hill Country in the spring of 1846, settling outside Austin (Figure 2.1; Bitton 1969). They had been on a journey from Wisconsin for two years, where they had been lumbering, because Wight had refused to accept Brigham Young's leadership of the church following Joseph Smith's murder in 1844. Wight saw in Texas a field for proselytizing among the Indians and perhaps the slaveholders of the South. While Mormon politics surrounding this schismatic group held national significance, the group's appearance in the Hill Country added a further competitive vision regarding the region's future character. Essentially cut off from the momentous events unfolding for the Mormons in the upper Mississippi Valley, Wight's group pursued its holy grail for over a decade within the Texas Hill Country.

The group built a gristmill on the Colorado River, which proved beneficial to the residents of Austin. When the water supply turned unpredictable, Wight searched for a new site, and the following year relocated his colony to the Pedernales River four miles southeast of Fredericksburg and sixteen miles north of Sisterdale. This spot, which he named Zodiac (Figure 2.1), held them for three years. Here they built another mill, a meetinghouse, and thirty-five houses, and raised corn and cattle. Here the colony gained a more solid foothold in the region, absorbed a small contingent from another wandering Mormon group, and made a few converts, though not among the Indians. With a temple built in 1849, they practiced ritual foot washing, body washing, annointings, baptism for the dead, and some polygamy. "Soon," as one chronicler has written, "they could start sending missionaries throughout Texas, to other parts of the South, to neighboring Indian tribes, and to Mexico, South America, and the Pacific islands. Then, as in the past, the Saints would start to gather" (Bitton 1969: 19). Such a grandiose vision for Zodiac as the center of a revived hemispheric Mormon host in the Hill Country did not come to pass. In 1850, high water swept away their mill dam, if not their dreams, and they moved once again. The next venue was Mormon Mill Colony on Hamilton Creek, but only months later the colony moved to Bandera (1854), and finally the same year to Mountain Valley on the Medina River (Figure 2.2).

The Mormon colony in the Hill Country seemed throughout most of its existence barely above the threshold for survival. It provided milling services to neighboring districts and thereby created a minute niche within the region's economy. It favored living in the frontier zone, partly to be close to the "lost tribes of Israel," as it regarded the Indians, and partly from desire to distance itself from slaveholders. While the group's geographical strategy, if risky, was at least easy enough to carry out, its demography certainly failed it, and its theology proved of little interest among Hill Country folk, Indian or European.

The Slaveholding Planter Vision

Slavery as a modern system was imported to Texas by Americans moving west from Louisiana and Arkansas. Whether profitable or not—and much of the time it was—slavery came shackled to cotton farming and other forms of Southern agriculture. Americans settled the coastal plain of Texas by moving up the major riverways from the Gulf of Mexico or west from the Sabine River along the immigrant road through Nacogdoches. Farms gradually multiplied in the 1830s and 1840s in the large zone covered by the Austin and DeWitt grants, bringing cotton culture and slave labor. By 1850, Austin County contained about 276 Southern white farmers, about 60 percent of whom owned slaves, and 101 German farmers of whom only 10 owned slaves. Ten years later, the proportions of slaveholders had diverged in the two groups: nearly three-quarters of the American farmers held slaves, whereas less than one in twenty Germans did (Jordan 1966: 61). The tendency for Germans to hold slaves was uniformly far less than their Southern neighbors throughout the Texas cotton belt, but this should not be explained by moral claims. Class, rather than ethnicity, among whites accounted for attitudes toward slavery. Many artisans and bourgeois Germans, transferring from a Germany in which the Industrial Revolution was thoroughly redefining the social order to a Texas where they hoped to recreate their erstwhile comfortable positions, were glad to avail themselves of the amenities of slavery (Küffner 1994: 110–13). It was this more complex social relationship which cotton culture and its supporting institutions were pushing northward and westward up the lower Colorado and Guadalupe rivers into the Hill Country during the 1840s and 1850s.

As settlers filed into the Hill Country counties during the 1850s, few had slaves, regardless of background, perhaps more through lack of capital than principled opposition (Campbell 1989: 56–57; Heinen 1985). In large part it was because the Germans in this stream inclined toward mixed farming and not cotton monoculture, and the Anglos who also began filtering in tended to favor ranching, neither of which demanded heavy investment in slave labor. Nevertheless the Peculiar Institution was gaining a foothold in the Hill Country. Gillespie County tax rolls in 1855 recorded 63 slaves, those of Burnet County 150 slaves, Comal County 126 slaves, and Hays County (largely non-German) 517 slaves. If we add one-quarter of Bexar County's slave count for that year (its territory straddles hill country and piedmont), giving 244 slaves, and likewise one-quarter of Travis

County (517 slaves), then the Hill Country as a region likely contained in excess of 1,600 slaves (Campbell 1989: 264–65).

If anything can be said about the balance among the "utopias" competing for territorial representation in the Hill Country up to the mid-1850s, it is that most had ample room to take root and to flourish without necessarily tripping over each other. Without question, the indigenous Apaches and Comanches were in the weakest position, having everything to lose and the disability of small numbers, susceptibility to European diseases and drink, and cultural values most removed from those of the remainder. Nevertheless, they did not concede ground without a fight, and through frontal assault or guerrilla action imposed on the American "civilization" the necessity of advancing its settlement frontier only through the agency of an ongoing, costly military operation.

Of the Euro-American social systems introduced to the Hill Country in the late 1840s, only the communitarian vision had disappeared completely by the mid-1850s. It is tempting to suggest that ideology and human nature were to blame. The social theory behind the Bettina experiment was crude, incomplete, romantic, and naive. It made no place for class conflict, and arguably this was its undoing. Certainly the pre-Marxian communists on the Hill Country frontier had no oppressive government to blame for their dissipation, for they had ample room, physically, for maneuver. Classically, the veterans of Bettina, Tusculum, and Darmstadter Creek chose to experiment on the very edge of their world, rather than in its heart (as Marx urged), and consequently their failures went largely unnoticed and unmourned by the mass of population struggling with change.

STRESS AND FRACTURE, 1854–1865

The Hill Country passed a quiet milestone in the mid-1850s. From "wild frontier" before, it passed into provisionally "settled" status thereafter. Germans had moved into the region with and without the Verein's help, and slowly their numbers grew and a generalized German culture solidified (Jordan 1991; Lich 1976). But Anglo-Texans also sought toeholds in the inviting hills, and everywhere English-speaking Americans showed up to represent government institutions and to conduct business. Over time, the region functioned far more as a zone to procure raw materials than as an odd refuge for self-sufficient idealist experiments. As a regional component of a modernizing polity seeking to find its locus within the American commonwealth, the Hill Country could no more escape the growing forces of political orthodoxy over slavery in Texas than it could free itself geographically from its embedded position within the heart of the state. Such conditions sharply colored the competition in the Hill Country over alternative utopias in the years leading up to the Civil War.

Communism Dissipated and Theocracy Contained

If the "true socialists" had by 1850 drifted back to mainstream settlements, it would not be long before the Mormon colony in Texas, too, would fold its tent, for it suffered from similar structural—and visionary—problems.

Wight's colony, unable to make substantial converts by force of example, was left to drift until internal pressures broke it up. By 1854, the group numbered about sixteen families, and the leader was now fifty-eight years old. Defections had occurred, some in answer to Wight's authoritarian ways. One group left in 1855 or 1856, reducing the colony by more than forty souls. In 1858 Wight prophesied a civil war and urged his now pathetic band to travel again, which they did, only to see their leader die a few miles outside San Antonio. His followers carried his body to Zodiac, where they buried it, and promptly dispersed, some finding their way to Utah, others to other parts of the state (Bitton 1969: 25). "No colonists in Texas were ever more thoroughly under the domination of one man," wrote one historian. "[Wight's] colony was a 'common stock' proposition, in which he was the absolute dictator" (Banks 1945: 243). Arguably, the Mormons lasted longer in the Hill Country than the communists because a strong leader exerted control over the collective, something a theocracy can offer more simply than a democracy.

Slavocracy Ascendant and Ethnocide Rampant

It is a commonplace that the number of slaves in Texas increased markedly during the 1850s and through the early years of the Civil War. Using the same areal definition for the Hill Country in 1864, slaves just exceeded 2,800 (compared with 1,600 in 1855). The slave counts for individual counties show significant variations (Figure 2.3). Kendall County contained 89 slaves in 1864, while neighboring Bandera County enumerated only 14. Without accounting for slaveholding at the individual farm level by ethnic background for the whole region, generalizations about the reach of slavery in the Hill Country must be cautious. However, there is ample evidence that the slavocracy, which had taken control of most of settled Texas by the outbreak of the war, showed a conspicuous lapse in the Hill Country. Here was an island of overt support for the continuation of the American Union. Nineteen Texas counties voted against secession in 1861, six of them in the Hill Country, with Mason and Gillespie returning majorities of over 95 percent for the Union. Blanco County, which at that time included the German-heavy part that the following year would split off as Kendall County, recorded a two-thirds majority for the Union. Had the future Kendall area voted separately, it would undoubtedly have produced as similarly lopsided a vote as did neighboring Gillespie County. In fact, the county split came precisely to give Germans in the Guadalupe and Cibolo valleys more political voice (Zelade 1983: 165). Sisterdale, in the heart of Kendall County, is widely recognized as the epicenter of abolitionist sentiment in Texas German areas (Campbell 1989: 215–16).

Figure 2.3 Texas Hill Country, 1864–1870: Triumph of Free Individualism.

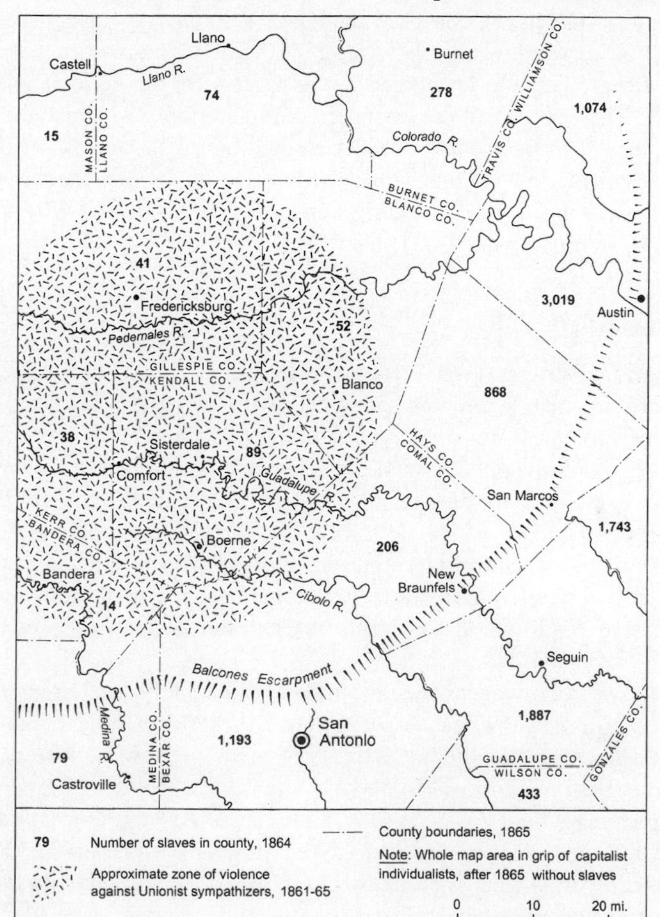

Slaveholding expanded areally through Texas between 1855 and 1864 and intensified in established areas. Even in localities such as the Hill Country where cotton was hardly grown, mostly without slave labor and largely by Germans, slaveholding as a domestic fact made quiet headway. Little wonder that the general political climate became progressively less tolerant of divergent opinions. In essence, proslavery sentiment became so pervasive that it could rule statewide public discourse, and "infect" areas, such as the Hill Country, where it was not welcome, with its presumptive elevation of any dissent to the level of treason. Geographically, slavery ruled the Hill Country through fear of the consequences of dissent, even though, materially, it did not fully possess the region. As a legal vision of society in Texas, and under the urgency of war, slavery could even nudge reluctant citizens into slave ownership as a defense against charges of disloyalty. At its worst, slavery begat thuggery, violence, and ethnocide, as Hill Country Germans came to be singled out as targets for hate crimes and general lawlessness. Their speech, dress, manners, and geographical concentration in a distinctive terrain made them especially vulnerable to the excess zeal of state military authorities and vigilante bands. Many Germans lost their lives in the worst periods of anti-German hysteria between 1862 to 1864 (Williams 1907: 336–37, 340; Duaine 1966: 30–31; Herff 1973: 57).

Dead Poets Society

Perhaps the shrill struggle over loyalty to the slavery cause across Texas would not have been driven to fever pitch had the strongest-willed abolitionists in Sisterdale held their tongues. But seven years before the outbreak of a civil war they could scarcely have imagined possible, it seemed more important for sincere men to speak out in the hope of influencing the general course of events for the better. Here was a consequence of the advanced character of Sisterdale as a "Latin" settlement: rationalist sensibilities spurred by idealism encouraged a leadership role in promoting political goals based on humanitarian principles, when, in fact, the surrounding Anglo world was in no mood to debate the merits of the abolitionist case.

The origins of controversy were obscure enough. In July 1851, a group of men formed "The German Political Association of West Texas." The signatories to the declaration were generally liberal thinkers, including several doctors and merchants, and at least two former members of the Bettina Colony (*Allgemeine Auswanderungszeitung*, Sept. 6, 1851: 419). Little emerged from this action, but two years later, following the first German-Texan Singing Festival in New Braunfels, a tavern meeting established a pan-Texas umbrella organization to coordinate the activities of all local German cultural organizations, and in particular to lobby against any state legislation that might outlaw the teaching of German in the public schools. Ottmar Behr chaired the meeting and was named to the Committee of Correspondence. Just one month later, in November 1853, the "Freie Verein" was organized at Sisterdale, with Ernst Kapp as president and August Siemering, a private tutor and schoolteacher brought to Sisterdale by a prominent resident, Eduard Degener, as secretary (*Neu Braunfelser Zeitung*, Oct. 21, 1853: 4; Biesele 1930: 196–97). The first public action came in December when the group petitioned the Texas legislature not to pass any bill modeled on the so-called Maine Liquor Law.[6]

The following March the Sisterdale Verein blindly lit the fuse to a political time bomb. The society issued a statewide call to sister organizations to send delegates to a German political convention to be held after the Singing Festival in San Antonio in May. On the agenda would be debate leading to a unified political stance in the upcoming presidential election so that German sentiment would be taken seriously in Texas. The convention was duly held and a platform of ideas approved, which contained a controversial plank. "Slavery is an evil whose ultimate removal is, according to Democratic principles, indispensable; but as it affects only individual States, we demand: That the Federal Government refrain from all interference in affairs of Slavery; but that, when any single State shall resolve on the removal of this evil, the aid of the government may be claimed" (Kapp and Siemering March 24, 1854: 2; *Neu Braunfelser Zeitung*, June 2, 1854: 2). This was widely understood to mean that federal involvement toward the abolition of slavery in states was being advocated. That prior action by a state in this direction was implied before any such involvement could commence was a subtlety immediately lost in the ensuing debate. Even with the subtlety recognized, this was a direct challenge to the states'-rights position of proslavery defenders. The records of the event make clear that Degener and Siemering were critical to the passage of this resolution and dominated much of the debate, and that Kapp did not even attend the meeting. Here was the first sign of a rift in the political front of Sisterdale's intellectuals. But whatever local discord this proposition sowed in Sisterdale, it unleashed a verbal firestorm in the press across the state. The net result was to set most Anglo-Texans against the Germans for meddling in an issue over which they were considered too recent arrivals for their opinions to carry any weight. It also set many Germans against each other, not necessarily over the principle of slavery, but over the wisdom of antagonizing the surrounding Anglo-Texan majority (Buenger 1979).[7]

This event constituted the defining moment in Texan–German political relations in the years leading up to the Civil War. It fixed the potentially false impression in Anglo minds that Germans in Texas could not be trusted to support the Southern cause. As the national conflict grew ever more intense, this mistrust reached deeper into excitable minds. The Hill Country would pay a heavy price for this mistrust (Figure 2.3), and Sisterdale in particular would be ravaged by marauding state militia detachments seeking out German abolitionists, and left in virtual ruins. Most poignant of all, following a deplorable army action in 1862 in which Degener lost two of his sons, he was seized and taken to San Antonio in chains for trial on charges of treason. Found guilty on some of the particulars, he was heavily fined and banned from ever returning to Sisterdale (Fröbel 1890–91, vol. 1: 477–78; Anonymous 1996; Schem 1873). As a Latin settlement, Sisterdale had been losing its luster years before the outbreak of the Civil War. Beginning around 1854, individuals who contributed materially to the social and cultural life of the community began to die, leave for the Far West, or pursue opportunities in the region's growing cities, principally San Antonio.

Yeomanry Questionable

An event as tumultuous as the Civil War naturally disrupted the orderly development of regional society in west-central Texas and exposed many social fault lines. But it cannot obscure the fact that the Hill Country, despite all the stresses and strains appearing between 1854 and 1865, was well adapted to the spread of an essentially flexible and resourceful yeomanry, and that land and stock changed hands at reasonable enough prices to support its vigorous extension as a social system over this kind of country. Slaveholding had made small gains there, but the region would never serve as the locale for vast cotton plantations or any other staple crops. The Hill Country would serve best as a ranching venue, with small pockets of improved pasture on the valley floors here and there. This well suited a yeomanry accustomed to individual operations dispersed over the hills, rather than a plantation or communal system of farming.

What queered the pitch for the rapid growth of successful yeoman settlement in the Hill Country was precisely the tensions and dislocations that led up to and came with the war. Ethnic groups were set against each other, and enmities could hardly subside except with the slow passage of time.

Indians in Retreat

If the Civil War proved an overwhelming distraction for Euro-Americans in west-central Texas, its effect on the region's Indians was no more helpful. It did result in the withdrawal of federal troops from frontier forts and their partial replacement with state troops. This provided opportunity for bolder raids deeper into ceded territory, but the effects, while disruptive to the families directly affected, had little long-term influence on the steady tide of white Euro-American advance (Wilbarger 1967: 660–61). Sisterdale's recorded experience with Indian visitations from 1854 seems fairly typical of a location slowly growing remote for the receding horsemen: September 1854: Kapp's horses taken; May 1855: Sisterdale horses taken; October 1856: Hermann Runge murdered; August 1857: All Sisterdale horses and mules taken; January 1862: Ludwig von Donop murdered. State ranger companies often provided superior defense, but at times the quality of soldiers and their leadership degenerated until their own activities threw more fear into the population being defended than did the Indians (Holden 1928).

AFTERMATH, 1865–1867

With the defeat of Confederate forces in fields far from Texas came the end of slavery. It took an overwhelming outside power to bring down slavery in Texas. Slavery did not exclude yeomen from its own economy, but it so distorted the equation that its removal thoroughly revamped the rural labor market for decades to come. In the Hill Country, former slaves found less opportunity to stay in the area than on the coastal plain, because the sharecropping principle was hardly adaptable to a ranching regime. Suddenly, the yeoman vision could operate without hindrance. Even agricultural mechanization would have a muted effect on the modernization of life under the yeoman principle in the Hill Country.

All the other visions for the Hill Country except slavery petered out quite early in the Euro-American settlement experience. Continued Indian occupance was incompatible with the ambitions and power of white society to use the land more intensively. Not all Europeans and Americans sought to extinguish the Indian populations, but willful settler incursions made peace treaties temporary instruments of accommodation, and so Indians fought white men to save first their domains and then simply their lives. Through American eyes, the Texas Hill Country was far too attractive to bypass and leave to the Indian as a reservation. For this the Germans are responsible, for the settlers only too eagerly gave up the trans-Llano tract obtained by the old Texas Verein for the green swells and comfortable nooks of the Hill Country to the south.

In light of the industrial-strength, class-obsessed Marxian communism that swept large portions of the globe during the twentieth century, the "communist" settlements of the Texas Hill Country seem laughably quaint, minute, and irrelevant. Yet such experiments were precisely those which convinced Marx and his peers to work for mass movements in society and to expose class-based labor exploitation as the key to political awareness. That neither approach has been fully successful is a testament only to human nature. The Bettina experiment probably never would have suceeded, yet remarkably its members seem to have opted in their disillusion for only one of the alternative lifestyles on offer in the Hill Country before the Civil War—that of independent yeomanry. It might seem inappropriate to bracket the Mormon experience in Texas with that of the communists, but the similarities of size, frontier isolation, lack of major infusions of further like-minded people and resources, and restlessness invite such a juxtaposition. Had Lyman Wight's colony made substantial local converts or had other Mormon groups who found their way to Utah joined Wight in Texas instead, the story of his group might have produced a less limpid ending. Certainly, neither group left any legacy in the Hill Country other than a rousing good story to tell and an odd grave marker or two.

And so we return to Sisterdale. One of the fortuitous consequences of Sisterdale's history of intellect is the rich array of recorded thought about the nature of the community sustained there during the 1850s. Sisterdale's later history became prosaic: "In the years following the end of the war, only a few of the first settlers in Sisterdale were still resident, none of whom had most contributed to the fun times, for their farms had passed into other hands," wrote a local chronicler. "The later residents of the charming valley," he goes on to say, "held no singing festivals and produced no 'earthshaking' political movements, but they were capable farmers and created a flowering settlement of well-to-do residents, such as perhaps the old 'Latin farmers' could never have accomplished. The period of the romantics had passed and that of the sustainers, practical though progressive, had begun" (Lafrentz 1929: 29–30).

Perhaps the best way to regard the six "utopias" in the Hill Country in light of their geographical propinquity is to note that no way of living has proven perfect; there are only better and worse alternatives, and when one mode becomes too oppressive, it likely generates at some point a countervailing "corrective." Whether excessive or corrective, each of these Hill Country visions oscillated between arrogance and embarrassment. Even so, the Hill Country saw a very rich panoply of experiments in lifeways play out over its comparatively small territory. In the age of nation-states and modernity, no six societal visions as diverse as those in the Hill Country, however ambitious, could coexist for long. Their variable sustainability, as well as their vulnerabilities were products of well-defined geographical as well as historical contingencies.

NOTES

1. All quotations originally in German were translated by the author.
2. Heinrich Gemkow has just completed the first full-length biography of Edgar von Westphalen, which is being prepared for publication.
3. It might be argued that the term "yeoman farmer" ill fits most American conditions, as it smacks of too much romantic attachment to traditional farming ways. But if it is accepted as a signifier of a relatively independent decision maker operating in the context of largely free market forces, then the term has value in contrast to the other "visions" examined in this chapter.
4. Society for the Protection of Immigrants in Texas.
5. Examples include Heinrich Ostermeyer in 1848, W. Steinert in 1849, John Russell Bartlett in 1850, Friedrich Kapp in 1852, Julius Fröbel in 1853, Frederick Law Olmsted in 1854, Prinz Paul Wilhelm von Württemberg and Jacob Stillman in 1855, Carl Sintenis in 1859, and Adolph von Wrede in 1860.
6. Passed by the Maine legislature in 1851, the law curbed sales of liquor and conferred extensive search and seizure powers upon the authorities. Several other states passed similar laws based on the Maine model as this early prohibition movement spread.
7. The most famous exchanges in the German press occurred between the abolitionist Galveston editor Ferdinand Flake and the accommodationist New Braunfels editor Ferdinand Lindheimer (Buenger 1979). Buenger errs, however, in presuming Kapp was a prominent abolitionist. Local evidence indicates a very different stance.

REFERENCES

Allgemeine Auswanderungszeitung 5 (105) Sept. 6, 1851: 419.

Anonymous. 1848. "An Ottmar Behr in New Braunfels: Wie gehts mit der Kolonie?" *Der Urwähler: Organ des Befreiungs-Bundes* (ed. by Wilhelm Weitling, Berlin) 1 (October): 8.

Anonymous. 1996. "Nueces, Battle of the." In *The New Handbook of Texas,* vol. 4, ed. Ron Tyler, 1054. Austin: Texas State Historical Association.

Banks, C. Stanley. 1945. "The Mormon Migration to Texas." *Southwestern Historical Quarterly* 49 (2): 233–44.

Biesele, Rudolph L. 1930. *The History of the German Settlements in Texas, 1831–1861.* Austin: Boeckmann-Jones.

Bitton, David. 1969. "Mormons in Texas: The Ill-Fated Lyman Wight Colony, 1844–1858." *Arizona and the West* 11 (1): 5–26.

Bollaert, William. 1956. "The Valley of the Guadalupe." In *William Bollaert's Texas*, ed. W. Eugene Hollon and Ruth Lapham Butler, 242–60. Norman: University of Oklahoma Press. [Reprint of 1843 ed.]

Buenger, Walter L. 1979. "Secession and the Texas German Community: Editor Lindheimer vs. Editor Flake." *Southwestern Historical Quarterly* 82 (4): 379–402.

Campbell, Randolph B. 1989. *An Empire for Slavery: The Peculiar Institution in Texas, 1821–1865.* Baton Rouge: Louisiana State University Press.

Der Volks-Tribun (New York; ed. Hermann Kriege) 1 (4) Jan. 24, 1846: 4.

Der Volks-Tribun (New York; ed. Hermann Kriege) 1 (17) April 25, 1846: 3.

Douai, Adolph. 1876. "Eine Communisten-Farm." *Vorwärts: Zentral-Organ der Sozialdemokratie Deutschlands* 21, Nov. 17: 1–2.

Dresel, Julius. 1943. "Diary, 1848–1850." Trans. Johanna Dresel (granddaughter), unpublished manuscript. New Braunfels: Sophienburg Archives.

Duaine, Carl L. 1966. *The Dead Men Wore Boots: An Account of the Texas 32nd Volunteer Cavalry CSA 1862–1865.* Austin: San Felipe Press.

Esselborn, Karl, ed. 1937. *Zwei Erbacher in Texas: Lebenserinnerungen von Ernst Dosch und Eberhard Ihrig.* Darmstadt: Selbstverlag des Herausgebers für die Buchhandlung H. L. Schlapp, Hessische Volksbücher, nos. 95–97.

Fischer, E. G. 1980. *Marxists and Utopias in Texas.* Burnet, TX: Eakin.

Fröbel, Julius. 1890–91. *Ein Lebenslauf: Aufzeichnungen, Erinnerungen und Bekenntnisse,* 2 vols. Stuttgart: Cotta.

Haas, Oscar. 1968. *History of New Braunfels and Comal County, Texas, 1844–1946.* Austin: The Author.

Heinemann, Hartmut. 1994. "Wo der Stern im Blauen Felde eine neue Welt verkündet: Die Auswanderung der Vierziger aus Darmstadt nach Texas im Jahr 1847 und ihre Kommunistische Kolonie Bettina." *Archiv für hessische Geschichte und Altertumskunde* 52: 283–352.

Heinen, Hubert. 1985. "German-Texan Attitudes toward the Civil War." *Yearbook of German-American Studies* 20: 19–32.

Herff, Ferdinand Peter. 1973. *The Doctors Herff: A Three-Generation Memoir.* San Antonio: Trinity University Press.

Holden, W. C. 1928. "Frontier Defense in Texas During the Civil War." *West Texas Historical Association Year Book* 4: 16–31.

Jordan, Gilbert J. 1991. "The German Settlements in the Texas Hill Country." *Journal of the German-Texan Heritage Society* 13 (1): 34–39.

Jordan, Terry G. 1966. *German Seed in Texas Soil: Immigrant Farmers in Nineteenth-Century Texas.* Austin: University of Texas Press.

Kapp, Ernst, Dr. and Mrs. 1936. "Briefe aus der Comalstadt 1850." *Jahrbuch der Neu-Braunfelser Zeitung fuer 1936* 31: 36–38.

Kapp, Ernst, and August Siemering. 1854. "To the Germans in East and West Texas." *Neu Braunfelser Zeitung* 2 (18): March 24: 2.

Kavanagh, Thomas W. 1996. *Comanche Political History: An Ethnohistorical Perspective, 1706–1875.* Lincoln: University of Nebraska Press.

Küffner, Cornelia. 1994. "Texas-Germans' Attitudes toward Slavery: Biedermeier Sentiments and Class-Consciousness in Austin, Colorado and Fayette Counties." Master's thesis, University of Houston.

Lafrentz, L. F. 1929. "Deutsche Ansiedlungen in Comal County nach der Gründung von Neu-Braunfels, mit Benutzung einiger Notizen des Herrn Hermann Seele." *Jahrbuch der Neu-Braunfelser Zeitung fuer 1929* 24: 29–30.

Lich, Glen. 1976. "Germans in the Hill Country: A Pictorial Essay on Immigration in the Nineteenth Century." *Journal of German-American Studies* 11 (3–4): 49–69.

Marx to Engels, May 20, 1865. Letter in *Karl Marx/Frederick Engels: Collected Works,* vol. 42: 160. New York: International Publishers, 1987.

Neu Braunfelser Zeitung 1 (48) Oct. 21, 1853: 4.

Neu Braunfelser Zeitung 2 (28) June 2, 1854: 2.

Ostermeyer, Heinrich. 1850. *Tagebuch einer Reise nach Texas im Jahr 1848–1849.* Biberach: The Author.

Petermann, Gerd Alfred. 1988. "Friends of Light: Friedrich Münch, Paul Follenius, and the Rise of German-American Rationalism on the Missouri Frontier." *Yearbook of German-American Studies* 23: 119–39.

Prinz Paul Wilhelm, Herzog von Württemberg. 1986. *Reisen und Streifzüge in Mexico und Nordamerika, 1849–1856,* ed. Siegfried Augustin. Stuttgart: Thienemann, Edition Erdmann. [reprint of 1856 ed.]

Ransleben, Guido. 1974. *A Hundred Years of Comfort in Texas: A Centennial History,* rev. ed. San Antonio: Naylor.

Reinhardt, Louis. 1899. "The Communistic Colony of Bettina." *Quarterly of the Texas State Historical Association* 3 (1): 33–40.

Richardson, Rupert. 1933. *The Comanche Barrier to South Plains Settlement.* Glendale, CA: Arthur H. Clark.

San Antonio Zeitung 1 (8) Aug. 20, 1853: 3.

Schem, Alexander J. 1873. "Degener, Eduard." In *Deutsch-Amerikanisches Conversations-Lexicon* 3: 573. New York: Commissions-Verlag von E. Steiger.

Schilz, Thomas F. 1987. *Lipan Apaches in Texas.* El Paso: University of Texas at El Paso, Southwestern Studies 83.

Siemering, August. 1878. "Die lateinische Ansiedlung in Texas." *Der Deutsche Pionier* 10 (2): 57–62.

Smithsonian Institution Archives, 1849–1875. Meteorological Project Records, 1849–1875, Record Unit 60, Box 19, Day Book, "Meteorological Instruments, 1851 to 1870."

Tiling, Moritz. 1913. *History of the German Element in Texas from 1820–1850.* Houston: Rein & Sons.

Westphalen, Edgar von. 1887. "Erinnerungen aus Texas." *Das Ausland* 60 (11) March 14: 208–10.

Wilbarger, J. W. 1967. *Indian Depredations in Texas.* Austin: Pemberton Press. [reprint of 1935 ed.]

Williams, R. H. 1907. *With the Border Ruffians: Memories of Far West, 1852–1868.* London: Hazell, Watson & Vinney.

Zelade, Richard. 1983. *Hill Country: Discovering the Secrets of the Texas Hill Country.* Austin: Texas Monthly Press.

3

The Struggle for Urban Public Space: Disposing of the Toronto Waterfront in the Nineteenth Century

Peter G. Goheen

For the sympathetic reader who regards the application of the concept of cultural landscape to a disorganized nineteenth-century waterfront as stretching beyond reasonable limits the spirit of the words contained in *Readings in Cultural Geography* (Wagner and Mikesell 1962), Marvin Mikesell's subsequent ecumenical acknowledgment of the study of "local environments or localized associations of culture" may relieve anxiety (Mikesell 1978: 15). Persuasive contributions have broadened the idea of cultural landscape from its early association with grand synthesis. David Sopher (1979) sought to domesticate the concept in his essay in the influential book *The Interpretation of Ordinary Landscapes*, edited by Don Meinig, which turned our attention to "ordinary landscapes." Peter Jackson (1989) creatively scuffled with traditional cultural geography in a pathbreaking effort to enrich urban and broaden cultural geography. He argued for a materialist approach that would explore everyday practice as a source for understanding the power of symbols and for interpreting ideology. The landscapes to be explored should be regarded as the product of "political and economic forces . . . demonstrating . . . close material links between past and present, private and public, the visionary and the pragmatic . . ." (Jackson 1989: 179).[1] New prescriptions have created plenty of territory on which to debate, but they have also invigorated significant lines of work.[2] Recent works demonstrate the naturalization of the idea of culture with the study of cities and illustrate the comfortable identification of cultural landscapes with ordinary landscapes (Zukin 1995; Groth and Bressi 1997).

This chapter will examine the role of the citizenry in defining and contesting rights to the Toronto waterfront during the second half of the nineteenth century.

This battle for the control and development of public land was waged between powerful private interests and an often inarticulate public. From the perspective of the urban citizen the struggle took form in relation to two crucial elements. First, the participation of the inhabitants defined the process. The slow process by which they became articulate, formulating and exerting their collective interest as a public, influenced all of the players and defined the politics of public space. Second, the private interests—especially the railways—vigorously pursued their interests. Their strategy was highly successful because they were closely attuned to opinion, while assiduously working to manipulate it and to influence decision-making bodies in every way possible.

THE NINETEENTH-CENTURY TORONTO ESPLANADE

> Whereas under and by virtue of the Act sixteenth Victoria, chapter two hundred and nineteen, the Mayor, Aldermen and Commonality of the City of Toronto, have contracted with the Grand Trunk Railway of Canada, for the building and construction of the Esplanade in front of the said City . . . Therefore Her Majesty, by and with the advice and consent of the Legislative Council and Assembly of Canada, enacts as follows:
>
> 1. It shall and may be lawful . . . to locate the roadway of the said Grand Trunk Railway and other Railways to the width of forty feet thereon . . . locating the different lines . . . along the frontage of the said City in such manner . . . as shall be most conducive to the public interests. (Canada, *Statutes*, 20 Victoria, Cap. 80)

The public interest received many salutes in the rhetoric of official proclamations, of which legislation was not the least practiced form, in nineteenth-century Canada. It was invoked as a rationale for many plans: these words from the 1857 bill granting the Grand Trunk Railway access to the Esplanade along Toronto's waterfront were penned to suggest that advancing the purposes of the railway would likewise serve the public. Such formal statements reflect the conviction in nineteenth-century society that the public counted (Ethington 1994; Goheen 1997, 1995, 1994; Ryan 1997). A public, according to Habermas (1989), forms when a significant portion of the population self-consciously asserts its interests which it has come to understand as distinct from those of the state or institutions privileged by the state. The public thereby attains recognition and standing in the political process by which contesting claims are adjudicated.[3]

This chapter investigates the process by which legally designated public space became socially significant public space in nineteenth-century Toronto. It was a contentious process which was joined when a citizenry, increasingly self-conscious of its rights, began to campaign for the protection of its interests against the claims of the mightiest corporations—the railways—in the nation. The public waterfront, known as the Esplanade, became symbolically important as public space, and this change in attitude by the population influenced the nature of the landscape which took shape there.

The earliest designation of the waterfront as public space occurred in 1818 when the Crown declared by Letters Patent that the Esplanade should be reserved "for the use and benefit of the inhabitants . . . as and for a public walk or mall in front of the . . . Town . . ." (quoted in Canada, *Statutes*, 16 Victoria, Cap. 219). The decree may have served the immediate purposes of government, but had little meaning to most of the small number of inhabitants in the still unorganized settlement which was to become Toronto. The inhabitants in 1818, lacking both institutions of local government and any recognized vehicles for corporate action, were without the tools to define their collective will (Firth 1966: lxviii). Toronto's first public space was created by a government eager to dispose of some of the abundant land in its possession (Figure 3.1). By its decree the government nominated several leading citizens who were to hold the Esplanade lands in trust for the use and benefit of the inhabitants. Such an arrangement immediately created the prospect of a conflict of interest: it would work to protect the public interest only if those citizens who were nominated as its guardians were to act to do so. Available evidence does not suggest that they did. Private interests soon enjoyed entrenched privileges on these public lands (Mellen 1974: 5–7). By its action the government both declared the Esplanade to be public space and made possible the usurpation of public lands for private purposes.[4] In this confusion between stated principle and permitted practice were sown the seeds of the subsequent bitter conflict.

Private interests quickly recognized the economic value of the waterfront; in the absence of an effective public they found no obstacles to appropriating much of it for their own purposes. Some clarification of the public nature of the land was provided in 1837 when the government issued Orders in Council granting to the recently incorporated City of Toronto all the water lots in front of the developed portion of the city (Canada, *Statutes*, 16 Victoria, Cap. 219). This grant was confirmed in Letters Patent of Upper Canada issued on February 21, 1840. Under its terms the city was to assure that within a fixed period of time the land would be filled 350 feet into the bay, providing thereby an Esplanade. City Council promptly passed a bylaw to parcel the land into units suitable for leasing. The initial term was for 42 years on these water lots; the requirement was that "the purchaser shall cause to be made or constructed, all the improvements required by and specified in the said patent deed . . . within the time" specified (By-Law No. 43, *By-Laws* 1870). City Council, the guardian of the public interest, encountered no opposition in disposing of leases to private interests. Its attitude, in regarding the waterfront as a resource to be exploited by private enterprise, seemingly reflected community values.[5]

Walter Shanly, chief engineer of the Toronto and Guelph Railway, harbored no doubts as to the value of the waterfront to the success of his business. In May 1852, when he addressed his Board of Directors, Toronto was on the verge of

Figure 3.1 Toronto, 1834. Source: 1834 City of Toronto by H. W. J. Bonnycastle, litho by S. O. Tazewell. Royal Ontario Museum accession no. 955.87.2. Photograph courtesy of the Royal Ontario Museum.

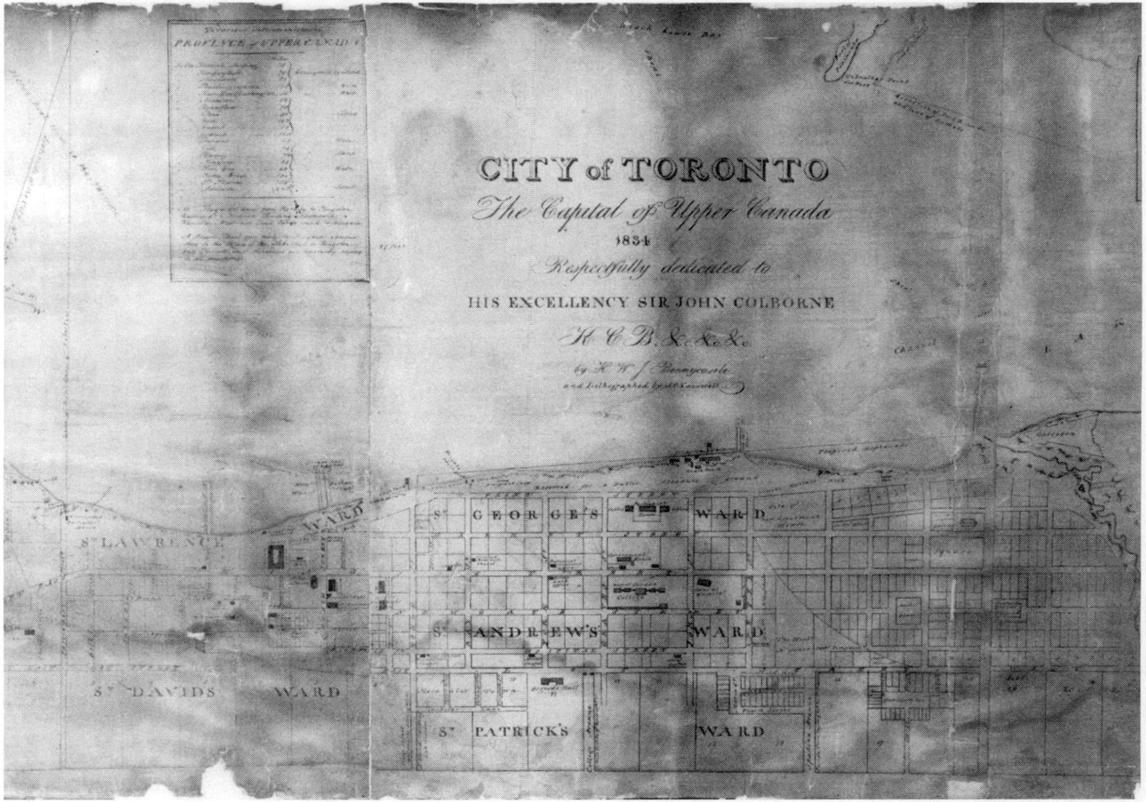

entering the railway age. "The direct connection of the Track with the Lake" was, he wrote, "indispensable." The "railroad terminus should be designed merely as a part of a future Marine Depot of vast extent, taking in, I should say, the whole navigable front of the City" (Toronto and Guelph Railway 1852). He proceeded forthwith to try to realize his plan, with the approval of his board. In his second annual report, in June 1853, he noted his considerable success in negotiating access to the city's waterfront (Toronto and Guelph Railway 1853). Shanly's firm, soon to be incorporated into the Grand Trunk system, and its competitors proceeded to make their claims and execute their designs. A pattern of appropriation of public rights for private interests would be set, with the concurrence of the City Council and senior levels of government.

A CHALLENGE TO THE PUBLIC

Governments—municipal and colonial—negotiated away public rights on the waterfront to private concerns by conflating the one with the other. How did the citizenry of Toronto respond? One commentator approved. Writing in the inaugural volume of the *Canadian Journal* he declared that "the interests of the public and the Railway Companies are one in the most important particulars . . ." (A Member of the Canadian Institute 1853: 16). A wider canvas of opinion from the 1850s can be retrieved through a survey of opinion, official and unofficial, engendered by a major scandal associated with the construction of the Esplanade. The episode attracted the attention of the articulate citizenry.

The issue of the public interest arose quite suddenly when it became urgent to undertake the long-delayed construction of the Esplanade because of the impending arrival of the railways. This was the judgment of the Select Committee of City Council appointed to investigate circumstances surrounding the issuance of the contract for the construction of the Esplanade. It concluded that "the passage of the Esplanade Act of 1853, was undertaken chiefly for the benefit of the Grand Trunk Railway Company . . . rather than for the peculiar benefit of the City of Toronto."[6] Many aspects of the complex scenario—negotiations over the construction contract, volatile relations of the City Council with the railways and the senior levels of government, the abrogation of the contract, and the consequences for the city of breaking the contract—can be quickly passed over here. The public nature of this protracted issue offers an ideal opportunity to examine the formulation and expression of the public interest on a matter which was central to the political agenda and held public attention for several years.

The interests of the public had been acknowledged in one proposal for constructing the Esplanade and accommodating the railways on the waterfront. On behalf of the Ontario, Simcoe, and Huron Railroad Union Company, its chief engineer, Frederick W. Cumberland, proposed a plan for an Esplanade incorporating a plaisance stretching the entire front of the city to be "devoted to pedestrians, and interrupted only at intervals and at right angles by the streets from the north projecting to the wharves." Below this, at water level, would be laid the railway tracks. Cumberland's Esplanade would not be "subservient" to the railways, a situation which he suggested would be "unwise in the last degree. The Esplanade . . . would be ruined . . . the city would be cut off from its water frontage on the bay, except under permanent hazard" (*Daily Leader* July 30, 1853). Cumberland's design received little editorial support from the press. The issue, one editorialist wrote, was for the railways to secure a right of way through the city, whereas Cumberland's plan simultaneously to develop the public space for the public's enjoyment was unnecessary and extravagant (*Daily Leader* August 1, 1853). Another editor supported a plan that would render the Esplanade useless as a promenade. The "very important object" was to bring the railway to the waterfront (*Globe* Oct. 20, 1853).

The contract for the construction of the Esplanade called for the filling of the waterfront fundamentally to accommodate the laying of railway tracks. Council accepted the tender of Messrs Gzowski and Company at its meeting of October 14, 1853, signing the contract on January 4, 1854 (*Leader* Oct. 17, 1853; *Leader* May 3,1854). In May, when construction was getting under way, the leaseholders raised questions of the legality of, and the city's ability to enforce, the contract (*Leader* May 3 and 16, 1854). By autumn yet another troubling element was in play. A bill, a "monstrous proposition" in one editor's opinion, before the legislature called for "the citizens of Toronto . . . to be taxed to any amount which may be named, for indemnifying the very parties to be benefitted by the Esplanade" (*Leader* Oct. 14, 1854). The public responded to this prospect of taxation as they had to no other issue raised by the Esplanade matter. They organized a public meeting for November 6, 1854. The "Esplanade business" was declared to be "one villainous transaction": charges of conflict of interest over the award of the contract were alleged; the rights of leaseholders to indemnification questioned; and the motivation of changes in proposed legislation scrutinized. Two further meetings to discuss taxation were promptly held on November 13 and 14, attracting a broad representation of the people. Significantly, the mayor took the chair for the final public meeting (*Leader* Nov. 8, 10, 14, and 15, 1854).

The problems multiplied until they became a scandal. When the agreed-upon plans were altered to permit the railways to occupy far more of the Esplanade than originally contemplated, one editor declared this deal to be "the most degrading part of the business" (*Leader* Dec. 9, 1854). Soon, when charges of corruption in the awarding of the contract filled the air, a special committee was appointed to investigate the affair. City Council voted on April 16, 1855, to annul the contract with Messrs Gzowski and Company.

The citizens, having evinced little interest in the charges surrounding the Esplanade construction, responded with alacrity when they realized the jeopardy of their private interests. This situation arose out of the stalemate on the Esplanade which followed the cancellation of the contract for its construction. The Grand Trunk Railway, having become frustrated in its efforts to have its tracks laid on the waterfront, threatened to implement a new plan and to place them on Queen Street, one of the city's principal avenues (*Globe* April 25 and June 19, 1855). As everyone knew, it was no idle threat. The Grand Trunk had succeeded in persuading the House of Assembly to pass the necessary enabling legislation. It was, consequently, now "in a position to thwart the wishes of the great majority of the citizens, as to the location of the line . . ." (*Leader* June 13, 1855). The city could not guarantee the protection of its interests: were it to refuse the railway permission for the alternate route, the company could appeal to a sympathetic agency with power to overrule the city. An aroused populace, as was customary when it wished its voice to be heard, organized public meetings. The first of these, held on June 23 in one of the wards most directly affected by the proposal, drew an overflow crowd. Those present expressed "the greatest alarm" over the proposal (*Globe* June 25, 1855).

The interests of the whole city were threatened, as the population well understood. A second meeting, to which inhabitants from throughout the city were invited, was called for the evening of July 3. Its purpose was nothing less than to offer the public the opportunity to advise, or instruct, Council on a course which might avoid the disaster of a railway line on Queen Street—described by the mayor in his introductory address as incurring "the greatest possible injury, not only to those living on that particular street, but to the whole city." The speakers were clear in their opposition to the Grand Trunk's proposal. They declared that Council's duty was to assure that no railway line was constructed on Queen Street. The resolution which was duly passed urged Council to allow the ratepayers themselves to decide the matter. The particular issue was whether the Council's decision to refuse to issue the Grand Trunk a contract for construction of the Esplanade should be overturned. The meeting was attended by a wide representation of the citizenry. Notable public figures, including a member of Parliament, addressed the meeting (*Globe* July 4, 1855). The weight of public opinion, forcefully expressed, had its impact. The City Council reconsidered its course of action, and agreed to proposals that the Grand Trunk be granted a contract to construct the railway track on the Esplanade. The accord ended a concurrent, serious threat to the Esplanade: the railway had proposed that the government revoke the city's rights to the waterfront held under letters patent. The railway and the city agreed that the city was entitled to obtain a patent from the Crown "of all lands covered with water, the City now holds or claims title to, under the license of occupation issued to the City in March, 1853, as soon as the contract shall be executed between the City and the Grand Trunk Railroad Company, of and in respect of the matters now under negociation [*sic*]" (*Globe* Aug. 13, 1855; *Leader* Aug. 13, 1855).

The struggle over building a line for the Grand Trunk through the city had forged a strong public reaction which in turn had influenced the decision making of Council. There were few if any precedents in the previous record of Toronto politics for an articulate public forcing the reconsideration of a decision of City Council. Toronto's citizenry thereby proved it could define and argue effectively for its own interests. But the issue on which the public had spoken was the threat to private property and not the threat to its public lands, the Esplanade.[7] The Esplanade lay open to the railways. By 1858 three railways had laid their tracks on the Esplanade. Shanly's vision was being realized (Figure 3.2).

THE PUBLIC DECLARES ITS INTERESTS IN THE ESPLANADE

Neither the aims nor the strategy of the Canadian Pacific Railway (CPR) during the 1880s as it sought to claim the Esplanade for its railway purposes was very original. The company approached the Toronto waterfront and the officials with whom it must deal in much the same way as had other railways over the past thirty years. What resulted, on the other hand, was unprecedented: the company found its claims contested by an aroused and articulate public. The episode stands apart because of the public response which it engendered. The dispute was unusually bitter and protracted. It provides an ideal opportunity to ask several questions. What was the process by which the public defined and contested its perceived rights to public land? Who was speaking for the public? What were the claims being advanced on its behalf concerning the waterfront? What were the contending claims? What recognition was the public interest accorded, and by whom?

When Hugh Blain of the Citizens' Association declared in February 1890 that the attitude of the railways toward the Toronto waterfront "simply means a declaration of war between the railways and Toronto," he spoke for an aroused and angry public (*Telegram* March 1, 1890). His remarks followed an extraordinary meeting held as part of continuing negotiations between the railways and the city. What was extraordinary was the involvement at this time of the public—through such organizations as the Board of Trade, the Citizens' Association, and the Trades and Labor Council—along with the City Council and Harbor Board (*Globe* March 1, 1890). When, on March 13, the editor of the *Globe* introduced his report of a special meeting of the Citizens' Association with the words: "[T]he citizens' rights are menaced with danger," he acknowledged the public perception that the citizens had a legitimate interest in the disposition of the Toronto waterfront (*Globe* March 13, 1890). The public had, at last, forcefully asserted its interest in the Esplanade and had been recognized. A formerly quiescent public had found a strong and articulate voice and was determined to defend its perceived rights. No agreement could now be concluded with the Canadian Pacific Railway which ignored the public's interest.

The forceful expression of public opinion which marked the early months of 1890 as a special time followed a prolonged period in which the CPR sought to

Figure 3.2 Toronto Waterfront, 1862. Source: Detail. Plan of the City of Toronto showing the government survey by H. J. Browne under direction of J. O. Browne. 1862. Archives of Ontario C-295-1-163-0-29.

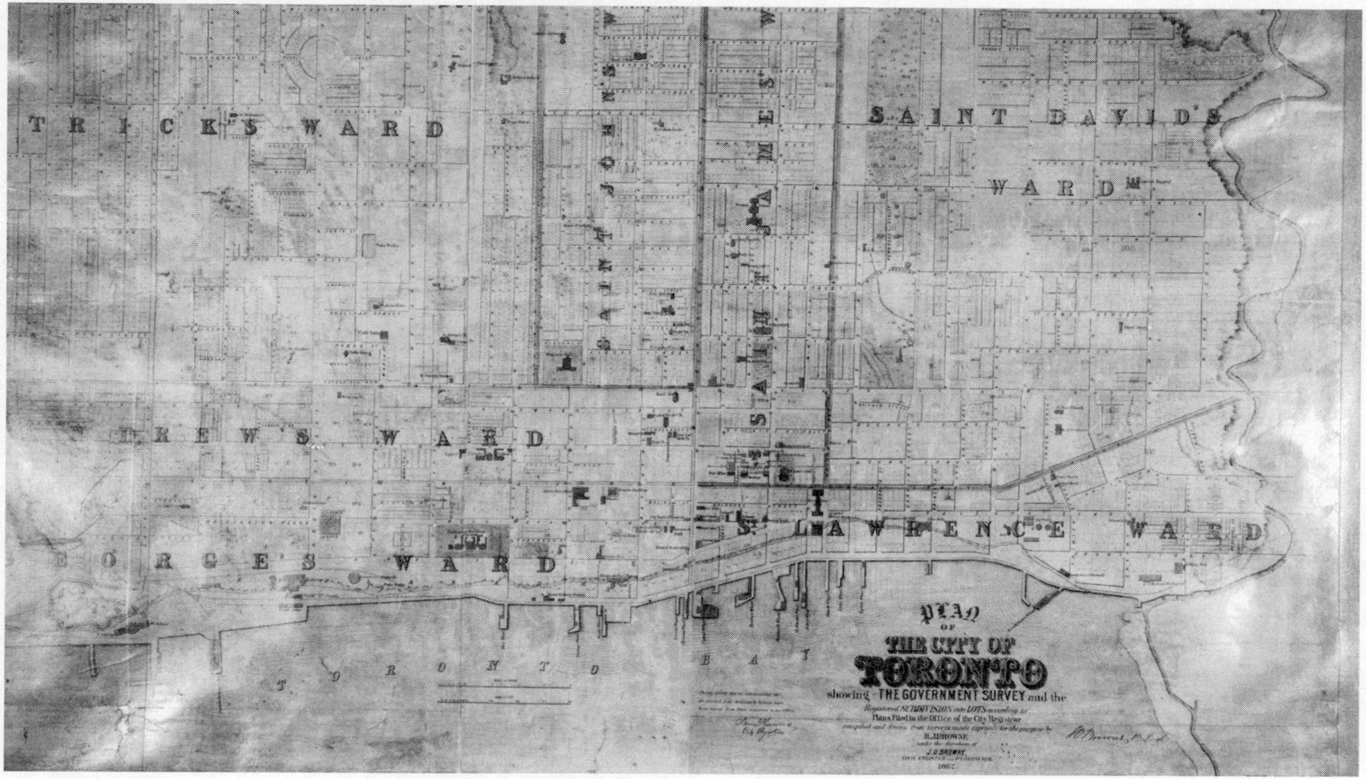

secure the privileges it wanted on the Esplanade without regard for the public interest. An episode in 1886 illustrates the railway's tactics and disregard for the interests of the people. It equally highlights the unconcern for public opinion by City Council and its officials. The railway, through its vice president, W. C. Van Horne, attempted to pressure the city by writing to complain that the accommodation offered his company was insufficient to meet its needs. Council learned of this letter at the same meeting which brought the news that the CPR had succeeded in placing a private bill before the Legislative Assembly of Ontario containing proposals that would restrict the powers of Council to deal with its public lands "except with the agreement and consent of all of the railways companies interested in the Esplanade."[8] At its meeting of April 27, Council received news of the CPR's intention to expropriate the land it wanted on the waterfront.[9] Because railways had the right under law to expropriate lands if necessary to meet their operating requirements this was no idle threat. City Council, facing the necessity of further negotiation, established a special committee to confer with the railways.[10] Council gave no consideration to involving any representatives of the public interest, aside from elected politicians, in these negotiations. Nor were public rights to the Esplanade considered by the Board of Trade when it became party to the negotiations in November 1887. The Board's only objective was "to assist the Mayor and Corporation in their efforts to further the interests of increasing trade and commerce of the City . . . by affording equal accommodation to all railways entering the City."[11]

The signing of an agreement in Montreal on April 26, 1888, between the city and the two principal railways marked the end of this phase of negotiations. The Montreal Agreement, as it came to be known, had been reached behind closed doors in a Montreal hotel. Representatives of the two principal railways, the mayor and several city officials, together with the president of the Board of Trade, conducted the discussions, which, in the mayor's words, were designed to avoid "protracted litigation and conflict" between the city and the railways. The draft agreement contained no reference to the rights of the public on any of the Esplanade lands and even reduced access to the streets by agreeing to close many of them. A proposed new street was to be "subject to such . . . regulations" as the railways dictated.[12] Some members of Council objected to the deal, considering it impossible "to ascertain how much the City is granting to the Railway Companies."[13] Council, promptly served with an injunction restraining its action, could not ratify the agreement.[14]

The Canadian Pacific Railway openly threatened to expropriate the waterfront public lands in the hands of the city.[15] More quietly, it continued its efforts to purchase leases of water lots from those who held them. In February 1889 Council was informed by a surprised city solicitor that the company had taken possession and, without permission, was laying rails across all of the slips over much of the city's water frontage (*Globe* Feb. 15, 1889). By June the city was arguing its case against the railways before the highest Canadian tribunal with authority in the matter, the Railway Committee of the Privy Council.[16] In the words of the *Globe*, the CPR had become "the great gobbler" in what was now a "war on the waterfront . . ." (*Globe* June 25, 1889).

At the June hearings the members of the Railway Committee listened to representatives of several groups appearing to protect interested parties. They included: members of City Council and city officials, members of the Board of Trade, and representatives of owners of private property whose interests were directly affected.[17] No persons who might have spoken on behalf of the public interest had been invited to attend. This would soon change. The Board of Trade, meeting immediately after the hearings in Ottawa, "admitted the gravity of the situation, and manifested the utmost anxiety to be furnished with the views of the citizens generally . . ." (*Globe* June 29, 1889). The citizenry, too, was becoming aroused as the railway lay tracks at night on disputed land (*Globe* July 11, 1889).

The Board of Trade was not alone in beginning to appreciate that in this crisis the voice of the citizenry needed to be heard if the railways were to be prevented from securing absolute control of the whole city waterfront. City Council took the unusual step of inviting citizens to aid in drafting a petition to be sent to Ottawa (*Globe* July 12, 1889). A Toronto alderman emphasized that the citizens, when they "understand the value of their rights on the Esplanade and can appreciate what it would mean to surrender these in perpetuity to railroad companies" will be able to prevent the railways from winning the contest. The New York City consulting engineer whose advice the Board of Trade sought regarding railway accommodation on the waterfront was of the same view: the force of public opinion would be crucial in overcoming the excessive demands of the railways for control of the Esplanade.[18]

An aroused public called on the mayor to announce a public meeting. Held on July 26, it brought together members of Council, the Board of Trade, labor organizations, business interests, and the general public. The distinguished senior judge, a former mayor, who proposed the first resolution hoped it "would result in reinvesting the citizens with their old rights." Another speaker moved that a permanent organization, to be known as the Citizens' Association, comprised of a representative group of ratepayers, be formed "to watch carefully the interests of the city." The meeting was "highly successful" in the opinion of the writer for the *Globe* (*Globe* July 27, 1889).

The Citizens' Association promptly held a large and enthusiastic public meeting, and presented a report to Council on September 2. At the same meeting the Toronto Trades and Labor Council presented its resolution endorsing the preservation of public access to the waterfront.[19] A broad and well-organized public was expressing itself and defending its perceived rights vigorously. The press was paying close attention.[20] So were public officials. In a report of September 17 prepared for Council, the assistant city engineer and the city surveyor, for the first time in a city document, championed the rights of citizens. "No one who has the slightest acquaintance with the social conditions of this place will deny that the Bay and Island are becoming yearly of more importance and value to Toronto. . . . And in order to maintain and preserve these water rights and enjoyments, it is necessary that the citizens of Toronto should carefully watch and control the disposal of the water front. To allow the railways to come in and . . . to purchase or acquire the water front . . . with the ease that land may be acquired back in the country, would be, for Toronto, a suicidal policy."[21] The editor of the *Globe* clearly registered the change in key of the debate in his editorial of December 26, 1889, when discussing a scheme then under consideration. The first of "certain necessities" for the city was a plan which "would bring the water front under civic control . . ." (*Globe* Dec. 26, 1889).

THE INFLUENCE OF THE PUBLIC

The press redefined the story; it became much more than a record of meetings and official statements. Railway representatives energetically argued their case in its columns.[22] The public kept the matter alive, and the newspapers followed it closely. During March and April no fewer than eleven public meetings were held throughout the city.

Perhaps the most important of the meetings was held on March 21, 1890. For the editor of the *Telegram* a "promise of success spoke in the enthusiasm that prevailed at the great meeting of citizens on Friday night." The meeting's mood constituted "a sure sign of the popular unanimity against the outrageous claims of the CPR." The public's attitude should assure the railway that "the struggle will be prolonged" and should "force the representatives of the taxpayers [the City Council] to fight" (*Telegram* March 24, 1890). The body of final appeal, the Railway Committee of the Privy Council, in the view of the editor "will not defy public opinion and outrage every principle of justice" by granting the CPR its demands for expropriation (*Telegram* March 25, 1890). Goldwin Smith, perhaps the most renowned public speaker in the city, was accorded the privilege of moving the first motion. He spoke of "the right of the public to safe and free access to the water front, and to proper facilities for the transaction of their business." He stressed the role of the lake as a place of resort for men, women and children and appealed to "the patriotism and public spirit of the citizens." He spoke the public mood when he suggested that while nobody wished to fight the railways they, the public, "were perfectly justified in looking after the great interest placed in our charge . . ." (*Globe* March 22, 1890).

The political process engineered by the public enhanced the sense of cooperation among diverse interests, permitting the broad coalition to exert more effective pressure on City Council. The cooperating groups bombarded Council with memoranda and petitions, eroding its long accustomed freedom of action. Council discovered that it could no longer ignore so forcefully expressed a public opinion. Recognition of the legitimacy of the public's input into city governance came when representatives of the Citizens' Association gained a place on the Esplanade Committee of Council.

The spring of 1890 presented the city with a series of challenges. Under a recent agreement new lands were being created by another round of filling in the harbor.[23] Legal title to this new land being created remained in dispute. Meanwhile, the CPR proceeded with its construction, action which its legal counsel would subsequently acknowledge before the Railway Committee of the Privy Council to have been "not legal." He would offer in justification of the company's actions that it was "not of vital importance to the City that this matter should be stopped."[24] The *Telegram* described the actions of Council surrounding the matter as "wavering, lame and impotent" (*Telegram* May 9, 1890).

The enduring dispute over the rights of the railways to expropriate public lands on the Esplanade reached its climax in the autumn of 1890 when the protagonists again argued their cases before the Judicial Committee of the Privy Council. This occasion differed in one principal regard from the others when the city and the railways had presented their cases: the public was now represented as never before. The Citizens' Association and the Toronto Trades and Labor Council had been granted standing to speak for the public. The city and the Board of Trade as well as the two railways were represented. In contemplating the issue, the editor of the *Globe* was keenly aware of the importance of public opinion. The CPR, he opined, should modify its proposals out of consideration of the "feeling of hostility" that would be aroused were the Privy Council to agree to the expropriation "against the wish of the citizens" (*Globe* Sept. 16, 1890).

Mr. Kingsmill, speaking for the city, referred to the recent Order in Council recognizing that much of the new land to be created by filling the harbor "should be devoted to the purposes of a Park and for the public purposes of the city of Toronto, and that no railway track should be placed upon it."[25] Mr. Biggar, on behalf of the city, explained its wish to accommodate the CPR but not without protecting its citizens' interests: "when it was discovered in the spring of 1889, for the first time that the CPR claimed to have taken possession of the whole of the water front under the Order-in-Council of June, 1886 on allegations which to their knowledge were false and behind the backs of the citizens of Toronto, and by that Order in Council proposed to close every thoroughfare from Yonge street to the western end of the city, I do not wonder that there has been a change in the views of our citizens."[26] Speaking as with one voice, the representatives of the Citizens' Association and the Board of Trade claimed that citizens were entitled to the protection of their public interests. The representative of the Trades and Labor Council agreed, calling the waterfront "the only lung that we can lay claim to . . . We are deeply interested in this question, perhaps more so than any other portion of the community represented in this deputation, because we cannot, as a class, go away from the city to breathe the fresh air for any length of time . . .

We trust the Committee will kindly consider that fact in the interests of the working people."[27]

In his summation on September 27, Mr. Clark stated the case for the CPR bluntly. "The city," he declared, "have not any water front. They parted with it long ago. They leased it to different parties and those leases have now an average of 14 years to run and they are obliged by law, as our Company has secured those leases, to sell the reversion to us for railway purposes. There can be no question at all as to our title to the land down to the Old Windmill Line [former waterfront]." The CPR, he said, were "simply getting from the Crown something which everybody else has got."[28]

The CPR, having denied the city's claim to the waterfront, then acknowledged that they must take account of public opinion. Their spokesmen offered revised proposals wherein the railways acceded to the public's demands for control of some of the Esplanade at the heart of the city and for improved access to the waterfront. These proposals provided the basis for an expeditiously negotiated agreement. On June 12, 1891, the Joint Esplanade Committee of City Council presented its report containing a draft agreement between the railways and the city, the basis for the document that all parties formally ratified on July 26, 1892 (Figure 3.3).[29]

CONCLUSION

This agreement would not preclude future protagonists from raising the question of public and private rights to the Esplanade. By 1892, however, the city and the private interests concerned with public space had recognized the public's legitimate influence and role. By the late twentieth century the railways have almost completely abandoned their yards and shops which once occupied much of this land, which is again the subject of protracted negotiations.[30]

Meaningful urban public space is created not by legislation but as a result of the interest which people take in it. It is their prerogative, and their duty, to define and defend their interests in it, often against the concerted opposition of powerful institutions (Rosenzweig and Blackmar 1992). The legal status defining a particular public space may be material to its management and thus influence the contention over its use; it is seldom the definitive criterion in adjudicating the rights of the public in that space. Habermas emphasized the importance of articulating and promoting a broadly shared opinion in arguing for the concept of a public (Habermas 1989). This illustration from the long contest between private and public interests on the Toronto waterfront demonstrates the ability of a self-conscious and determined public to influence the terms in which the dispute was understood. The built environment reflected the powerful position of the railways, but they could not ignore an aroused public opinion or prevent it from making its mark on the land. Before the mid-1880s, during which period the fundamental

Figure 3.3 Toronto Waterfront, 1899. Source: Detail. J. G. Foster & Co. [Toronto] 1899. Special Collections, Metropolitan Toronto Reference Library.

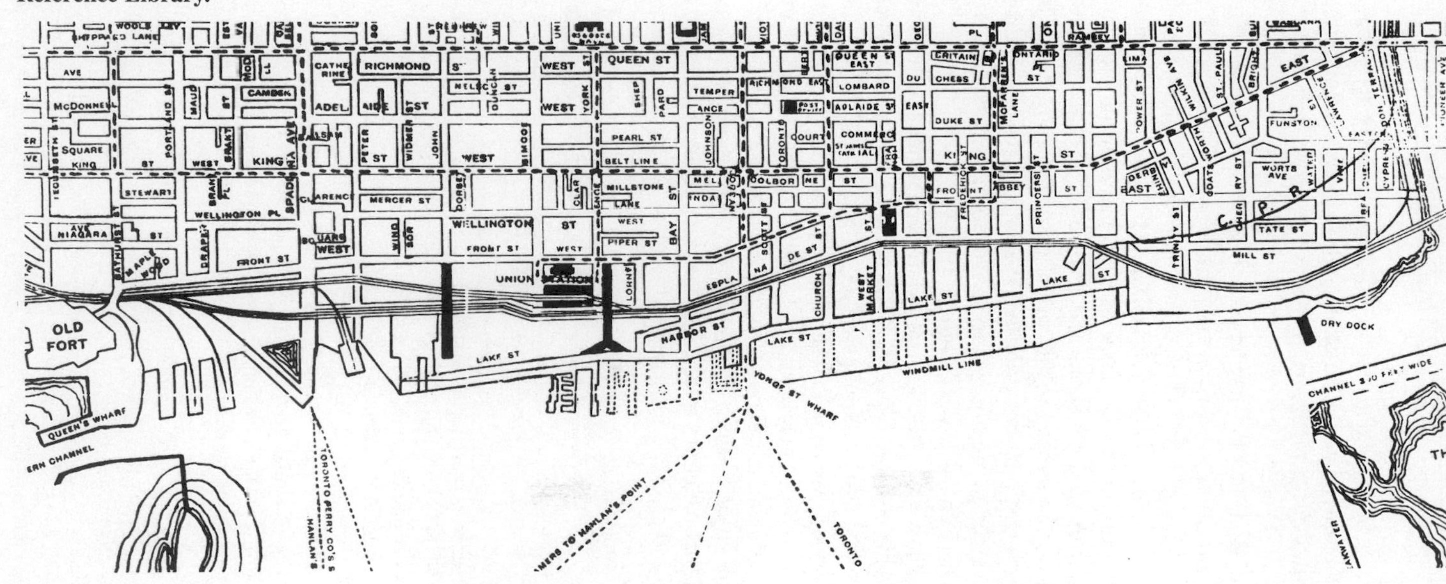

nature of the Esplanade as public space had not legally changed, Toronto City Council, the official custodian of the public interest, dealt with railways without serious regard for public opinion. This changed when a public insisted on being heard.

The Esplanade in 1885 may have been, as Mr. Van Horne of the CPR described it to the prime minister, Sir John A. Macdonald, "a disgrace to any city. Sewage water was floating in and around the wharves, old tumble down buildings were there; there were coal yards, piles of telegraph poles, fence posts and a collection of unsightly objects."[31] Nevertheless, an interested public valued it. It came to reflect the expressed intentions and interventions of the people and of powerful institutions; it comprised a cultural landscape (Groth 1997). It was as much a product of intentional behavior and the implanting of cultural values as landscapes designed to signify religious and political authority or refined aesthetic sensibilities (Duncan 1990; Cosgrove 1993).

The Esplanade was transformed from space which the citizens collectively had failed to value into a significant public place to which they attached social meaning. The social and economic conflict which their changing attitudes helped to inspire created a valued place (Hayden 1995 and 1997). The very ordinariness of Toronto's waterfront may enhance its value as a landscape. Peirce Lewis advises that in American landscapes we must look at everything, and especially their commonest objects, which possess the strongest cultural meaning (Lewis 1983). These cultural meanings were being defined and embedded on the Esplanade as the public and capitalist institutions contended to create a landscape symbolizing their values.

NOTES

1. Jackson uses these words to describe the classic work of Carl E. Schorske (1980).

2. Mitchell (1995a). For replies to Mitchell's argument and his riposte, see Cosgrove (1996); Duncan and Duncan (1996); Jackson (1996); Mitchell (1996a).

3. The political negotiation over rights in public space has been addressed by Don Mitchell in several recent papers (1995b; 1996b; 1997).

4. This illustration of the conflation of public rights and private interests is a footnote to the behavior of a privileged oligarchy with enormous influence in early Upper Canada (Ontario). For a recent discussion of a large literature, see Patterson (1991).

5. City of Toronto Archives (CTA), RG 1, City Council Minutes, meetings of March 23 and April 13, 1840.

6. CTA, RG 1, City Council Minutes, meeting of April 10, 1855. It should be noted that the Grand Trunk had by this time purchased the Toronto and Guelph Railway Company. See McIlwraith (1991–92).

7. Evidence of the continued conflation by the Grand Trunk Railway of its private with the public interest can be found in the correspondence of the company on this issue. The company wrote to the attorney general of Upper Canada on July 6, 1855, urging that

"on all grounds the public interests demand that a right of way should be provided along the front of the city." The passage is quoted in a memorandum, signed by John A. Macdonald, Attorney General for Upper Canada, Quebec, 16 July, 1855. It was printed in the *Globe*, August 13, 1855. It is significant that this claim, when published, elicited no comment in the press.

8. CTA, RG 1, City Council Minutes, March 15, 1886.

9. CTA, RG 1, City Council Minutes, April 27, 1886, and Appendix 36.

10. CTA, RG 1, City Council Minutes, May 17, 1886.

11. CTA, RG 1, City Council Minutes, November 21, 1887.

12. CTA, RG 1, City Council Minutes, 1888. Appendix 97, 504, 507.

13. CTA, RG 1, City Council Minutes, April 30, 1888.

14. CTA, RG 1, City Council Minutes, May 2, 1888.

15. CTA, RG 1, City Council Minutes, May 13, July 3, and October 8, 1889.

16. National Archives of Canada (NAC), RG 46, Railway Committee of the Privy Council, Vol. 779. Proceedings, June 18 and June 26, 1889.

17. CTA, RG 1, City Council Minutes, 1889. Appendix 186.

18. In a letter of August 21, 1889 to Mr. Wilkie [Chairman of the Esplanade Committee], Board of Trade, A. M. Wellington wrote that he was "very apprehensive" for the outcome of the dispute. "If the rys [*sic*] believe the people are in dead earnest they will yield and they really have everything to gain and nothing to lose by it—but their inertia will have to be overcome by steady and active pressure" (NAC, MG 28, III 56, Toronto Board of Trade, Vol. 328).

19. CTA, RG 1, City Council Minutes, September 2, 1889.

20. I have read two papers following generally different editorial positions; the *Globe* and the *Telegram*. Each provides detailed coverage of the story.

21. CTA, RG 1, City Council Minutes, 1889. Appendix 230.

22. President Van Horne of the CPR prepared a lengthy statement published in the issues of February 4, 1890 of the *Telegram* and the *Globe*. The editor of the *Globe* commented on the argument in an editorial of February 5, whereas the editor of the *Telegram* editorialized on February 5 and 6. R. M. Wells, a lawyer acting on behalf of the CPR, wrote lengthy letters published in the *Globe* on April 14, 21, and 28, 1890.

23. The agreement established a new line in the harbor, known as the New Windmill Line, for land fill. For details see: CTA, RG 1, City Council Minutes, 1887, Appendix 252; and City Council Minutes, 1888, Appendix 67.

24. The transcript is reprinted in: CTA, RG 1, City Council Minutes, 1890. Appendix 289, 1618.

25. NAC, RG 46, Railway Committee of the Privy Council, Vol. 697, file 4294, 29.

26. Ibid., 46.

27. These words were spoken by Mr. R. Glockling, president of the Trades and Labor Council. Ibid., 69.

28. Ibid., 11–12.

29. CTA, RG 1, City Council Minutes, 1891, Appendix 39; City Council Minutes, July 6, 1892, and Appendix C. The agreement was passed into law in 1893: Canada, *Statutes*, 56 Victoria, Cap. 48.

30. Among the burgeoning literature on Toronto's late twentieth-century waterfront redevelopment are several studies by David L. A. Gordon (1996, 1997). There is also a

substantial output from the Royal Commission on the Future of the Toronto Waterfront (1989 and 1992) and related documentation (Greenberg and Sicheri 1990; Lemon 1990; Merrens 1989).

31. NAC, RG 46, Railway Committee of the Privy Council, Vol. 697, file 4294. Arguments and Statements, September 19, 1890.

REFERENCES

By-Laws of the City of Toronto. 1870. Toronto: Henry Rowsell.

Canada. *Statutes*. 1853–1893.

Cosgrove, Denis. 1993. *The Palladian Landscape*. State College: Pennsylvania State University Press.

———. 1996. "Ideas and Culture: A Response to Don Mitchell." *Transactions, Institute of British Geographers* n.s. 21: 574–75.

Daily Leader and *Leader*. Toronto. 1853–1855.

Duncan, James S. 1990. *The City as Text: The Politics of Landscape Interpretation in the Kandyan Kingdom*. Cambridge: Cambridge University Press.

Duncan, James, and Nancy Duncan. 1996. "Reconceptualizing the Idea of Culture in Geography: A Reply to Don Mitchell." *Transactions, Institute of British Geographers* n.s. 21: 576–79.

Ethington, Philip J. 1994. *The Public City*. New York: Cambridge University Press.

Firth, Edith G., ed. 1966. *The Town of York, 1815–1834*. Toronto: University of Toronto Press.

Globe. Toronto. 1853, 1855, 1889–1890.

Goheen, Peter G. 1994. "Negotiating Access to Public Space in Mid-Nineteenth Century Toronto." *Journal of Historical Geography* 20: 430–49.

———. 1995. "Creating Public Space in Nineteenth-Century Toronto: The Semi-Centennial Celebrations of 1884." In *Espace et Culture — Space and Culture*, ed. Serge Courville and Normand Séguin, 245–52. Sainte-Foy: Les Presses de l'Université Laval.

———. 1997. "Honouring 'One of the Great Forces of the Dominion': The Canadian Public Mourns McGee." *Canadian Geographer* 41: 350–62.

Gordon, David L. A. 1996. "Planning, Design and Managing Change in Urban Waterfront Redevelopment." *Town Planning Review* 67: 261–90.

———. 1997. "Managing the Changing Political Environment in Urban Waterfront Redevelopment." *Urban Studies* 34: 61–83.

Greenberg, Ken, and Gabriella Sicheri. 1990. *Toronto's Moveable Shoreline*. Toronto: Canadian Waterfront Resource Centre, Working Papers 5.

Groth, Paul. 1997. "Frameworks for Cultural Landscape Study." In *Understanding Ordinary Landscapes*, ed. Paul Groth and Todd W. Bressi, 1–21. New Haven: Yale University Press.

Groth, Paul, and Todd W. Bressi, eds. 1997. *Understanding Ordinary Landscapes*. New Haven: Yale University Press.

Habermas, Jürgen. 1989. *The Structural Transformation of the Public Sphere*, trans. Thomas Burger. Cambridge: Massachusetts Institute of Technology Press.

Hayden, Dolores. 1995. *The Power of Place*. Cambridge: Massachusetts Institute of Technology Press.

———. 1997. "Urban Landscape History: The Sense of Place and the Politics of Space." In *Understanding Ordinary Landscapes*, ed. Paul Groth and Todd W. Bressi, 111–33. New Haven: Yale University Press.

Jackson, Peter. 1989. *Maps of Meaning*. London: Unwin Hyman.

———. 1996. "The Idea of Culture: A Response to Don Mitchell." *Transactions, Institute of British Geographers* n.s. 21: 572–74.

Lemon, James. 1990. *The Toronto Harbour Plan of 1912*. Toronto: Canadian Waterfront Resource Centre, Working Papers 4.

Lewis, Peirce. 1983. "Learning from Looking: Geographic and Other Writing About the American Cultural Landscape." *American Quarterly* 35: 242–61.

McIlwraith, Thomas. 1991–92. "Digging Out and Filling In: Making Land on the Toronto Waterfront in the 1850s." *Urban History Review* 20: 15–33.

Mellen, Frances N. 1974. *The Development of the Toronto Waterfront During the Railway Expansion Era, 1850–1912*. Ph.D. dissertation, University of Toronto.

A Member of the Canadian Institute. 1853. "Railway Termini and Pleasure Grounds." *Canadian Journal* 1 (10): 16.

Merrens, Roy. 1989. *A Selected Bibliography on Toronto: Port and Waterfront*. Toronto: Canadian Waterfront Resource Centre, Working Papers 1.

Mikesell, Marvin W. 1978. "Tradition and Innovation in Cultural Geography." *Annals of the Association of American Geographers* 68: 1–16.

Mitchell, Don. 1995a. "There Is No Such Thing as Culture: Towards a Reconceptualization of the Idea of Culture in Geography." *Transactions, Institute of British Geographers* n.s. 20: 102–16.

———. 1995b. "The End of Public Space? People's Park, Definitions of the Public, and Democracy." *Annals of the Association of American Geographers* 85: 108–33.

———. 1996a. "Explanation in Cultural Geography: A Reply to Cosgrove, Jackson and the Duncans." *Transactions, Institute of British Geographers* n.s. 21: 580–82.

———. 1996b. "Political Violence, Order, and the Legal Construction of Public Space: Power and the Public Forum Doctrine." *Urban Geography* 17: 152–78.

———. 1997. "The Annihilation of Space by Law: The Roots and Implications of Anti-homeless Laws in the United States." *Antipode* 29: 303–35.

Patterson, Graeme. 1991. "An Enduring Canadian Myth: Responsible Government and the Family Compact." In *Historical Essays on Upper Canada*, ed. J. K. Johnson and Bruce G. Wilson, 485–511. Ottawa: Carleton University Press.

Rosenzweig, Roy, and Elizabeth Blackmar. 1992. *The Park and the People: A History of Central Park*. Ithaca: Cornell University Press.

Royal Commission on the Future of the Toronto Waterfront. 1989. *Interim Report*. Toronto: Minister of Supply and Services Canada..

———. 1992. *Regeneration*. [Ottawa]: Minister of Supply and Services Canada.

Ryan, Mary P. 1997. *Civic Wars*. Berkeley: University of California Press.

Schorske, Carl E. 1980. *Fin-de-Siècle Vienna*. New York: Knopf.

Sopher, David E. 1979. "The Landscape of Home." In *The Interpretation of Ordinary Landscapes*, ed. D. W. Meinig, 129–49. New York: Oxford University Press.

Telegram. Toronto. 1890.

Toronto and Guelph Railway. 1852. *Chief Engineer's Report, May 21, 1852*. Toronto: Brewer, McPhail and Co. In Pamphlet Collection, Archives of Ontario.
———. 1853. *Second Annual Report, June 6, 1853*. Toronto: Henry Rowsell. In Pamphlet Collection, Archives of Ontario.
Wagner, Philip L., and Marvin W. Mikesell. 1962. *Readings in Cultural Geography*. Chicago: University of Chicago Press.
Zukin, Sharon. 1995. *The Culture of Cities*. Cambridge: Blackwell.

4

Place Your Bets: Rates of Frontier Expansion in American History, 1650–1890

Carville Earle

Embedded in American history is an assumption of an "ever-westering" frontier, of a more or less linear rate of settlement expansion from colonial times to the closing of the frontier in 1890 (Turner 1920). As astonishing as it may seem, a century of frontier inquiry scarcely has acknowledged this assumption let alone subjected it to empirical test. Only a handful of scholars have regarded settlement expansion as problematic, and fewer still have assessed its variability in time and space (Mood 1952; Clark 1960, 1972; Florin 1977; Warntz 1967; Otterstrom 1997). Frontier scholarship has been content with a rude impressionism (Turner 1920; Mitchell 1978; Zelinsky 1973; Meinig 1986; Mitchell and Groves 1987). Insofar as casual inspection has conveyed a misleading characterization of the frontier and rendered it an easy target for the harsh criticisms of the "new western historians," that methodology and its results invite reconsideration (Bernstein 1989).

This chapter takes a different approach toward American settlement history, one that was clearly enunciated four decades ago in Marvin Mikesell's (1960: 65) provocative essay on frontier history. "It goes without saying," Mikesell observed, "that the establishment of objective standards of comparability is a necessary prerequisite of any comparative [frontier] study." Mikesell thus spoke of the variable nature of frontiers, of frontiers as static or dynamic, as inclusive or exclusive. This chapter regards frontier expansion as a variable and therefore as problematic. Was it fast or slow? Linear or cyclical? Were there many frontiers or only one? Were its causes singular or plural? Determinant or stochastic? Unilineal or recursive? To ask these questions is to assert the quantitative proposition that settlement expansion rates are measurable and that variability in these rates is sufficiently large as to warrant reconceptualization of the frontier process.

Following from the proposition that settlement expansion is problematic are two subsidiary propositions. The first is the incontrovertible notion that *secular rates of settlement expansion* are more or less sympathetic with the long-run performance of demographic and economic systems. That said, we simply do not know whether in the long run settlement expands faster, slower, or at the same rate as the economy or the population.

The second proposition is that the assumption of a *linear rate of frontier expansion* constitutes a testable hypothesis. Does this test confirm the presence of a singularly steady rate of settlement expansion or of multiple rates? And if the latter, of what temporal form and structure (i.e., cyclical, step-like, random)? By infusing precision and a sense of the problematic into the marrowy center of American settlement history, we can entertain the possibility of cyclical rather than linear variation in the aggregate and regional rates of settlement expansion.

Early American history offers a case in point. In the half century that has elapsed since Herman Friis (1940) published his landmark essay on the population distribution of the mainland colonies of British North America, 1625–1790, his exquisite series of population maps has eluded rigorous analysis. Only one study (of the colony of Pennsylvania) has used his maps to measure long-term trends of settlement expansion (Florin 1977).[1] Scholars have preferred an impressionist methodology predicated upon visual inspection and verbal description of the Friis maps. The shortcomings of these methodologies exert their most insidious effects subliminally. By denying the possibility of falsification, they affirm through default the conventional wisdom of a linearly expanding frontier.

Impressionistic methods have their place. But our continuing reliance on these methods is mocked on one side by the scrupulous accuracy of Friis' map series, on the other by the precision of growth estimates produced by contemporary economic and demographic historians of early America. That historical geography persists in its impressionistic accounts of our closest analogue—the rate of settlement expansion—is regrettable indeed. Friis' enormous cartographic undertaking assuredly deserves more. His meticulously reconstructed dot maps, ten in all between 1625 and 1790, invite a more analytical assessment of the rates of early American expansion. This assessment is imperative if historical geography is to keep pace with the mainstream of historical social science.

In paying homage to Friis' monumental work, this study of American settlement expansion eschews casual description in favor of precise measurement. Part one measures the rates and tempos of frontier expansion on Friis' dot maps during the first two centuries of Anglo-American experience. Part two carries the analysis forward from 1790 through 1890—the date when, according to Frederick Jackson Turner, the American frontier was finally closed. Measurements of settlement expansion between 1790 and 1890 are based on the population density maps in Paullin (1932). This analysis of the American frontier over nearly two and a half centuries (1650 to 1890) yields three conclusions: (1) the *secular lag* in settlement expansion rates vis-à-vis economic and demographic rates; (2) the *cyclical tempo* in the aggregate and regional rates of settlement expansion—findings which have important implications for interpretations of frontier society and policy, historical periodization, long-wave cyclical interpretations of early American history and macroregional divergence and convergence with respect to that cycle; and (3) the effective closure of the frontier in the 1840s, a half century ahead of Turnerian schedule, and the ensuing involution of American settlement processes.

With this precis we turn to the colonial period and an argument that unfolds in five parts: (1) source and definitions; (2) absolute measures of settlement expansion; (3) aggregate rates of expansion; (4) regional rates of expansion; and (5) some reflections on landscape evolution and macrohistory.

MEASURING FRONTIER EXPANSION, 1650–1790: SOURCE AND DEFINITION

A word is in order about the Friis maps. Friis' mapping project rests on a carefully conceived and rigorously executed methodology of three parts: (1) compilation of population data and estimates from a variety of fugitive and often intractable local sources including tax lists, militia lists, poll lists, and early censuses; (2) location of these data in their respective townships, parishes, counties, and colonies; and (3) portrayal of colonial population distributions in a time series of ten high-resolution dot maps (at 200 persons per dot).

It is ironic that Friis may have done his work too well. The accuracy and reliability of his maps, now attested to by several generations of scholarship, convey a degree of closure and finality so complete that scholars have been content to use them principally for purposes of context or ornament. In our admiration we have overlooked the enormous utility of these maps for the quantitative analysis of settlement expansion. Indeed Friis' maps are nearly ready-made for testing the hypothesis of a linearly expanding American frontier. All that is required for that test is: (1) measurement of changes in the settled areas evident on Friis' maps of the American colonies; (2) calculation of annual rates of colonial and regional settlement expansion; and (3) charting the rates of expansion over time. Applying elementary arithmetic to Friis' maps suggests a fundamental revision in our linear interpretation of American settlement history.

Before testing and reinterpreting the hypothesis of linear settlement expansion, we must give some attention to definition. In this analysis, settlement is defined as that area circumscribed around proximate dots on the Friis maps; that is, areas wherein Friis' dots of population are sufficiently close to warrant a characterization of settlement. For operational purposes, the circumscribed area of settlement includes all dots whose nearest neighbor lies at a distance of 20 miles or less.

Translated into the metric of population density, this definition of settled area amounts to a lower-bound density threshold of about two-thirds of a person per square mile. This threshold is deceptively conservative since it assumes that population is evenly distributed around Friis' dots—an unlikely assumption on the edge of the frontier. Adopting the more realistic assumption that arrays frontier population between the dots rather than evenly around them has the effect of halving the inclusive area around the most distant point and increasing the density threshold from two-thirds to 1.3 persons per square mile. My settlement density threshold is considerably lower than Hart's (1974) six persons per square mile, slightly lower than Turner's (1920) two persons per square, and probably more akin to Preston James's (1959: 5–13) notion of an "effectively settled territory."

MEASURING FRONTIER EXPANSION: AREA

Early American settlement expansion (Figures 4.1 and 4.2) defies easy description. Perusal of these maps underlines the difficulty of drawing conclusions about the rate of frontier expansion from visual inspection alone. The sheer quantity of graphic information contained in these maps presents a daunting task for intellectual assimilation. The variable time intervals in the map series—intervals of 25 years before 1700, 20 years between 1700 and 1760, and 10 years between 1760 and 1790—play tricks on our capacity for differentiating temporal and regional variations in settlement expansion.

First impressions that settlement expanded at a relatively steady rate between 1625 and 1790 are misleading. On closer examination of Friis' maps, it is apparent that settlement expanded more rapidly at some times and in some regions than in others. But how much more rapidly it spread is almost impossible to say from visual observation. The best one can do is to speak about coarse changes in the spatial pattern of settlement, noting the change from a series of discrete frontiers before 1700 to a more continuous frontier of settlement after 1700 (Mood 1952). Given the frailty of conclusions based upon inspection and imprecise language, the logical step is to do what a century of frontier scholarship has steadfastly and inexplicably avoided: measure the rate of settled-area expansion.

The initial round of measurements appears in Table 4.1. The table lists the absolute areas under settlement at the various dates specified on Friis' maps and in the various regions in the colonies. The time frame for these estimates is defined by Friis' nine cross-sectional years between 1650 and 1790, while the macroregions constitute an adaptation of Donald Meinig's (1986) sociocultural regions of colonial America circa 1750. Graphing these measurements in a time series offers a preliminary perspective on spatial and temporal variabilities in early American settlement expansion. The two graphs in Figure 4.3 suggest that settlement expansion in the colonies divides into three phases: (1) 1650–1720; (2) 1720–1770; and (3) 1770–1790. In the first phase, settled area increases 3.5-

Figure 4.1 Macroregions on the Atlantic Seaboard of North America, circa 1750.

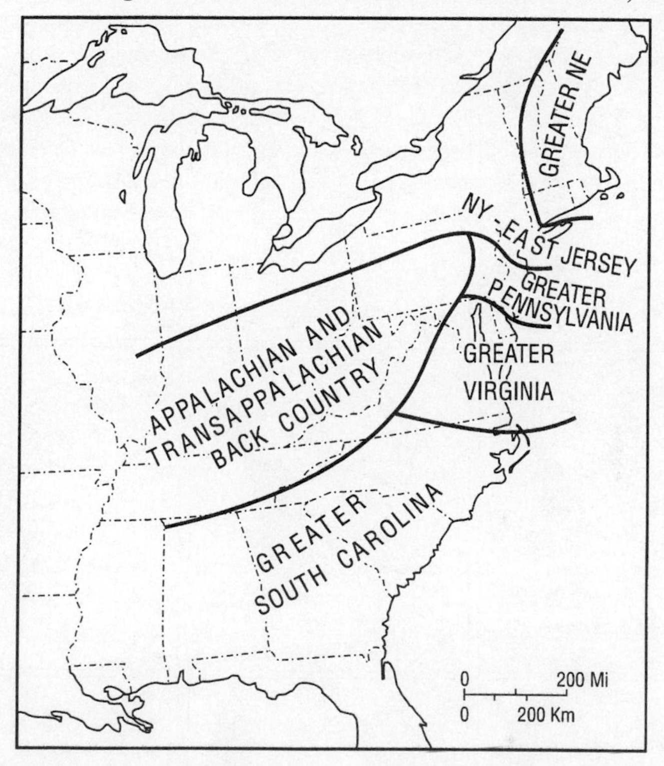

Source: Adapted from Meinig (1986: 244–54).

fold; in the second phase by 3.2-fold; and in the third phase by 1.6-fold. The variable length of these phases obscures the secular acceleration of frontier expansion. When the time series is partitioned into two equal periods centered on 1720, settlement expanded 3.5-fold before and about fivefold after 1720.

If the big picture of settlement expansion revealed in these numbers and graphs seems reasonably clear, the regional details are hazier (Figure 4.3). Amid the variability in regional trends of expansion one pattern stands out: the inverse relationship with the age of settlement. The older the region, the slower and steadier the pace of expansion. Pennsylvania and New York fit the pattern precisely, as do Virginia and New England with two exceptions: the punctuated bursts of settlement growth between 1720 and 1740 in the former and after 1760 in the latter. In the newer regions of South Carolina and the Back Country, settlement expands rapidly after 1720, particularly in the Back Country between the Revolution and 1790.

The net effect of these regional variations is a long-run inversion in the spatial

Figure 4.2 Settled Areas, 1650–1790.

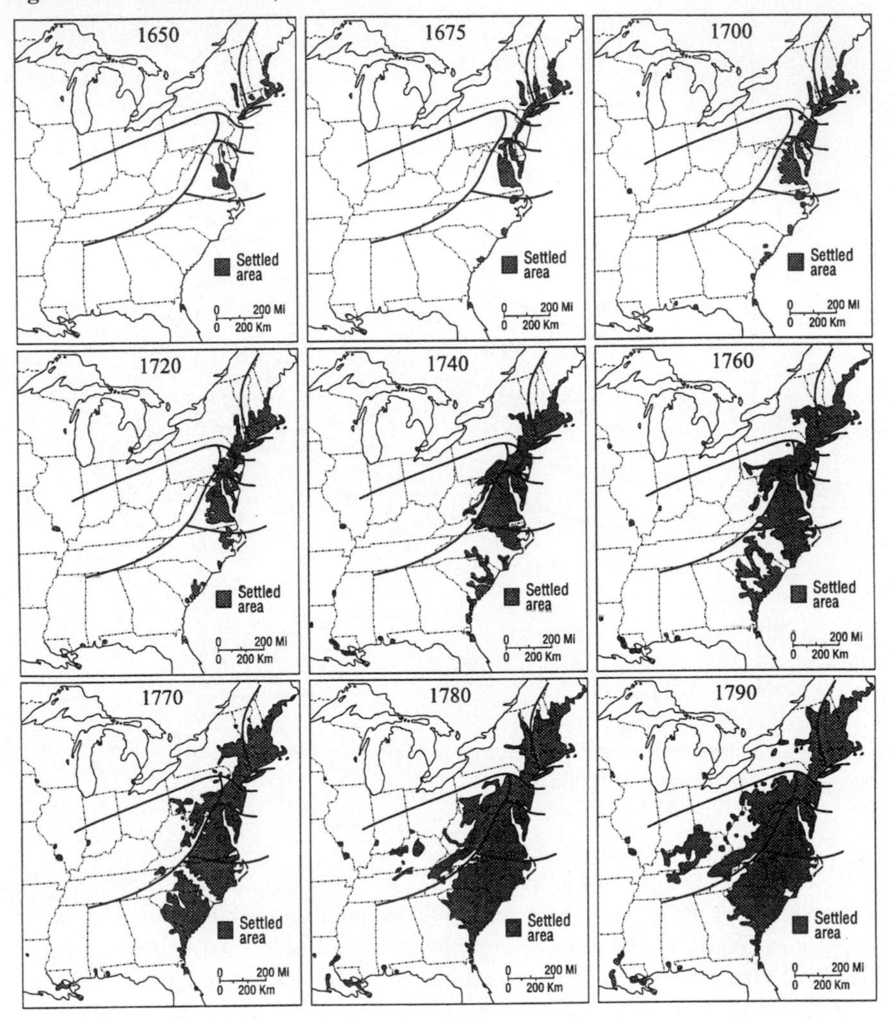

Note: See text and notes for definition of "settled area."

Source: Adapted from the dot-density maps of population in Friis (1940).

distribution of early American settlement. In 1650, the eldest regions account for 80 percent of the settled area of colonial America. Thereafter New England's and Virginia's dominance recedes steadily, falling to 62 percent by 1720, 38 percent by 1770, and 28 percent by 1790. Displacing them from top rank are the rapidly expanding regions of South Carolina and the Back Country, which to-

Table 4.1 Early American Settled Areas (in square miles)

Year	*Nation*	*Greater New Eng.*	*Greater N.Y.*	*Greater Pa.*	*Greater Va.*	*Greater S.C.*	*Back Country*
1650	14,561	6,011	2,498	479	5,572	0	0
1675	30,220	7,793	4,791	1,811	14,518	1,308	0
1700	39,511	11,370	5,730	2,447	17,913	1,958	92
1720	52,068	13,260	8,710	7,513	19,183	3,133	269
1740	99,014	16,810	12,600	8,594	32,890	18,176	9,944
1760	137,484	18,940	15,850	9,300	33,750	42,828	16,817
1770	167,567	27,390	16,850	10,310	35,750	60,200	17,067
1780	223,211	32,020	17,140	11,490	36,750	76,650	49,161
1790	261,954	36,400	21,543	11,490	36,861	80,890	74,770

Sources: Herman R. Friis, *A Series of Population Maps of the Colonies and the United States, 1625–1790,* American Geographical Society Mimeographed Publication No. 3 (New York: American Geographical Society, 1940), and Figure 4.4.

gether accounted for 43 percent of the settled area in 1760 and 59 percent in 1790.

It is easy to imagine the strains that this spatial inversion of settlement imposed upon British frontier policy. Imagine them, however, when compounded by sizable fluctuations in the temporal rates of settlement expansion. It is to these vola-

Figure 4.3 Expansion of Settled Areas, for All Colonies (left) and Macroregions (right), 1650–1790.

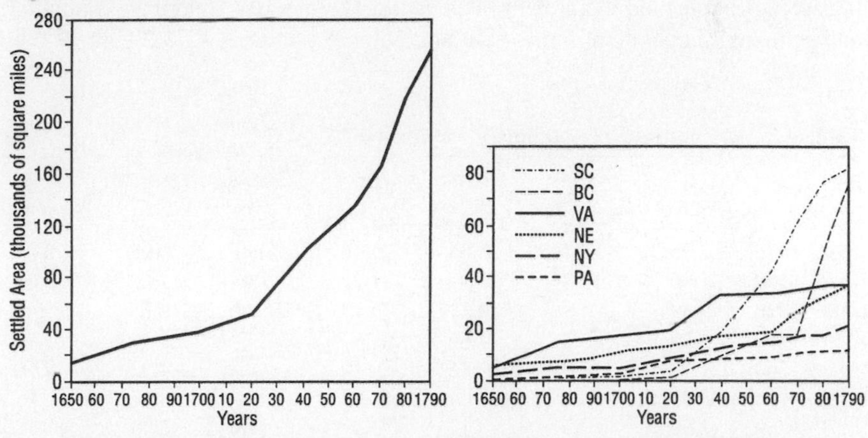

Note: Regional abbreviations: (BC) Back Country; (NE) Greater New England; (NY) Greater New York; (PA) Greater Pennsylvania; (SC) Greater South Carolina; (VA) Greater Virginia.

Source: Table 4.1.

tile changes, secular and cyclical, and their policy consequences that we now turn.

AGGREGATE RATES OF SETTLEMENT EXPANSION

Settlement mensuration is a first step toward a test of the hypothesis of a linear rate of settlement expansion. The full test requires conversion of these absolute measures of settlement expansion into commensurable rates of expansion. In that connection, Table 4.2 introduces greater exactitude. The table reports colonial and regional growth rates in settled area, compounded annually, for each of Friis' sequential map pairs and for the secular trend between 1650 and 1790. These time series and kindred ones are graphed in Figures 4.4 and 4.5. The story they tell casts considerable doubt on the conventional wisdom that American settlement expansion was more or less progressive and linear. The new story divides into two parts: the first compares secular trends in the rate of settlement expansion with trends in early American demographic and economic growth rates; the second describes the cyclical trends in expansion rates.

THE SECULAR TREND IN AGGREGATE RATES OF SETTLEMENT EXPANSION

Over the long run, early American settlement expands at the fairly rapid rate of about 2.1 percent per annum (compounded) between 1650 and 1790. That rate lags somewhat behind the growth rates estimated for early American population and economy. Population in the mainland colonies increases at 3.2 percent over

Table 4.2 Annual Rate of Settlement Expansion (%)

Year	*Nation*	*Greater New Eng.*	*Greater N.Y.*	*Greater Pa.*	*Greater Va.*	*Greater S.C.*	*Back Country*
1650–1675	2.96	1.05	2.64	5.46	3.90	0.00	0.00
1675–1700	1.06	1.52	.72	1.22	.84	1.62	0.00
1700–1720	1.39	.78	2.11	5.77	.34	2.37	5.48
1720–1740	3.26	1.20	1.86	.67	2.37	9.19	19.78
1740–1760	1.64	.60	1.15	.40	.13	4.38	2.66
1760–1770	2.05	1.37	.61	1.04	.58	3.46	.15
1770–1780	2.80	1.60	.17	1.07	.28	2.44	11.15
1780–1790	1.60	1.30	2.31	0.00	.03	.54	4.28
1650–1790	2.08	1.30	1.55	2.29	1.35	3.65	4.90

Source: Table 4.1.

Figure 4.4 Increase of Population and Settled Areas, for the Thirteen Colonies and Their Macroregions, 1620–1790.

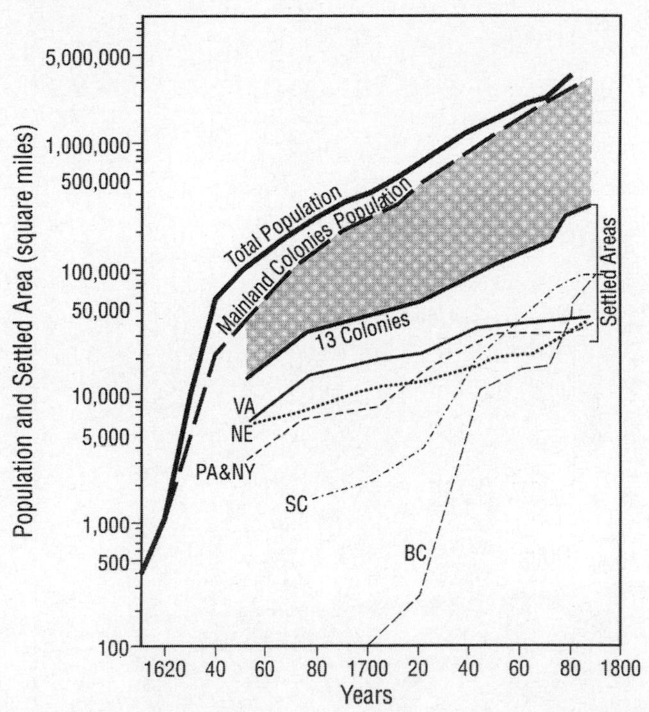

Note: The shaded area between the semilogarithmic curves serves to highlight the divergence in the rates of growth of population and settled areas after 1700. For regional abbreviations, see Figure 4.3.
Sources: United States Bureau of the Census (1975) and Table 4.1.

the same period, and national product estimates imply a growth rate on the order of 3.5 percent per annum between 1650 and 1774 (McCusker and Menard 1985). Even allowing for error in these estimates, it seems likely that the rate of settlement expansion lagged a third or more behind the rates of demographic and economic growth.

Doubtless the reasons for the lag are complex. One important factor is the sympathetic bond between settlement expansion and demographic and economic growth in newly settled regions. Settlement expansion is contingent on the achievement of a threshold of socioeconomic intensification in the region as a whole. Specifying the level of that threshold could go a long way toward explaining the one-third lag in the secular rate of early American settlement expansion. Lending credence to this interpretation of settlement lags is the striking parallelism in the semilogarithmic curves of demographic growth and settlement expansion (Figure 4.4). Their relatively fixed trajectories after 1700 imply the estab-

Figure 4.5 Variability of Macroregional Rates of Settlement Expansion, 1650–1790.

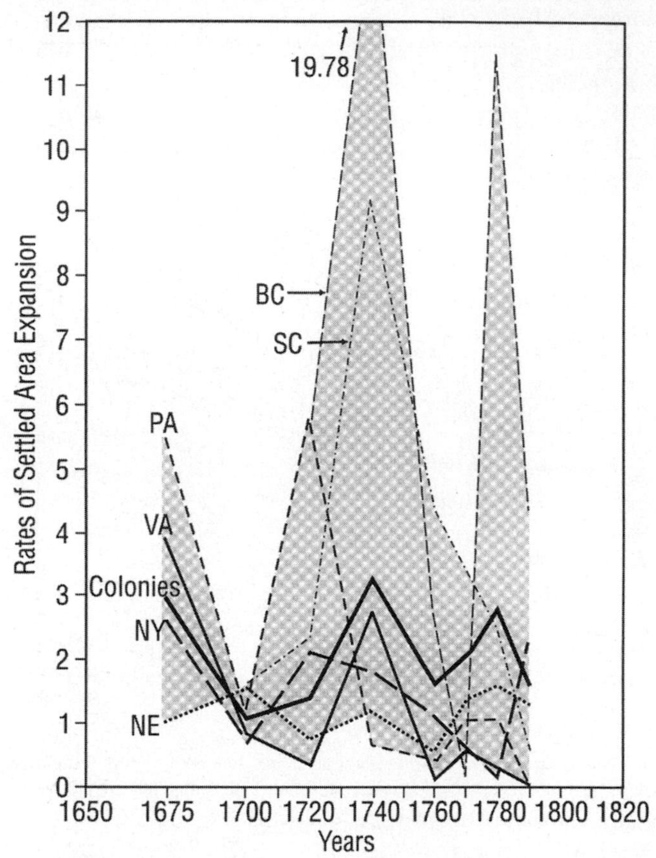

Note: Expansion rates equal percentage increases compounded annually. For regional abbreviations, see Figure 4.3. The shaded area delineates the range of macroregional rates at given dates.

Source: Table 4.2.

lishment of an equilibrium population density (the intensification threshold) that modulated the upper bound on the rate of settlement expansion. While it is generally true that settlement and population advanced in unison, the trends in their curves before and after 1700 are not identical. Sharp convergence of their slopes before 1700 indicates that seventeenth-century expansion outpaced the growth of population. Divergence of their slopes after 1700 implies an increase in the intensification prerequisite for eighteenth-century settlement expansion. The year 1700 therefore marks a critical turning point in the achievement of a new equilibrium density threshold—an achievement that seems to have its sources in a dramatic and deliberate compaction of early American settlement.

The "great compaction" of American settlement in the late seventeenth century altered once and for all the quantitative relationship between settlement expansion and population density. In a mere twenty-five years, early American society experienced a sudden increase in the customary level of population density. Prior to 1675, American settlement expanded at a dizzying pace. Settlement sprawled over a relatively large territory and colonial population density stood at just over three persons per square mile. A quarter of a century later, the implementation of a policy of settlement compaction had trebled the density of early American population, pushing it up from just three persons per square mile in 1675 to over ten—an equilibrium level that the colonies maintained more or less for the remainder of the eighteenth century.

What caused this sharp rise in population density between 1675 and 1700? One factor is problems created by the exceptionally fast pace of settlement expansion in the 1660s and 1670s. These were turbulent times in which a rapidly expanding frontier incited widespread Indian unrest and hostility during the mid-1670s. Bacon's Rebellion in Virginia, a frontier rebellion which grew out of real and perceived threats of Indian attack, and King Phillip's War in New England drove home the message that overzealous frontier expansion carried enormous social, demographic, and political costs (Morgan 1975; Washburn 1957).[2]

Policymakers and common sense subsequently insisted on a more deliberate pace for settlement expansion at about two-thirds the rates of growth in population and economy. Effecting that balance between settlement and population monopolized the agenda of the Crown and colonial governments for the rest of the seventeenth century. Their policy initiatives repeatedly appealed for intensification and its corollaries: the need for towns in the colonial Chesapeake, for orderly administrative bureaucracies in New England, and for carefully planned and modulated settlement expansion in Pennsylvania and South Carolina. When policymakers in the Chesapeake abandoned the generous land system of headrights in favor of cash payment and assiduous quit-rent collection, they reinforced the process of compaction through the imposition of an artificial scarcity value upon otherwise abundant supplies of land (McCusker and Menard 1985: 34–50; Rainbolt 1969, 1974; Sirmans, 1966; Nash 1968).

It is in this context of density-equilibrium maintenance that history should judge British frontier policy in the eighteenth century. After suffering the military consequences of another phase of expansion between the 1720s and 1740s, imperial policymakers can hardly be blamed for instituting a restrictive series of frontier policies prohibiting settlement west of the Appalachians. The proclamation of 1763 and Quebec Act of 1774, although poorly timed and perhaps clumsily applied, cannot be regarded as unreasonable in their logic nor ill-tempered and mean-spirited in their intent. British policymakers sought to restore the old equilibrium between settlement and population, an equilibrium which had ensured at least a respite of tranquility on the early American frontier (Egnal 1988; DeVorsey 1966).

The great compaction of the late seventeenth century ameliorated but did not eliminate the dilemma of Anglo-American frontier policy. The tensions between population growth and the frontier persisted for the simple reason that the inherent cyclicity of settlement expansion during the eighteenth century repeatedly destabilized the density equilibrium that policymakers so desperately attempted to maintain. In a vast and sparsely settled land subject to the cyclical rhythms of capitalist markets, achieving a dynamic equilibrium was easier said than accomplished.

CYCLICAL TRENDS IN AGGREGATE RATES Of SETTLEMENT EXPANSION

The rate of early American settlement expansion followed a series of rhythmic long waves that were closely attuned to the changing fortunes of the Atlantic economy (Table 4.2 and Figure 4.5). In three sweeping oscillations of about a half century, settlement raced forward at rates of 2 to 3 percent per annum in times of prosperity and then stumbled into a slow trot of 1 to 1.6 percent in times of economic crisis. Take first the case of long-wave depressions. Expansion rates bottomed out at 1.6 percent per annum or less on three occasions: 1675–1700, 1740–1760, and 1780–1790. These intervals encompassed the most severe and protracted economic depressions in early American history. The rate of expansion was least (1.1 percent) during the great compaction of the late seventeenth century and only slightly higher (1.6 percent) in the depressions of the 1740s and the 1780s. At no other times in the colonial period does the aggregate rate of expansion fall to comparably low levels (Earle 1992; Wallerstein 1974, 1980; McCusker and Menard 1985; Harris 1996).

Conversely, settlement expansion quickened on the resumption of prosperity. Following the troughs in frontier expansion during times of economic depression, the rate of settlement expansion routinely accelerated for three to four decades before cresting at about 3 percent per annum toward the end of the cycle. Peaks in frontier expansion during the 1660s, 1720s, and 1760s–1770s thus coincided with moments of legendary prosperity. Shortly thereafter, rates of settlement expansion fell off precipitously (to less than 1.7 percent) as economic depression recurred.

The rates of early American settlement expansion progressed in long logistic waves. Oscillating within a period of about fifty years, these waves troughed amid the economic depressions of the 1680s–1690s, the 1740s, and the 1780s–1790s and crested in the ensuing thirty to forty years of economic recovery. For imperial policymakers, charged with maintaining some semblance of equilibrium between population and settlement and managing the social strains and tensions arising from a cyclical economy periodically pushing settlement beyond the limits where it was safe to go, the situation was maddening indeed.

MACROREGIONAL RATES OF SETTLEMENT EXPANSION

Regional compliance with the expansionary cycles of colonial America might be expected under certain centralized political economies, but not in the loose-knit bureaucracy that prevailed in the British Empire. Imperial and colonial policymakers confronted enormous regional differences in settlement expansion rates. The regional scene, if less orderly than the colonial aggregate, was not entirely chaotic in its secular trends and cyclical rhythms.

Secular rates of frontier expansion for the period 1650 to 1790 grew slowly in some regions and expanded at a breakneck pace in others (Table 4.2), the rate varying inversely with the longevity of settlement. In the older coastal regions, save for South Carolina, expansion rates were the slowest, varying from as little as 1.3 percent per annum in greater New England to 2.29 percent in greater Pennsylvania—all of which compared favorably with the colonial rate of 2.08 percent. Expansion proceeded much more rapidly in the more recently settled Back Country and in South Carolina, the former expanding at a rate of 4.9 percent per annum, the latter at 3.7 percent. These regional trends indicate moderate expansion rates in the northern and middle colonies flanked by exceptionally rapid growth on both the southern and western frontiers.

The settlement histories of early American regions are simultaneously distinctive and generic. The outcome is determined by the variance of regional expansion rates around the colonial mean in conjunction with long waves in the economy. Regional rates of expansion diverge widely during times of prosperity and then converge sharply during times of protracted economic depression (Table 4.2 and Figures 4.5 and 4.6).

During the economic depressions of the 1680s, 1740s, and 1780s, convergence prevailed; virtually every macroregion experienced a deceleration in the rate of settlement expansion with the onset of economic crisis. Of the 16 regional cases presented in Table 4.2, the rate of expansion slows down in 14. In the depression of the 1680s, three of four regional rates trend downward (only New England's rose); in the 1740s depression, six of six trend down; and in the 1780s, five of six do the same (New York's alone trends up). This descent is often quite swift. In one case the regional expansion rate drops from 20 percent per annum to 2 percent and in another from 9 percent to 4 percent.

Severe depressions were a powerful force in American settlement history, causing sharply decelerating rates of settlement expansion in the several regions of early America as well as in the colonies as a whole. This remarkable regional convergence suggests that hard times constituted formative moments in the rhythm of early American history.

In good times, regional rates of expansion are erratic in their behavior, disorderly in their pattern, and divergent in their trends. Regional rates go their own way, accelerating in half of our macroregions, and decelerating in the rest (Figure 4.5). In that respect, some regions buck the colonial trend just once (Pennsylvania

Figure 4.6 Pattern of Settlement Expansion by Cycle, Phase, and Macroregion, 1650–1790.

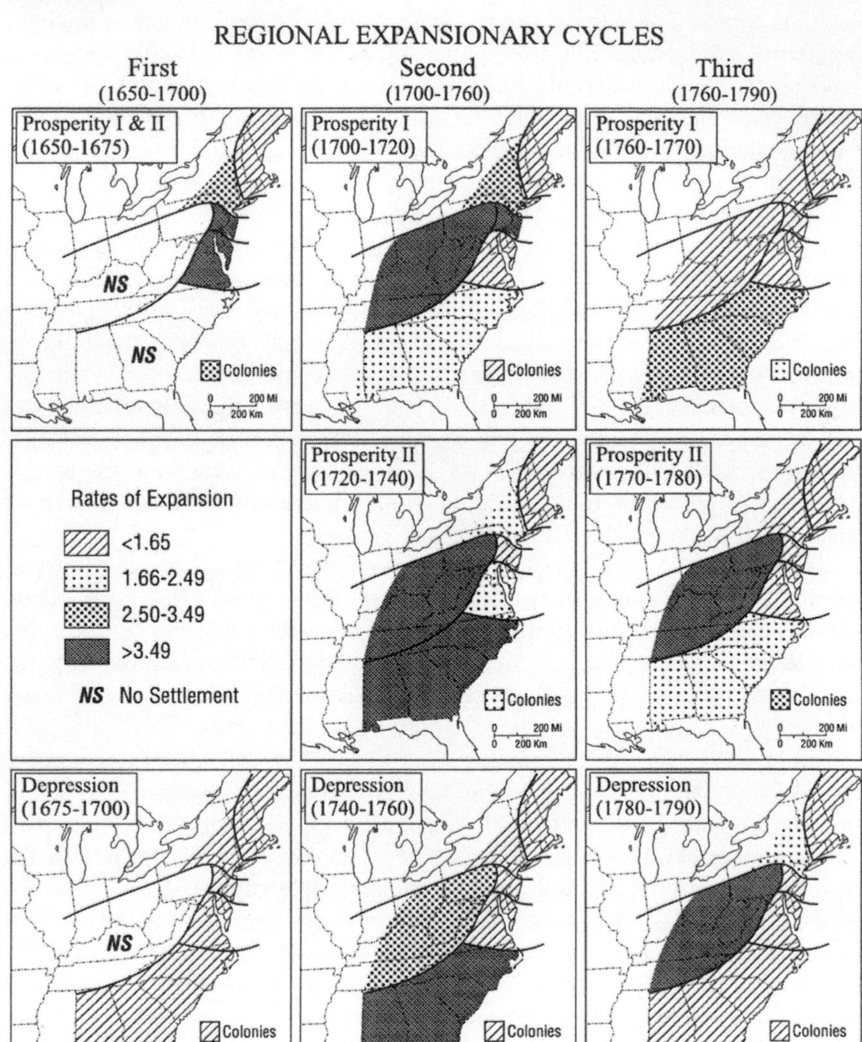

Note: The pertinent rate for the colonies/states as a whole is given by the symbols in the box located just east of South Carolina.

Source: Table 4.2.

and the Back Country); some twice (New England, the Chesapeake, and the Carolinas); and one thrice (New York). The immense variability of regional settlement during long periods of economic prosperity hints that outcomes were contingent more upon local resources, changing markets, immigration policies, individual initiatives, and luck than long waves in the colonial economy.[3]

Yet there is some consistency in this otherwise confusing scene. The regions of greater Pennsylvania and the Back Country (often settled by Pennsylvanians) duplicated most closely the national trajectory of colonial settlement expansion during periods of economic prosperity. These two macroregions, long regarded as the hearths of American society, occupied a deep channel in the mainstream of early American history.

I do not presume to explain the erratic and particularistic behavior of regional expansion rates during phases of economic prosperity. That task is best left to local and regional scholars of early American history and geography. Here I underline my central points: (1) early American settlement expansion was cyclical and its rate covaried with long waves in the colonial economy; (2) mean expansion rates varied from lows of 1 to 1.6 percent amid severe depressions to highs of around 3 percent toward the end of a three- to four-decade phase of economic prosperity. Aggregate and regional settlement experiences were more nearly alike in depression than in prosperity—except in Pennsylvania and its back country extension.

The traditional view of the frontier as a continuously expanding domain is, in light of the evidence presented here, misleading if not fallacious, a sort of historico-geographical half-truth. Turner's confident declarations of continuous frontier advance notwithstanding, early American frontier expansion was far from steady. Actually, the frontiers of settlement pulsed forward in long logistic waves, their rhythm regulated by larger structural changes in the Atlantic economy and an emergent capitalist world system. Because early American settlement expansion was cyclical rather than linear in trend and fast frontiers alternated with slow frontiers, we may conclude that the frontier was not one but many varying in macroregional space and macrohistorical time.[4] It is toward this plurality of frontier societies and landscapes that we now turn.

FRONTIER AND SOCIETY IN MACROHISTORICAL PERSPECTIVE

Frederick Jackson Turner erred in proclaiming that what he saw in his brief glimpse into the complex process of settlement expansion—a frontier of mostly males, fast profits, self-reliance, violence, and political opportunity—was in fact the whole of the matter. Historians and historical geographers have been proving him wrong ever since.

The critique of Turner has been an exercise in negation, and in that sense it has proven insufficient as a guide for inquiry. Nor do the disparate findings of countless studies of particular frontiers add up to a fruitful synthesis comparable to Turner's in scope or power. They fail because these interpretations, unlike Turner's, are insensitive to macrohistory. A proper critique should share Turner's vista overlooking long spans in the American past, for there alone can we survey the extraordinary variability and the dynamic qualities of the early American frontier.

The Frontier Mosaic

Turner was wrong about the frontier. Its true nature is dynamic rather than static, plural rather than singular, and cyclical rather than unilineal. It consists not merely of one Turnerian frontier experience endlessly replicated but of several swaying in rhythmic oscillation with long waves in the early American economy. To the extent that the early American frontier runs in half-century cycles, observers will experience a recurrent sense of déjà vu. Society and landscape on the slow frontier of Pennsylvania in the 1680s probably had more in common with the frontiers of Carolina and the Back Country in the 1740s or with frontier Kentucky in the 1780s than with Pennsylvania's fast frontiers in the 1720s or the 1760s.

These periodic refrains in the history of settlement expansion suggest a taxonomy of frontiers flanked at either end by the polarities of slow and fast, gradualist and volatile, gentle and robust, depressed and prosperous, permanent and evanescent. The need for a frontier taxonomy is elementary for purposes of analysis, but our exercise has the added virtue of sharpening the periodic resemblances among frontiers and developing the interpretive muscularity for a comparative American history.

Let us begin on the gentler end of our taxonomy, sketching in the ways in which the gradual passage of the frontier effects enduring transformations in landscapes and societies. When the frontier of settlement expands slowly, as it did during the depressions of the 1680s, 1740s and 1780s, landscape transformations are more likely. On slow frontiers, settlers are afforded the luxury of time to establish a settled way of life, nurture rudimentary socieconomic and political structures, and preserve ethnocultural traits that rapidly expanding frontiers too readily wash away. Slow frontiers afford a "breathing space" for the establishment, intensification, and perpetuation of ethnocultural communities. These were not the fast frontiers of Turnerian mythology, bustling with hunters, trappers, graziers, and other restless young men on the make. They consisted rather of a more stolid sort of citizenry, of families in flight from the trials of economic depression and the tribulations of ethnic and religious persecution and in search of a frontier refuge for the practice of their ethnocultural and communitarian traditions. Such were the motives that drove westward the "great migration" of Puritans in the 1630s and 1640s, the Quakers and Huguenots of the 1680s, the profusion of sectarian groups in the Great Awakening of the 1740s, and the "new lights" of the 1790s—not to mention the Mormon, utopian, and communitarian migrations of the 1840s.

For these adventurous souls, the term "frontiersmen" is a misnomer for settlers who came not alone but in groups, families, and communities. This fact explains the balanced sex ratios on frontiers as diverse as New England in the 1630s, Pennsylvania and the Carolinas in the 1680s, and Kentucky in the 1790s. In these hard economic times, "family" frontiers advanced slowly, thus allowing for the gradual thickening of ethnocultural geographies. Slow frontiers, peopled by groups seeking peace and salvation rather than by footloose males lusting after fame and fortune, were more nearly "sacred" than secular or profane (Earle 1992: 52–87).

A world apart from these gradualist, ethnocultural and sacred frontiers were the more volatile and robust frontiers of prosperity and rapid expansion. Here social change proceeded at a furious, almost inassimilable pace. On these "fast frontiers," the pace of expansion literally overwhelmed local institutions in a ceaseless change of personnel. Men came and went, rarely having time to settle down, marry, have children, and join in local life. For those who wanted to do these things, the scarcity of women precluded it. What a curious social matrix this was, with its imbalanced sex ratios (six males per female in early Virginia), migratory flux, "get-rich-quick" mentality, propensities toward violence, institutional instability, and transient landscapes. It is hard to imagine that these fast and robust economic frontiers and the slower ethnocultural frontiers were derived from the same historical stock.[5]

In these extremes of frontier expansion lie the seeds of a more dynamic interpretation of landscape evolution. This interpretation envisages not one landscape but many, a mosaic of historically specific landscapes whose residual elements subtly reflect the cyclical conditions prevailing at the precise moments of their formation.

This mosaic of frontier landscapes is composed of graduated shadings on the map of frontier expansion. In this frontier chiaroscuro, the darker shades symbolize the slow and gradualist frontiers of American history. They signify the frontiers of hard times, of enduring landscapes deeply etched by family, ethnocultural institutions, religious purpose, and communitarian value. The lighter shadings signify those ephemeral frontiers of outrageous prosperity and hell-bent settlement expansion. Transitory and opportunistic, created by the milling about of men in atomistic motion, these landscapes have a thinner and more feeble feel about them, born as they were out of haste and economic exploitation. These landscapes retain very little of their ephemeral frontier origins and a great deal more of the staple economies which eventually commandeered and reconfigured these rather vacant spaces.

THE NINETEENTH-CENTURY TRANSITION: FROM EXTENSIVE TO INTENSIVE SETTLEMENT EXPANSION

When Frederick Jackson Turner proclaimed the closing of the American frontier in 1890, he erred by perhaps as much as half a century. The evidence is fairly compelling. Following one last cycle of settlement expansion between 1790 and 1840, the rate of expansion plunged below 2 percent per annum during the 1840s and remained there for the rest of the century. The secular rate of expansion that had risen to a historic high of 2.47 percent per annum for the period between 1790 and 1840 fell to a historic low of 1.14 percent between 1840 and 1890 (Table 4.3). Frontier expansion after 1840 never again rose above 1.8 percent per annum in any single decade. On one occasion it dropped nearly to zero (0.2 percent). These rates hitherto had been associated only with the worst periods of protracted long-wave depression (Figures 4.7 and 4.8). If the frontier after 1840 was not completely closed, it was nonetheless a very different phenomenon than the frontier Americans had experienced during the preceding two hundred years.

Preoccupied by the pressing affairs of a new nation, most Americans scarcely noticed the midcentury closing of the frontier. A buoyant American economy and a growing population amply compensated for the declension of the frontier. For a people whose numbers mounted by 2.99 percent per annum before 1840 and 2.46 percent between 1840 and 1890, the freefall in frontier expansion was of modest consequence (Table 4.3).

The pivotal nature of the mid-nineteenth century deserves more notice than it has received. These were unusual times for the American settlement system.

Table 4.3 Settlement Expansion Rates, Population Growth, and Their Ratios, 1650–1890

Period	*Rate of Settlement Expansion (S) (% per annum)*	*Population Growth (P) (% per annum)*	*P/S*
1650–1700	2.02	3.26	1.61
1700–1760	2.10	3.13	1.49
1760–1790	1.93	3.05	1.58
1790–1840	2.47	2.99	1.21
1840–1890	1.14	2.46	2.32

Sources: C. O. Paullin, *Atlas of the Historical Geography of the United States,* ed. J. K. Wright (Washington, D.C.: Carnegie Institution of Washington and American Geographical Society, 1932), and U.S. Bureau of the Census, *Historical Statistics of the United States: Colonial Times to 1970,* Part 1, 8–9 (Washington, D.C.: Government Printing Office, 1975).

Note: Settlement expansion rates are calculated from digitized estimates of U.S. counties settled at a population density greater than or equal to 2 persons per square mile (Paullin 1932). Population growth rates are calculated from U.S. population counts in U.S. Bureau of the Census (1975).

Figure 4.7 Settled Areas and Populations of the Colonies and the United States, 1650–1890.

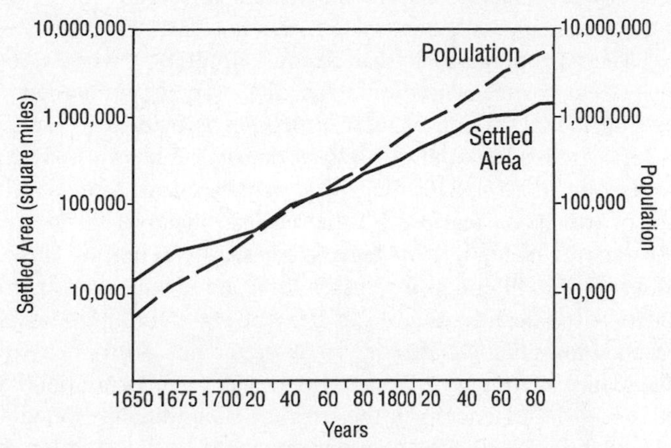

Sources: See text, notes, and United States Bureau of the Census (1975).

Their uniqueness is documented by changes in the ratio of population growth and settlement expansion (Table 4.3). During the three settlement cycles prior to 1790 that ratio varies within a narrow range (1.49–1.61). Thereafter the ratio converges dramatically toward parity (1.21) in the fourth cycle between 1790 and 1840 and diverges toward its historic extreme (2.32) between 1840 and 1890. The ratio's rapid convergence between 1790 and 1840 and its equally rapid divergence after 1840 hints at a fundamental restructuring of the American settlement system midway through the nineteenth century.

More precisely, the system shifted from extensive expansion to intensive development in previously settled regions (Earle and Cao 1994). American settle-

Figure 4.8 Annual Rates of Settlement Expansion, 1675–1890.

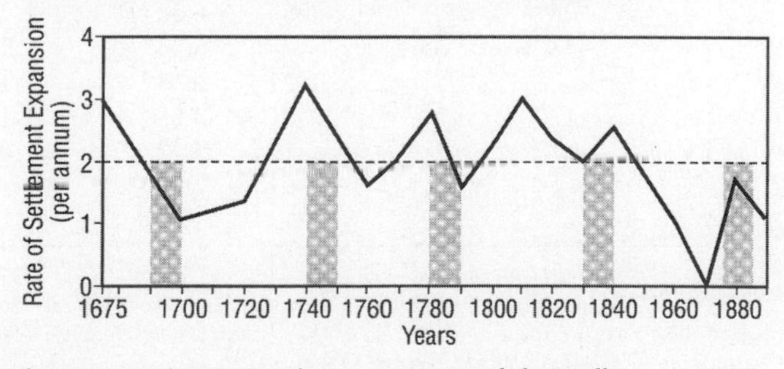

Note: Expansion rates equal percentage increases compounded annually.

Sources: Text, Table 4.2, and Earle and Cao (1994).

ment gravitated from the rural frontier toward the concentration of Americans in an emergent urban, industrial, and metropolitan core (Figure 4.9). This transformation in the American settlement system was not as discontinuous as it might first appear. The transition actually had begun before 1840, when vigorous frontier expansion and spatial concentration in older regions proceeded simultaneously. The transitional mixing of these processes is evident in rates of spatial expansion for various classes of population density (Table 4.4). Of 25 possible cases (five decades, 1790–1840, times five density classes), 17 density classes report rates of expansion that exceed the national mean of frontier expansion. Four of these cases occur in low-density classes (2–6 and 6–18 persons per square mile) and 13 in high-density classes. In a period when the American frontier expanded at the healthy rate of 2.5 percent per annum, high-density areas also increased at rates that equaled or surpassed the rate of frontier expansion.

The coincidence of frontier expansion and spatial concentration came to an abrupt end after 1840. Henceforth the frontier's significance faded away, and processes of spatial concentration assumed command of the American settlement system. Settlement expansion ratcheted upward into higher-density classes. In the decades between 1840 and 1890, expansion rates in low-density classes exceeded the national mean in just one of ten cases (Table 4.4) as compared to 14 of 15 cases among high-density classes.

Increased spatial concentration was a continuous historical process. A closer analysis of settlement expansion reveals a curvilinear transition from extensive

Table 4.4 Above-Average Rates of Settlement Expansion by Density Class, 1790–1890

	Density Class (persons per square mile)					
Decade	*2–6*	*6–18*	*18–45*	*45–90*	*>90*	*Mean (all classes)*
1790–1800	x	3.53	**4.45**	2.60	x	2.29
1800–1810	3.80	x	3.18	**6.71**	5.62	3.07
1810–1820	x	x	2.80	**2.84**	2.40	2.36
1820–1830	x	2.28	2.06	4.28	**6.54**	2.00
1830–1840	x	3.18	2.85	x	**8.06**	2.64
1840–1850	x	x	2.51	1.82	**6.33**	1.80
1850–1860	x	x	2.93	2.18	**3.08**	1.04
1860–1870	x	x	0.62	1.41	**1.72**	0.02
1870–1880	x	x	x	**7.96**	3.81	1.74
1880–1890	**2.15**	x	1.67	1.71	1.69	1.10

Sources: C. O. Paullin, *Atlas of the Historical Geography of the United States,* ed. J. K. Wright (Washington, D.C.: Carnegie Institution of Washington and American Geographical Society, 1932), and U.S. Bureau of the Census, *Historical Statistics of the United States: Colonial Times to 1970,* Part 1, 8–9 (Washington, D.C.: Government Printing Office, 1975).

Note: Rates in boldface type denote the highest rate of expansion in each decade; x equals below-average rate of expansion.

Figure 4.9 Cumulative Settlement Area by Population-Density Classes, 1790–1890.

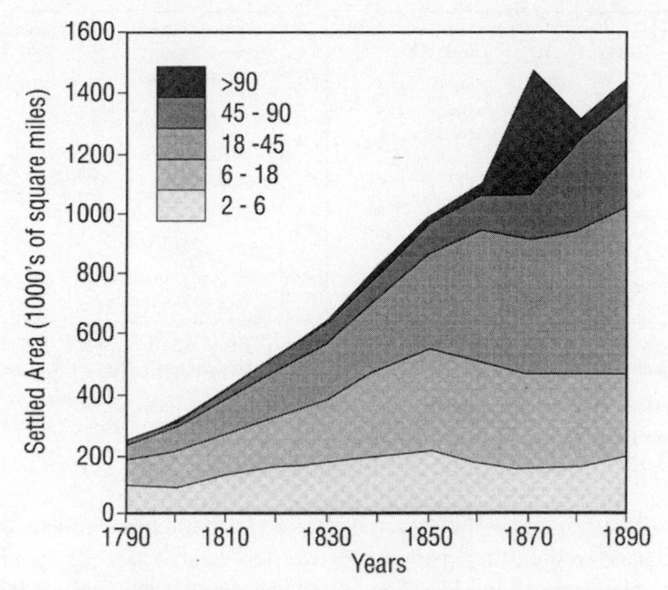

Note: Shaded areas represent the indicated levels of population density per square mile.
Sources: See sources cited in Table 4.3.

to intensive settlement processes (Figure 4.10). Tracking the leading edge of density-specific expansion discloses that peak rates of expansion rise and then fall, ascending from the density class of 18–45 persons per square mile between 1790 and 1800 to the 45–90 class between 1800 and 1820 and the greater-than-90 class between 1820 and 1870, before descending to the 45–90 class in the 1870s and the 2–6 class in the 1880s in what amounts to the frontier's last gasp. The graphs of density-specific rates reinforce the view of a curve that is concave downward in 1800, concave upward by 1880, and more or less linear (and positive) in between. These curves fix the temporal coordinates of the transition from spatial expansion to spatial concentration in the middle third of the nineteenth century. The transition began after 1820, reached its climax on or before 1870, and rose to an anticlimax in a brief and ironic return of frontier expansion during the 1880s.

Plotting the geographical coordinates of this transition requires a few methodological preliminaries. The first of these partitions the nation into a series of regional systems defined with respect to well-known hinterlands and functional patterns of interaction (Pred 1980; Conzen 1977). The second calculates rates of expansion for the five population-density classes in each of these regional systems. The resulting regional patterns document both the mid-century transition and the spatial realignment of frontier expansion and spatial concentration. In the case of frontier expansion between 1800 and 1840—a period of exceptionally

Figure 4.10 Population-Density Class Growth Rates, 1800–1890.

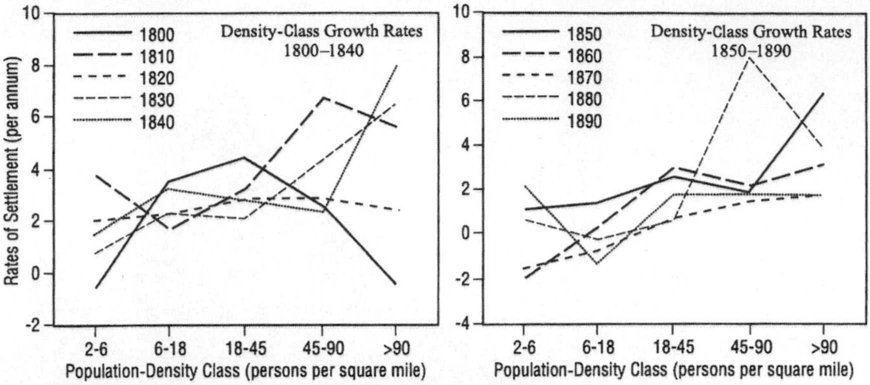

Note: Expansion rates equal percentage increases compounded annually.
Sources: See Table 4.3.

high rates of expansion—the regional data underline the prominent role of the South. Southern regions consistently register the highest rates of frontier expansion. Matters change after 1840. Thereafter, the South surrenders its leading role in frontier processes as the locus of expansion (of which there was very little in any case) shifts to the Midwest, and most notably to the urban systems of Chicago and St. Louis.

The regional realignments at midcentury due to settlement intensification are equally striking (Table 4.5). Between 1790 and 1850, the northeastern urban systems of Boston, New York, Philadelphia, and Baltimore consistently report the highest rates of expansion in the highest population-density class (greater than 90 persons per square mile). Unremarkably perhaps, New York leads the way in three of the six decades—including 1830–40 and 1840–50. After 1850 the locus of high-density expansion shifts toward the upper Midwest, and more particularly to Chicago and Minneapolis–St. Paul.

The closure of the American frontier and the concurrent transition from spatial extension to spatial intensification turned on three geographical changes: (1) the termination of rapid expansion on the southern frontier after 1840; (2) the concentration of population in the incipient northeastern megalopolis between 1790 and 1850; and (3) the enlargement of the megalopolitan core to the Midwest after 1850.[6] All are problematic. In the first instance, why the collapse of the southern frontier after 1840 precisely when the region's economy enjoyed its highest level of performance? In the second and third instances, why the concentration of population in the northeast before 1850 and the enlargement of the urban core to the Midwest after 1850 when ample supplies of land were available in the west? Is it possible that the impacts of the industrial revolution on American settlement came into play a good deal earlier and with considerably greater effect than is

Table 4.5 Regions with the Highest Rate of Settlement Expansion by Density Class, 1790–1890

	Density Class (persons per square mile)					
Decade	*2–6*	*6–18*	*18–45*	*45–90*	*>90*	*All Classes*
1790–1800	NY	Appal	SC	NY	Balt	Appal
1800–1810	Appal	SC	Appal	NY	NY	SC
1810–1820	Mobile	NO	NO	NE	Balt	Mobile
1820–1830	SC	Mobile	Up.MW	NO	Phil	Va
1830–1840	Up.MW	Up.MW	Phil	Up.MW	NY	Up.MW
1840–1850	NO	NO	Up.MW	Va	NY	Up.MW
1850–1860	[a]	OhioV	Up.MW	NO	Up.MW	Up.MW
1860–1870	Mobile	Mobile	Balt	NY	NY	Balt[b]
1870–1880	Chic	Minn	NO	StL	Minn	StL
1880–1890	Minn	Minn	Chic	Minn	Chic	Chic

Sources: C. O. Paullin, *Atlas of the Historical Geography of the United States,* ed. J. K. Wright (Washington, D.C.: Carnegie Institution of Washington and American Geographical Society, 1932), and U.S. Bureau of the Census, *Historical Statistics of the United States: Colonial Times to 1970,* Part 1, 8–9 (Washington, D.C.: Government Printing Office, 1975).

Notes:

(a) No region reported positive rates of expansion in this density class in the 1850s.

(b) Baltimore's leadership in frontier expansion is an artifact of the city's incorporation of the Virginia regional system rather than any new net gains.

Regional system abbreviations: Appal-Appalachian South; Balt-Baltimore; Chic-Chicago; Minn–Minneapolis–St. Paul; Mobile-Mobile; NE-New England; NO-New Orleans; NY-New York; Phil-Philadelphia; StL-St. Louis; SC-greater South Carolina; Up.MW-Upper Midwest; Va-greater Virginia.

usually allowed (David 1967)? What is clear is that the rate of American frontier expansion had largely collapsed after the 1840s. The fast frontiers that invariably accompanied prosperous times had become, after the 1840s, a dead letter. Henceforth the involutional processes of spatial concentration and intensification had displaced the centrifugal processes of frontier expansion as the prime movers of the American experience.

AFTERTHOUGHTS

The study of the American frontier has fallen on hard times; Frederick Jackson Turner's frontier thesis has been subjected to a withering assault by various representatives of the "new western history." Turner's thesis, they have argued, papered over the unseemly sides of American frontier history: the destructive occupance of new lands; the exploitation of soils, forests, waters, and air; the expulsion of Native Americans; and the subversion of Native American ways of life and ecological harmonies—all in the pursuit of the short-run profits of an unfettered capitalist economy (Kearns 1998).

What is remarkable about this debate is how little either side knows about the dynamics of American settlement expansion. In an intellectual version of Gresham's law, the cheap money of speculative interpretation has driven out the good of empirical evidence. Neither side is aware of the half-century cyclicity of frontier expansion prior to 1850; nor of the regular alternation of fast and slow frontiers in conjunction with long waves in the transatlantic economy; nor of the coaxial associations of economic crises, slow frontiers, and ethnocultural landscapes; nor of frontier closure after the 1840s and the concurrent spatial involution of American society.

Neglecting empirical evidence of the rates of American settlement expansion has profound implications for both old and new perspectives on the frontier. Proponents of Turner's thesis must come to grips with the key role of economic forces, with the primacy of cyclical over linear processes, and with the enormous variabilities of frontier expansion in space and time. Much needs to be explained, not least frontier closure and the involution of American society in the middle of the nineteenth century.

Proponents of the "new western history" have their work cut out for them as well. Indeed the implications of frontier variability are more profoundly troubling for them than for Turnerians. The new western historians must come to grips with the fact that the robust era of frontier expansion was coming to an end precisely when the occupation of the western half of the United States was just beginning. From a geographical perspective, these western regions constituted a slowly expanding periphery of marginal interest relative to the dynamic concentrations of population and economy then under way in the Northeast. From a comparative perspective, the settlement processes in the United States east and west of the one-hundredth meridian bore increasingly less relation to one another after the 1850s. The west was the tail end of the American dog—all of which goes a long way toward explaining the defensiveness of the new western historians and the bitterness of their assaults on Turner's thesis, the "eastern historical establishment," and the marginalization of western history in the American past.

NOTES

1. Neither Florin's (1977) study of the Pennsylvania frontier nor trend-surface models of settlement expansion pay much attention to rates of expansion (Norton and Smit 1977; but see the novel estimates of American frontier expansion in Warntz 1967: 207).

2. Political constraints on frontier expansion after 1680 are evident in the proliferation of regulatory policies and sharp increases in land prices (Rainbolt 1974; Webb 1979; McCusker and Menard 1985: 34–50). Nearly a century later, following the French and Indian War, British policymakers introduced similar constraints on frontier expansion (DeVorsey 1966).

3. Interpretation of these regional variations may be found in my forthcoming volume *Space, Time and the American Way: Liberalism, Republicanism, and Geographical History, 1600–2000*. Lanham, MD: Rowman & Littlefield.

4. The plurality of frontiers in time and space has been obscured by the assumption that the frontier is all one thing or all the other. This has precipitated a series of sterile debates on whether the frontier experience was principally economic or ethnocultural (Newton 1974; Mitchell 1978; Bernstein 1989).

5. Turner's caricature of the frontier amalgamated the several behavioral frontiers noted here. His amalgam blended the notion of the self-sufficient farmer from slow frontiers and the notions of male predominance and social volatility from fast frontiers.

6. On the emergence of the American core region, see Bensel (1990) and Ward (1971). As for southern frontier closure on the eve of the American Civil War, this fits nicely with Ransom's (1989: 41–60) theoretical argument that spatial expansion was contrary to slaveholder interests, since increasing the supply of cotton decreased cotton prices and slave assets.

REFERENCES

Bensel, Richard. 1990. *Yankee Leviathan.* Cambridge: Cambridge University Press.

Bernstein, Richard. 1989. "Among Historians, the Old Frontier is Turning Nastier with Each Revision." *New York Times*, December 17: E5.

Clark, Andrew H. 1960. "Geographical Change—A Theme for Economic History." *Journal of Economic History* 20: 607–16.

———. 1972. "Some Suggestions for the Geographical Study of Agricultural Change in the United States, 1790–1840." *Agricultural History* 46: 155–72.

Conzen, Michael P. 1977. "The Maturing Urban System in the United States, 1840–1910." *Annals of the Association of American Geographers* 67: 88–108.

David, Paul. 1967. "The Growth of Real Product in the United States before 1840: New Evidence, Controlled Conjectures." *Journal of Economic History* 27: 151–97.

DeVorsey, Louis. 1966. *The Indian Boundary in the Southern Colonies, 1763–1775.* Chapel Hill: University of North Carolina Press.

Earle, Carville. 1992. *Geographical Inquiry and American Historical Problems.* Stanford, CA: Stanford University Press.

Earle, Carville, and Changyong Cao. 1994. "Frontier Closure and The Involution of American Society, 1840–1890." *Journal of the Early Republic* 13: 163–80.

Egnal, Marc. 1988. *A Mighty Empire: The Origins of the American Revolution.* Ithaca, NY: Cornell University Press.

Florin, John. 1977. *The Advance of Frontier Settlement in Pennsylvania, 1638–1850: A Geographical Interpretation*. University Park: Department of Geography, Pennsylvania State University, Papers in Geography, No. 14.

Friis, Herman R. 1940. "A Series of Population Maps of the Colonies and the United States, 1625–1790." New York: American Geographical Society, American Geographical Society Mimeographed Publication No. 3.

Harris, P. M. G. 1996. "Inflation and Deflation in Early America, 1634–1860." *Social Science History* 20: 469–505.

Hart, John Fraser. 1974. "The Spread of the Frontier and the Growth of Population." In *Man and Cultural Heritage*, ed. H. J. Walker and W. G. Haag, 73–81. Baton Rouge: School of Geoscience, Louisiana State University.

James, Preston. 1959. *Latin America*, 3rd ed. New York: Odyssey Press.

Kearns, Gerry. 1998. "The Virtuous Circle of Facts and Values in the New Western History." *Annals of the Association of American Geographers* 88: 377–409.

McCusker, John J., and Russell R. Menard. 1985. *The Economy of British North America, 1607–1789.* Chapel Hill: University of North Carolina Press.

Meinig, Donald W. 1986. *The Shaping of America: A Geographical Perspective on 500 Years of History: Volume I: Atlantic America, 1492–1800.* New Haven: Yale University Press.

Mikesell, Marvin. 1960. "Comparative Studies in Frontier History." *Annals of the Association of American Geographers* 50: 62–73.

Mitchell, Robert D. 1978. "The Formation of Early American Cultural Regions: An Interpretation." In *European Settlement and Development in North America*, ed. James R. Gibson, 66-90. Toronto: University of Toronto Press.

Mitchell, Robert D., and Paul Groves. 1987. *North America: Historical Geography of a Changing Continent.* Totowa, NJ: Rowman & Littlefield.

Mood, Fulmer. 1952. "Studies in the History of American Settled Areas and Frontier Lines, 1625–1790." *Agricultural History* 26: 16–33.

Morgan, Edmund. S. 1975. *American Slavery, American Freedom: The Ordeal of Colonial Virginia.* New York: Norton.

Nash, Gary B. 1968. *Quakers and Politics: Pennsylvania 1681–1726.* Princeton: Princeton University Press.

Newton, Milton. 1974. "The Cultural Preadaptation of the Upland South." In *Man and Cultural Heritage,* ed. H. J. Walker and W. G. Haag, 143–54. Baton Rouge: School of Geoscience, Louisiana State University.

Norton, W., and P. D. Smit. 1977. "Rural Settlement Surface Evolution: Cape Province, 1865–1970." *Geografiska Annaler* 56B: 43–50.

Otterstrom, Samuel Mark. 1997. "An Analysis of Population Dispersal and Concentration in the United States, 1790–1990: The Frontier, Long Waves, and the Manufacturing Connection." Ph.D. dissertation, 2 vols., Louisiana State University.

Paullin, C. O. 1932. *Atlas of the Historical Geography of the United States,* ed. J. K. Wright. Washington, DC: Carnegie Institution of Washington and American Geographical Society.

Pred, Allan. 1980. *Urban Growth and City-Systems in the United States, 1840–1860.* London: Hutchinson.

Rainbolt, John C. 1969. "The Absence of Towns in Seventeenth-Century Virginia." *Journal of Southern History* 35: 343–60.

———. 1974. *From Prescription to Persuasion: Manipulation of the Seventeenth-Century Virginia Economy.* Port Washington, NY: Kennikat Press.

Ransom, Roger. 1989. *Conflict and Compromise: The Political Economy of Slavery, Emancipation, and the American Civil War.* Cambridge: Cambridge University Press.

Sirmans, M. Eugene. 1966. *Colonial South Carolina: A Political History, 1663–1763.* Chapel Hill: University of North Carolina Press.

Turner, Frederick Jackson. 1920. *The Frontier in American History* New York: Henry Holt.

United States Bureau of the Census. 1975. *Historical Statistics of the United States: Colonial Times to 1970*, Part 1, 8–9. Washington, DC: Government Printing Office.

Wallerstein, Immanuel. 1974. *The Modern World-System: Capitalist Agriculture and the Origins of the European World-Economy in the Sixteenth Century.* New York: Academic Press.

———. 1980. *The Modern World-System II: Mercantilism and the Consolidation of the European World-Economy, 1600–1750.* New York: Academic Press.

Ward, David. 1971. *Cities and Immigrants: A Geography of Change in Nineteenth-Century America.* New York: Oxford University Press.

Warntz, William. 1967. "Macroscopic Analysis and Some Patterns of the Geographical Distribution of Population in the United States, 1790–1950." In *Symposium on Quantitative Geographical Research: Part I*, ed. W. C. Garrison and D. F. Marble, 195–218. Evanston, IL: Northwestern University Studies in Geography 13.

Washburn, Wilcomb. 1957. *The Governor and the Rebel.* Chapel Hill: University of North Carolina Press.

Webb, Stephen Saunders. 1979. *The Governors-General: The English Army and the Definition of Empire, 1659–1681.* Chapel Hill: University of North Carolina Press.

Zelinsky, Wilbur. 1973. *The Cultural Geography of the United States.* Englewood Cliffs, NJ: Prentice-Hall.

II

REMAKING THE ENVIRONMENT

5

Wittfogel East and West: Changing Perspectives on Water Development in South Asia and the United States, 1670–2000

James L. Wescoat Jr.

In a conversation on Karl Wittfogel's "hydraulic hypothesis" for the origin of the "Oriental state" in the mid-1980s, Marvin W. Mikesell reflected, "Why doesn't that idea stay dead?"[1] I want to try to answer his question by situating Wittfogel's arguments within a historical, cultural, and practical perspective on water development in two regions, South Asia and North America, over the past three and a half centuries. Water resource comparisons between South Asia and the United States have only occasionally involved independent case studies; more commonly they have involved long-term processes of interaction and exchange.

From the 1970s to the present, water experts in both regions have been debating the roles of the state, community, and private property in guiding water and environmental management (some recent works involving geographers include Chapman and Thompson 1995; Emel and Roberts 1995; Jacobs and Wescoat 1994; National Research Council 1996; Templer 1997; Wallach 1996; White 1997). Controversies have escalated over large dams, river-basin development, irrigation management, aquatic ecosystem degradation, desertification, flood hazards, water laws and pricing, and community-based water management (White 1997).

For much of the twentieth century, American water specialists have regarded themselves as exporters of ideas, technologies, and institutions of water development to regions like South Asia (TVA 1961; Wescoat, Smith, and Schaad 1992). A hundred years ago, however, U.S. scientists, engineers, and lawyers actively imported irrigation innovations from South Asia and other parts of the world for application in the Central Valley of California and elsewhere (Brown 1905; Davidson 1875; Hall 1886; Hilgard 1886; Norton 1853; Smith 1861; Wilson 1890–91, 1894). The roots of policy deliberations concerning water development, property rights, state power, and environmental degradation reach back even further, drawing upon some of the great debates in western social thought of the seventeenth through nineteenth centuries.

My aim in retracing this historical geography of ideas about water development in South Asia and the United States is to show that it is once again time for American water specialists to seek out water management innovations from other regions of the world such as South Asia and to seek a more equal exchange of expertise with such regions. I believe such historical and cultural geographic perspectives have a role to play in informing water policy debates.[2]

My story begins with Wittfogel's controversial "hydraulic hypothesis," how it has been used in recent research on water and power in the United States and South Asia, and why we need a longer-term macrogeographic perspective (Figure 5.1). The core of the chapter then traces the flow of western ideas about Asian irrigation from the mid-seventeenth century to the present, beginning with Francois Bernier's *Travels in the Mogul Empire AD 1656–1668* and continuing to the live policy debates of the present day. Although ideas about "Oriental Despotism" have had little direct impact on water policy, the section on "Wittfogel: Truth or Consequences?" outlines a chain of indirect influences that account, in part, for why it has not yet died.

WITTFOGEL'S LEGACY

Karl Wittfogel is a complex figure in the history of twentieth-century social and geographic thought (see *Antipode* 1985; Bernard and Reynolds 1992; Ulmen 1978; Wittfogel 1981/1957). Born in 1896 in Germany, he became a teacher and wrote prolifically. As a young member of the German Communist Party he wrote probing, polemical tracts on environmental aspects of geopolitics and social organization during the 1920s (Wittfogel 1985/1929). He developed a Marxist critique of Ratzel, Richthofen, Kjellen, Haushofer, and others, and a geographic critique of contemporary Marxist theory. He took a doctorate at the University of Frankfurt in economics in 1925, developing a special interest in Chinese economic history, and became a research associate at the Institut für Sozialforschung there. Imprisonment in 1933 and outrage at Nazism and Stalinism led him to flee to the United States, break with the communist movement in 1939, and swing sharply to the political right (including a highly controversial role in the U.S. congressional anti-communist hearings of the early 1950s). His substantive contributions focused on Chinese history, pursued through the Chinese History Project at Columbia University and the University of Washington.

These experiences reshaped Wittfogel's arguments about the relations between nature and society, as he sought to draw connections between environmental

Figure 5.1 Wittfogel East and West: Western Ideas about Asian Irrigation.

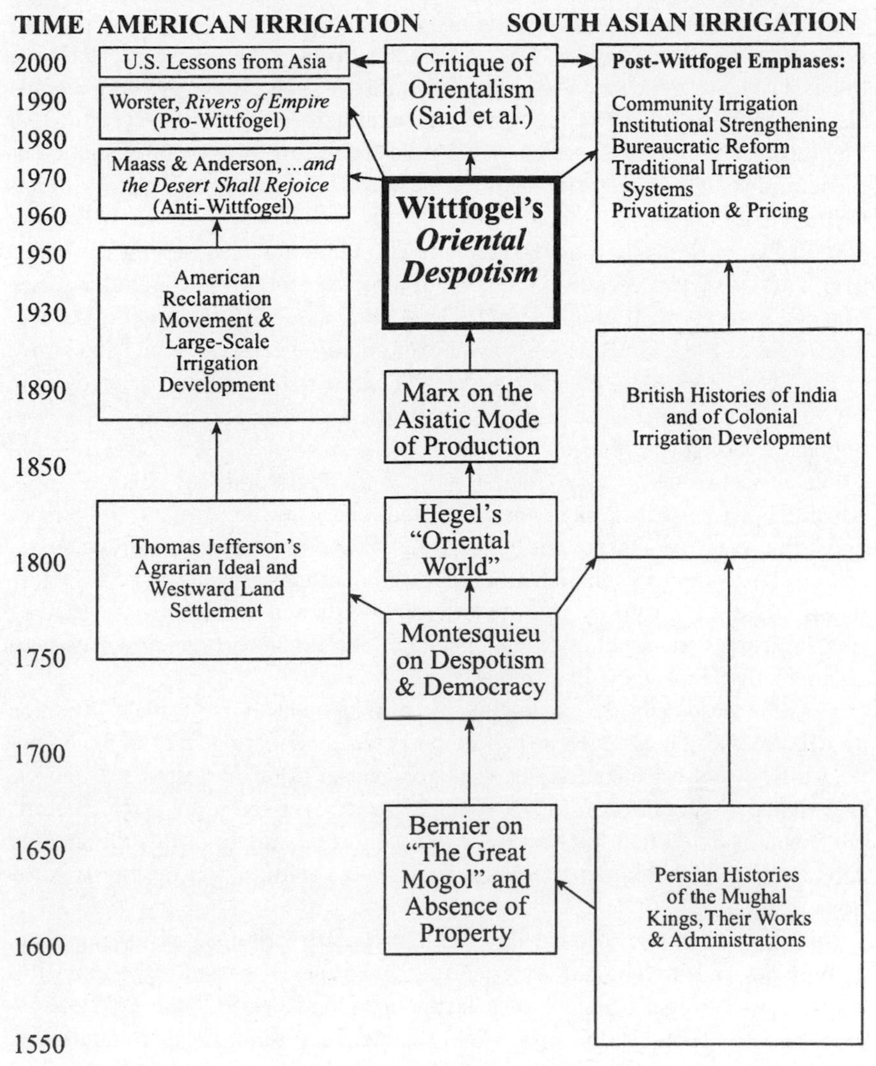

processes, resources management, and the exercise of social power. His canvas shifted from China to the world scale, inspired in part by his wife, anthropologist Esther Goldfrank, who studied Pueblo Indian cultures of the southwestern United States. His correspondence with Julian Steward (1955) indicates growing emphasis on irrigation agriculture and what Wittfogel termed "hydraulic societies" or "hydraulic civilizations," which referred as much to a pattern of despotic-bureaucratic social organization as it did to irrigation technologies and systems (Wittfogel 1955).

These strands of theoretical, substantive, and ideological writing coalesced over a period of two decades in a grand comparative work on irrigation agriculture and social organization titled *Oriental Despotism: A Comparative Study of Total Power* (1981/1957) which ranged over continents and millennia and strove to classify and evaluate the relationships between water, culture, and power in ways that spoke to the most controversial political debates of his times. Wittfogel published numerous extensions of the arguments in *Oriental Despotism* from positions at Columbia and the University of Washington until his death in 1988, which helped perpetuate the debate; but that one work became the benchmark and straw man for subsequent generations of social researchers on water resources. The story of its influence is interesting for many reasons, in part because it was initially engaged by cultural anthropologists and archaeologists, later by social scientists and water resources specialists. It is to this progression of debates about Wittfogel's legacy that we now turn.

Oriental Despotism sought to link the development of large-scale irrigation agriculture with the absence of private property, the emergence of complex bureaucratic social organization, the entrenchment of political despotism, and the experience of "total terror" (Wittfogel 1981/1957). Even more provocatively, Wittfogel asserted that the best twentieth-century example of this threatening conjunction of environment, technology, society, politics, and despair was the Soviet Union.

Oriental Despotism stimulated waves of scholarly criticism of its empirical, theoretical, and ideological errors. Some geographers tended to be more impressed by Wittfogel's breadth of investigation (Jones 1958; Spate 1959). However, Shabad (1959) and Gourou (1961) wrote sharp rebuttals that emphasized Wittfogel's abuse of concepts of the "Oriental," "despotic" and "hydraulic." Some water resources faculty read it, but it had no immediate impact upon water resources research or management.

In addition to his pre- and protohistoric errors about the causal role of irrigation in the origin of complex social organization, Wittfogel's global classification of irrigation systems failed to even mention the colonial irrigation programs of Asia and Africa, or the large water bureaucracies of the U.S. Bureau of Reclamation, the Tennessee Valley Authority (TVA), and the postcolonial water authorities of India, Pakistan, and Bangladesh—that is, the largest and most state-centered water systems in world history. These errors should have been enough to put the "hydraulic hypothesis" to rest, and for most scholars they have.

But Wittfogel is still invoked, reexamined, and often dismissed again (Bernard and Reynolds 1992; Sidky 1997; Ulmen 1978). In the western United States, Maass and Anderson (1978) drew upon Jean Bruhnes' ideas to test Wittfogel's propositions in six water case studies, concluding that powerful local irrigation organizations more often manipulated national reclamation agencies than vice versa. Focusing on large-scale water development in California, however, Worster (1985) adapted Wittfogel's perspective to what he termed a "state-capitalist mode of production." Although more recent authors seem to have set Wittfogel himself aside for the moment, research on "water and power" is growing, if anything, and often owes something to debates sparked by Wittfogel (Emel and Roberts 1995).

Outside the water resources field, Richard Peet organized an effort to reexamine Wittfogel's ideas in a special issue of *Antipode* (1985), arguing that attention should be directed toward Wittfogel's (1985/1929) early contributions to the fields of geographical materialism and the nature–society dialectic (also Chappell 1971; Peet 1988). Large literatures have retraced the flow of ideas about "Oriental despotism," "hydraulic civilization," and the "Asiatic mode of production" (O'Leary 1989).

But little has been written about the international flow of ideas and associated technologies, commodities, and institutions between irrigated regions of the world on the timescale of centuries. Aside from Maass and Anderson (1978) and Worster (1985), the influence of Asia and Europe on North American water development, and vice versa, in more recent decades remains a neglected subject. It is this story that I sketch out below.

EARLY ENCOUNTERS, 1660–1780

The story begins at the climax of the Mughal empire in India, an empire that expanded during the sixteenth and seventeenth centuries to cover much of South Asia, and that drew the keen attention of European travelers and traders who initiated trade of bullion from the West Indies for finished textile goods and spices from the East Indies (Figure 5.2).

They also trafficked in ideas about water resources, property rights, and political organization. A particularly influential travel account was written by the French physician-philosopher Francois Bernier, who attended the courts of the sixth Mughal emperor Aurangzeb Alamgir in Agra and Delhi in the 1660s.

Upon his return, Bernier (n.d./1670: 200) wrote a letter to Jean-Baptiste Colbert, the powerful finance minister to Louis XIV on "the principal cause of decline of the states of Asia." Bernier wrote that, "As the ground is seldom tilled otherwise than by compulsion, and as no person is found willing and able to re-

Figure 5.2 Early Encounters, 1660–1780.

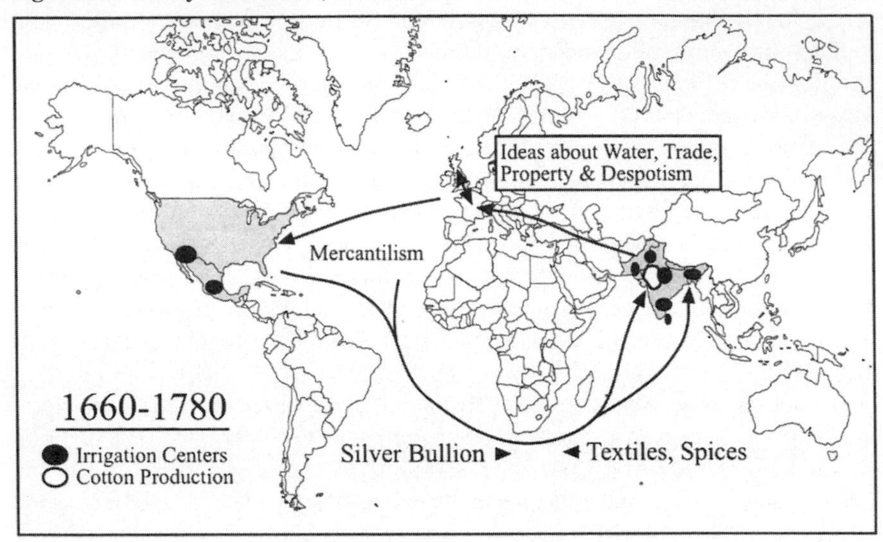

pair the ditches and canals for the conveyance of water, it happens that the whole country is badly cultivated, and a great part rendered unproductive from the want of irrigation" (Bernier n.d./1670: 226–27). He attributed these deficiencies to the absence of private property among the agricultural and aristocratic classes, excessive imperial taxation, and the escheat of personal property to the king.

Bernier thus associated despotism with *deteriorating*, not magnificent, irrigation works around Delhi and Agra. This rendering stood in sharp contrast with Bernier's descriptions of imperial monuments such as the Taj Mahal and Yamuna riverfront gardens that surpassed in beauty the views from the Pont Neuf (Bernier n.d./1670: 297ff.)! It is important to keep in mind that Bernier's letter was written at the height of French absolutism and monumental garden construction at home, and expansionism overseas. It was thus more a veiled and cautionary tale for the French leader than it was a purely objective account of India. He concluded the letter to Colbert with the general geographical assertion: "take away the right of private property in land, and you introduce, as a sure and necessary consequence, tyranny, slavery, injustice, beggary and barbarism . . . it is the prevalence or neglect of this principle which changes and diversifies the face of the earth" (Bernier n.d./1670: 238). Although Bernier did not discuss private property in water, his arguments would later be extended to water as well as land resources.

Interestingly, the irrigation works of northern India at this time were small in scale, local in effect, and simple in technology—as indicated by evidence from Mughal paintings of wells, tanks, and Persian wheels. There were a few large perennial canals on the Yamuna River, such as the one constructed by Firoz Shah

Tughluq in the late fourteenth century and renovated by the third Mughal ruler, Akbar, and a Persian noble, Ali Mardan Khan, in the late sixteenth to seventeenth centuries. A large map of Ali Mardan Khan's canal makes it appear monumental, but the map also depicts a large number of canal turnouts for labeled properties along its length (Gole 1989).

The predominance of small irrigation systems and complex diversions from larger canals invites the question: How did western scholars come to associate large-scale imperial irrigation works with political despotism in South Asia?

To answer this question, we need to examine Bernier's influence on subsequent currents of European social thought. His account appeared in Paris in 1670, in English in 1671, and was reprinted frequently in Europe—it was a "bestseller."

Bernier was cited by the most influential social theorists of enlightenment France and England, from John Locke, who was a personal friend of Bernier's to Encyclopedists like d'Herbelot. Utilitarians and Marxists carried these ideas further. This general chronology has been recounted in recent histories of ideas about Orientalism and the Asiatic mode of production (O'Leary 1989), but less attention has been given to the inclusion of ideas about water, a theme outlined below:

1. In his second treatise on government, published in 1690, John Locke repeated the connection between despotism and the absence of private property (Locke 1965/1690: 430–31; Locke 1812).
2. In 1748, Baron de Montesquieu drew upon Bernier and Locke to argue in *The Spirit of the Laws* that aridity was associated with, and helped explain, the occurrence of despotism and the absence of private property in Asia (Montesquieu 1949/1748: 57–65, 224, 226, 269).
3. In *The Wealth of Nations*, Adam Smith noted the importance of canals in the domestic economies of China and Hindustan. He appreciated Bernier's critical perspective on public works in Asia, stating that there were other "accounts of those works which have been transmitted to Europe . . . by weak and wondering travelers; frequently by stupid and lying missionaries" (Smith 1976/1776: vol. 2: 256). But Smith went on to speculate that Asian rulers might have placed greater emphasis on canals and roads to increase land revenues, upon which their wealth was based, but he concluded that such public works are better managed by local governments.
4. John Stuart Mill spent most of his career as an employee of the East India Company in London. He stated in his *Principles of Political Economy* (1965/1848: 13) that irrigation works were occasionally patronized in "the enlightened self interest of the better order of princes" in Asia. Finally,
5. Marx and Engels returned to Bernier to re-assemble all of the pieces of this puzzle in a new way that associated the absence of private property with large-scale irrigation works and political despotism in the arid continental

environments of Asia, which they characterized as a distinctively Asiatic mode of production. (Avineri 1969)

Before examining Marx and Engels, who influenced Wittfogel, in more detail it seems important to ask to what extent this chain of ideas accorded with or influenced water development patterns or practices in either Asia or the Americas.

FROM COMPARISON TO EXPERIMENTATION AND CRITIQUE, 1780–1860

There is little evidence that any of the ideas assessed thus far influenced water management in either South Asia or the United States (Figure 5.3). Some British officers studied Mughal and early travel accounts (Colvin 1833; Yule 1846). Montesquieu's ideas certainly influenced Jefferson, Madison, and Franklin on politics and property. And those ideas, among others, may have constrained federal involvement on river, harbor, and canal projects of the 1780s through 1820s. Although Bernier's direct influence on American social thought appears slight, and there was no direct connection between ideas about the ownership of water and land as yet, there was a broad flow of ideas about property, trade, and despotism to which the literature on India and the East India Company contributed.

American leaders took offense at comparisons by Abbe Raynal (1776) and oth-

Figure 5.3 From Comparison to Experimentation and Critique, 1780–1860.

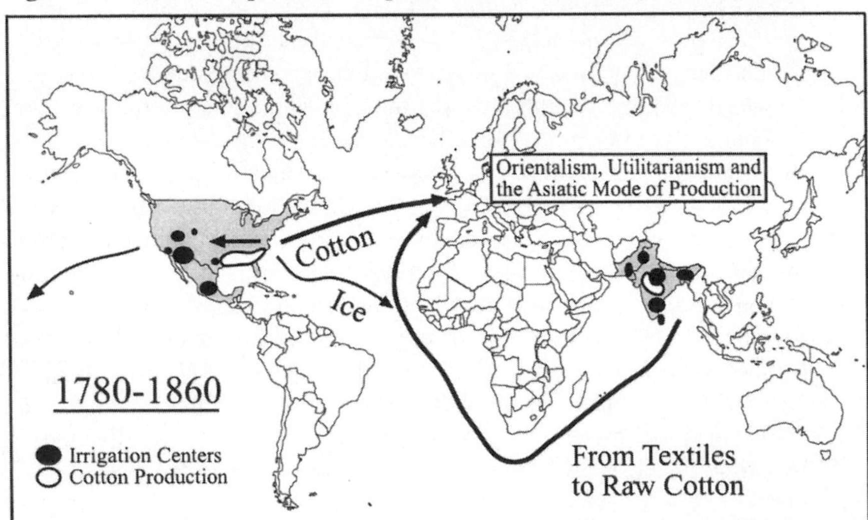

ers of the East Indies as "decadent" and the West Indies as "savage." European political leaders sought to draw lessons from the revolution in the American colonies and apply them in India. They engaged in heated comparisons and debates about colonial tax and land-tenure policies. Soon after his defeat in the American revolution, for example, Cornwallis became governor-general of the East India Company, where in 1793 he enacted sweeping changes in land-tenure laws that established a new *zamindari* propertied class. In London, Edmund Burke launched impeachment proceedings against the previous governor-general, Warren Hastings, on grounds of corruption. When Hastings tried to invoke the accounts of travelers like Bernier to justify continuation and incremental transformation of "Oriental" customs, Burke, a conservative Whig, disputed the entire notion of "Oriental Despotism" as unsound in theory and evidence (Burke 1981).

Back in India, battles arose between those whom historian Eric Stokes (1959) has called the "Paternalists," the "Utilitarians," and the "Evangelicals." The Paternalists patronized studies of customary law and resource use, partly to guide the renovation of historic waterworks like the Ganges and Yamuna canals, as well as to increase revenues. Utilitarians, by contrast, advocated radical legal reforms based on the ideas of Jeremy Bentham (1962/1838–43: 179–94). With James and John Stuart Mill in the employ of the East India Company, and high-ranking disciples in India, they had the muscle to force large-scale experiments in legislation, adjudication, and finance—in which private property was eschewed as a likely drain on company revenues (Mill 1965/1848)!

Both regions were surveyed with similar mapping projects that followed coastal harbors to major inland waterways such as the Ganges, Indus, Missouri, and Mississippi rivers into the continental interiors. Government-sponsored geographical surveys, such as those of James Rennell in India and Louis and Clark in the United States, followed and mapped river corridors (Edney 1997: 98–102; Rennell 1785). Freedom of navigation and commerce was invoked both in India, to gain access to the Indus River, and in North America to open the Mississippi River. The principle of free navigation was followed in each case by military conflicts and cessions of territory to England (Wescoat 1996).

Trade relations between the American colonies and South Asia remained limited through the late eighteenth century. Perhaps the earliest water resources connection was the export of ice from New England to Bengal as coolant as well as ballast. This situation would change, however, with westward expansion in America and colonial expansion in India in the mid-nineteenth century. To understand how those economic changes were related to intellectual and ideological debates, we need to return to Marx and Engels.

During the nineteenth century, criticisms of colonial experiments began to converge with ideas about Oriental societies. On June 2, 1853, Marx wrote to Engels that "on the formation of Oriental cities, one can read nothing more brilliant . . . than old Francois Bernier," noting in particular Bernier's comments on the absence of private property (Avineri 1968: 425–26). Engels replied that "Old Bernier's material is really very fine. It is a real delight . . . to read something by a clearheaded old Frenchman, who keeps hitting the nail on the head without appearing to notice it" (Avineri 1968: 429).

It was Engels who shifted the conversation and context from cities to irrigation and public works, a thesis that Marx picked up and published two weeks later in the *New York Daily Tribune* when covering parliamentary debates on renewal of the East India Company charter (Avineri 1968: 93–101).[3] Marx criticized British rule as a sickening yet necessary stage to pass beyond Oriental Despotism, which he caricatured as governments of plunder and public works resting upon a sea of unchanging, insular, superstitious village communities.

These arguments were overtaken by the 1857 rebellion that led to the transfer of control from the East India Company to the British crown; by Marx's shift in interest away from Asia; and by the American Civil War which blocked the flow of cotton from the southern United States to the mills in England, contributing to the expansion of canal irrigation in India and Egypt (Farnie 1979).[4]

INDIAN INFLUENCE ON AMERICAN IRRIGATION, 1860–1930

Expansion of perennial canal irrigation in India during the late nineteenth century represents one of the most dramatic transformations of land and water resources in the modern era—in terms of both the extent and rates of change in land cover, irrigation technologies, and institutions (Figure 5.4). Early irrigation projects in northern and southern India were dramatic and, for the most part, profitable (Wallach 1996; Whitcombe 1982). Colonial engineers and officers sought to combine customary and statute law, traditional and scientific practices, political and economic objectives (Ali 1988; Gilmartin 1994, 1995; Zafar 1985). British engineers looked to southern Europe, particularly Spain, France, and the Piedmont region of Italy, for practical irrigation lessons (Moncrief 1868; Smith 1849, 1855). They showed little interest in the fledgling irrigation systems of American irrigators in Utah or California, or in the older systems of Mexico and the middle Rio Grande Valley.

American engineers and agriculturalists, by contrast, gave close attention to water development in India. Acting on behalf of the U.S. Congress, the State of California, and irrigation investors, American engineers visited Indian works. They prepared reports on irrigation structures, economics, and institutions. The influence of Indian irrigation in the western United States involved both diffusion of innovations and differentiation of American from Indian approaches.

American engineers imported innovations in hydraulic engineering and earthworks technologies, for example, and they lamented the shoddy construction at home compared with structures built to last in India. The concept of large-scale water development diffused, *contra* Wittfogel and others, from colonial India and

Figure 5.4 Indian Influence on American Irrigation, 1860–1930.

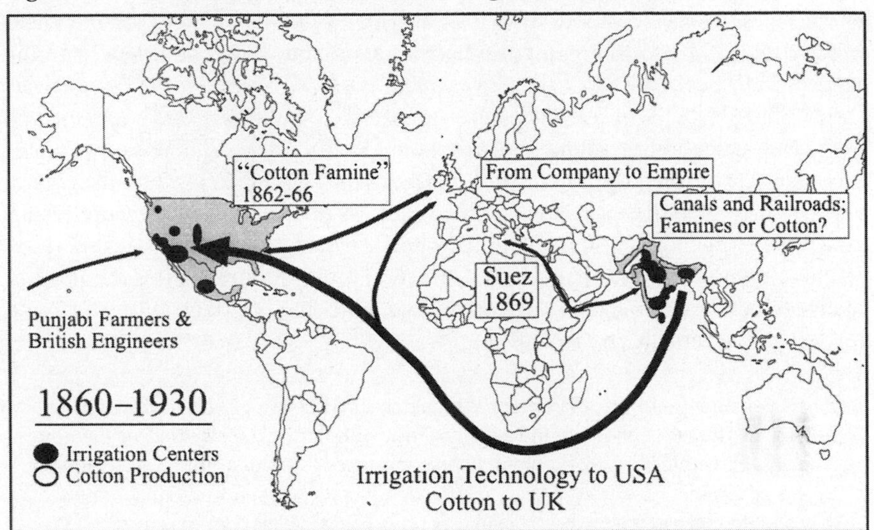

not from the precolonial empire of the Mughals. It is no exaggeration to say that colonial India demonstrated the large-scale possibilities and problems associated with state-sponsored water development in sparsely populated semi-arid regions, such as the Punjab. Although not all American scientists supported irrigation development—George Perkins Marsh (1874), for example, testified to the U.S. Congress on the "evils" of irrigation—the reclamation movement gained momentum in the late nineteenth century.

The diffusion of irrigation innovations was aided by the migration of British irrigation engineers and Punjabi cultivators to the Imperial Valley in the first decade of the twentieth century (Flynn 1892; Jensen 1988; La Brack 1988; Leonard 1992; Mazumdar 1984). Another possible, but as yet unexamined, connection may have involved a common set of British investors in American and Indian irrigation projects (Lee 1980).

American irrigation promoters also stressed differences between the United States and India. These included differences in labor supply and costs, explaining the scale and solidity of Indian irrigation works by low labor costs (Wilson 1894). American engineers were amazed at, and in some cases envious of, the level of state control and water ownership in India. They often cited the Indian Canal and Drainage Act of 1873, whose preamble runs: "Government is entitled to use and control for public purposes the water of all rivers and streams flowing in natural channels." It remains the law to this day. American legal treatises compared irrigation laws in India and other countries with those taking shape in the western United States (Kinney 1912).

A member of the Board of Commissioners for the Central Valley of California wrote in 1875 that "compensation for entry upon private lands is arbitrary and minimum. . . . It is utterly impossible that such conditions could exist in the United States, *except where the Government enter[s] new territory*" (emphasis added, Davidson 1875: 39; Jackson et al. 1990). "New territory" was exactly how some irrigation promoters regarded the Central Valley of California. Commissioner Davidson added that "As compared with the Italian system, the greater undertakings of India and their whole system of distribution seem more analogous to what is required in the United States" (Davidson 1875: 40). The Central Valley commissioners displayed not a trace of irony in applauding the government of India's combination of paternalism, on behalf of poor cultivators, and revenue enhancement, in these words:

> It was generally held that the property in water could not safely be intrusted to private hands; that the ignorant cultivators would, without the intervention of the government, be helpless against a powerful corporation . . . At this time it was thought by the government that the profits of irrigation were great and immediate, and that they should inure to the government and not to a corporation. (Commissioners 1874: 55)

> The full development of many of our broad valleys depends wholly and solely upon the adoption of some of these propositions. (Davidson 1875: 69)

I quote these influences of India on the United States at length because they are less well-known than, and yet are directly related to, subsequent influences of the United States on India.

AMERICAN INFLUENCE ON INDIAN WATER DEVELOPMENT, 1930–1980

By the 1930s, the balance of trade in water resource innovations had shifted from India to the United States (Figure 5.5). American water engineers and planners began to export new technologies of dam construction, hydropower engineering, and groundwater development. Significant numbers of South Asian students studied water resource engineering and planning in American universities, federal agencies, and consulting firms—with consequences that have yet to be fully appraised (Wescoat, Smith, and Schaad 1992).

The jewel in the American crown, however, was the Tennessee Valley Authority (TVA), which was exported around the world as a model for comprehensive state-sponsored regional water development even as its further application was rejected in the United States (Clapp 1955; Hargrove 1994; Lilienthal 1944). The Damodar Valley Corporation in India and the Gal Oya project in Ceylon served as a warm-up for the massive Indus, Mahaveli, and Mekong development pro-

Figure 5.5 American Influence on Indian Water Development, 1930–1980.

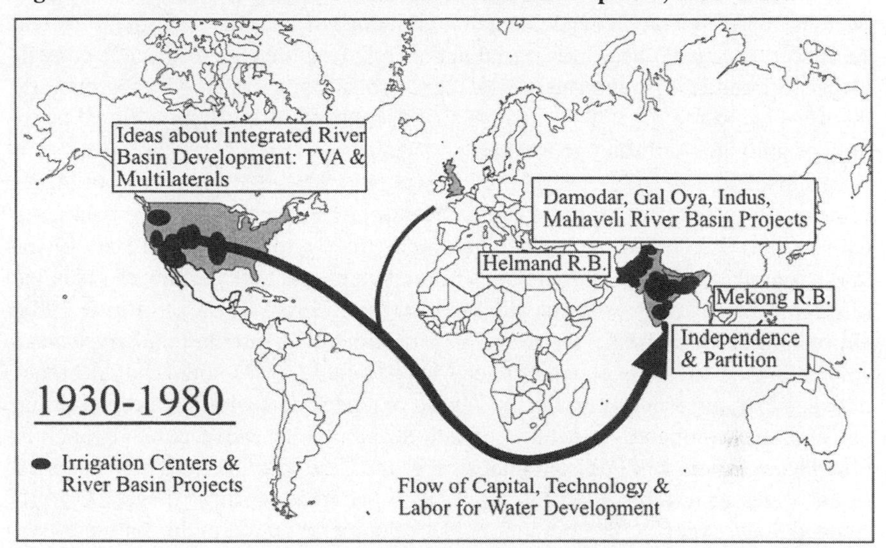

grams of the postcolonial era (Damodar Valley Corporation 1992; TVA 1961; Uphoff 1992).

Although these latter river-basin schemes were closely tied with U.S. Cold War geopolitical strategies, they were accompanied by vigorous promotion of principles for reforming public finance through project evaluation methods, water pricing, and water markets, reforms that made little headway in practice at home or abroad.

American water law had some influence in South Asia, though it and other branches of American law have been viewed as overly individualistic by Indian jurists (Baxi 1985; Beer 1979; Singh 1991; Vani 1992). American precedents on interstate streams, for example, were invoked in the early 1940s debates on the apportionment of the Indus River between Punjab and Sind—debates that were superseded by the partition of India and Pakistan in 1947 (Government of Sind 1944; Gulhati 1973; Michel 1967).

Legal frameworks for water-user associations, irrigation districts, and bureaucratic organization were promoted in U.S. development projects, but many were not sustained beyond the life of project funding (Radosevich and Kirkwood 1975). There is, however, increasing concern in India and Pakistan about the need for water-law reform, for which the U.S. experience is perceived to be relevant perhaps as much as a source of difference as possible diffusion. Geographic research on water law is not yet part of this exchange (Templer 1997).

The first U.N. Conference on Natural Resources at Lake Success, New York, in 1949 included a session on water resources that emphasized U.S. experience in river-basin development and in particular the Tennessee Valley Authority (United Nations 1950). A French participant stated, perhaps ironically, that the TVA had succeeded because it had "dictatorial authority," to which TVA officials convulsively responded with the rhetoric of "grassroots democracy." Unfortunately, this brief exchange did not stimulate a sustained discussion of the political economy of international water programs.

In subsequent decades, the United States became deeply involved in large-scale water development in the Indus, Mekong, Helmand, and Nile basins, to name a few. The Indus has special relevance for this discussion in light of the major role played by the United States in facilitating the treaty and then financing a massive program of water development (White House 1964; Lieftinck 1968; Duloy and O'Mara 1984). In one of the most detailed international river-basin studies carried out to date, geographer Aloys Michel (1967) noted the difference between U.S influence in Pakistan, which extended to technical expertise and construction, compared with India, which developed its own technical capabilities. The experience of Pakistan underscores the Cold War geopolitics of international water programs, as Pakistan grew to become the third largest USAID (United States Agency for International Development) recipient during the Soviet occupation of Afghanistan and then plummeted to no aid at all after the Soviet withdrawal and the selective resumption of U.S. nuclear nonproliferation policy. Curiously, these developments in the Indus River basin coincided with, but showed little influence from, academic debates about Karl Wittfogel's ideas about hydraulic societies.

Wittfogel: Truth or Consequences?

Oriental Despotism shook up the academic world. Rather than recite criticisms of the "truth" of Wittfogel's work, I want to concentrate here on its "consequences." The lasting scholarly contributions of *Oriental Despotism* were: (1) its identification of key social structures in water management; (2) its emphasis on the geographical context of water systems; and (3) its global and comparative perspectives on water development (Figure 5.6).[5] Wittfogel erred on many of the causal relations among these variables, but he was the first water-resource theorist to bring political, social, psychological, technological, and environmental variables together under one cover. At the same time, comparative geographic research on water-management systems remains, to this day, at a rudimentary level of development (Smith 1861; Wescoat 1994b).

What were the consequences? The first generation of Wittfogel's critics were leading historians, archaeologists, and social theorists who focused upon his ideas about either ancient societies or the Soviet Union (Butzer 1976; Habib 1961; Steward 1955). These early critics focused on long-term change in ancient irrigation systems.

A second generation of critics turned toward the social aspects of modern irrigation systems (Hunt and Hunt 1976). They examined the relationships between

Figure 5.6 The Structure of Wittfogel's Analysis.

Variables	Categories		
Climate	Core (arid)	Margin (mesic)	Submargin (humid)
Water Technology	Canal Irrigation	Hydroagriculture	Rainfed Agriculture
Hydraulic Density (Economic Role)	Compact (national hydraulic economy)	Loose (regional hydraulic economy)	Loose II (local hydraulic economy)
Property Relations	Simple (state property)	Semi-Complex (includes personal property)	Complex (includes real property)
State Functions	Hydraulic Works	⟶	Defense, Communications, Transport, and Trade
Human Experience	Terror	⟶	Freedom
Political Organization	Despotism	⟶	Democracy

bureaucracy, community, and property in irrigation systems. During this period, there was a shift from large-scale TVA approaches to innovative social and institutional experiments.

The "Management" Revolution, 1970–1980

These trends led to the present generation of development sociologists, anthropologists, and applied geographers who work with the Ford Foundation, multilateral development organizations, and bilateral agencies on institutional projects to improve water management in South Asia (Coward 1980; Freeman and Bhandarkar 1989; Merrey 1979; Uphoff 1992; Wade 1988). Although the point should not be exaggerated, these social scientists were to some extent influenced by theoretical debates sparked by Wittfogel in the 1950s and 1960s.

EMERGING EXPERIMENTS AND PROSPECTS, 1980–2000

Emerging relations between North America and Asian water management indicate some fundamental shifts that have special relevance for cultural geographers

(Figure 5.7). Struggles over monumental projects in the Narmada basin of India, the Three Gorges Dam in China, the Flood Action Plan in Bangladesh, and the Glen Canyon Dam on the Colorado River give increasing attention to cultural impacts and to cultural analysis of all aspects of water use (Chapman and Thompson 1995; Rogers, Lydon and Seckler 1989; Verghese 1990).

Several research trends and opportunities provide a fitting conclusion to this investigation of cultural encounter and exchange between Asian and American water systems. First, it seems time once again for American water managers and scientists to look overseas, particularly to South Asia, for experiments that may be instructive for the United States, as occurred 100 years ago. Cultural geographers from the United States who have participated in these Asian irrigation experiments have a special role to play (Wallach 1996), as do social geographers associated with the International Irrigation Management Institute (Bhutta and Vander Velde 1992).[6] Research on evolving cultural landscapes of water management seems especially promising. Building upon Francois Bernier's accounts of deteriorating canals outside the Agra of the Taj Mahal, we might reexamine the relations between beauty and efficiency in waterworks (Wallach 1996; Wescoat 1985). Bret Wallach (1996) has woven together aesthetic, emotional, and practical essays on irrigated landscapes of India, with some involvement with development organizations such as the Ford Foundation and International Irrigation Management Institute.

Cultural research in related fields has begun to integrate theoretical and practical inquiry in innovative ways. Stephen Lansing (1991) combines social theory

Figure 5.7 Emerging Experiments and Prospects, 1980–2000.

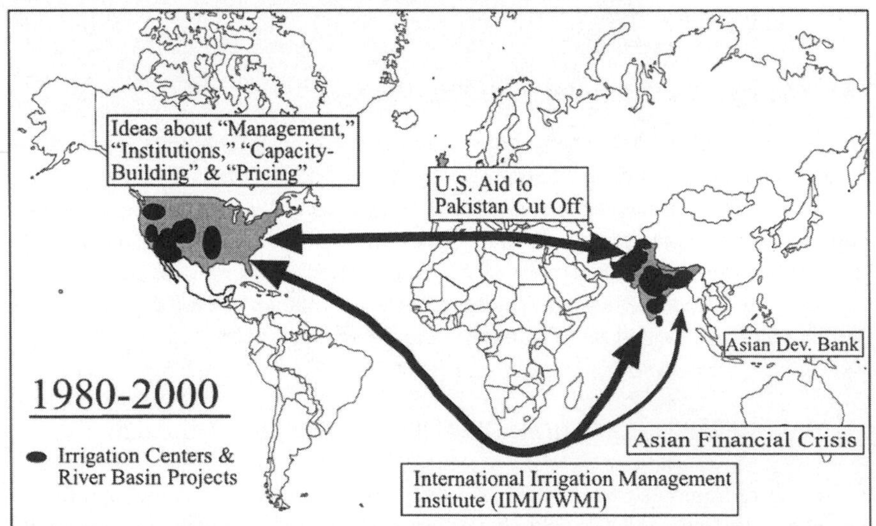

and development research on traditional and modern irrigation systems in Indonesia, leading to innovative hybrid water-management strategies. Norman Uphoff (1992) brings postmodern theory to bear on irrigation development planning in Sri Lanka, again with a commitment to going beyond critique to advance water-management systems.

In the United States, a report of the National Research Council (1996) on *A New Era for Irrigation* indicates that some of the most creative social research on irrigation has occurred in Asia by international teams of Asian and American social scientists. Beginning in the 1950s, American water plans and planners displayed an amnesia in which the U.S. Constitution, rather than international experience, was regarded as the historical foundation for national water planning (United States, President's Water Resources Policy Commission 1950).

Western ideas about South Asian water development reviewed in this chapter include numerous errors of cultural fact and judgment. To lessen such problems in the future, there is a need for advances in cross-cultural comparative theory and method (Wescoat 1994b). Geographic comparison remains largely subjective and unscientific, as Wittfogel argued, but also demonstrated, some forty years ago. We need more insightful comparisons, for example, of flood hazards problems and planning in the Ganges–Brahmaputra basin, which has just completed the largest multilateral flood investigations in history with the Indus, Mekong, and other river basins (Jacobs and Wescoat 1994). A Social Science Research Council project is comparing the environmental effects of property rights in different regions of the world (Wescoat 1994a). Schwartzberg (1992) has given a finely crafted perspective on historical and cultural geography's cartographic representation in South Asia, including "maps" of canals and water systems.

Research on long-term change in water management, of the sort Wittfogel pioneered, indicates that analogies may be as important as formal comparisons. Analogies translate experiences gained in earlier times and places to new situations and places. A team of Pakistani and American water scientists used analogies to assess the potential impacts of climate change in the Indus basin (Wescoat 1991). The study then sought to integrate modeled and unmodeled impacts and to combine high science and appropriate technologies.

Finally, recent work by new cultural geographers and colleagues is opening up promising lines of inquiry. Cosgrove and Petts (1990) drew together contributions from cultural, historical and physical geographers in a highly stimulating collection on *Water, Engineering and Landscape* that, with further collaboration and integration, could advance the field. Cosgrove et al. (1995) then contributed to the European Community Environmental Research Programme by convening British, Danish, Italian, and Swedish teams to focus on national perspectives on water regulation in the mid-twentieth century. These works incorporate themes of power and control with beauty and livelihood in ways that point toward new contributions for cultural geographers and new practical applications.

At present cultural geography plays little role in policy analysis and design. In my experience, water managers in India, Pakistan, Bangladesh, and the United States—even the most tough-minded water engineers—are fascinated by cultural geographic aspects of their field, including its conflicts and power relations, and would benefit from clear practical analyses of these issues. The pitfalls must not be discounted. Wittfogel's legacy rose through a commitment to contemporary political concerns, and fell through a peculiar conflation of ideology and science. The seeds for new syntheses of cultural geography and water resources geography are now germinating on that well-worked ground.

NOTES

This paper has been in formation since my early days as an M.A. student with Marvin Mikesell in the late 1970s. I am sure Marvin would encourage me to acknowledge debts to other colleagues at Chicago: Karl Butzer, Norton Ginsburg, Paul Wheatley, and Robert McC. Adams. Earlier versions of this paper were given in the lectures honoring Joseph Schwartzberg at the University of Minnesota and Edwin Hammond at the University of Tennessee.

1. Circa 1985. Marvin Mikesell also perpetuated Wittfogel's influence by regularly assigning Marx's article on "The British Rule in India" (Avineri 1968: 83–89) and Wittfogel's paper on "Hydraulic Civilizations" in *Man's Role in Changing the Face of the Earth* (1954) in Geography 313 (Cultural Geography) at the University of Chicago. For reflections on the persistence of Wittfogel, and a humorous comparison with Elvis, see Butzer (1996).
2. I have presented these arguments in projects sponsored by the U.S. Environmental Protection Agency (Wescoat 1991) and the National Research Council (1996).
3. This is the newspaper article Marvin Mikesell has used in his cultural geography course.
4. Habib (1983) provides a detailed critique of Marx's perceptions of India.
5. Geographers engaged in river-basin planning, such as Harlan Barrows and Gilbert F. White, were already engaged in integrating social and environmental factors in water development. Although they were less critical of relationships between civil society and the state, a more detailed comparison of their approaches and Wittfogel's is warranted.
6. Recently renamed the International Water Management Institute (IWMI) to broaden its scope beyond irrigation and drainage.

REFERENCES

Ali, Imran. 1988. *The Punjab under Imperialism, 1885–1947.* Princeton: Princeton University Press.
Antipode. 1985. Special issue. *The Geographical Ideas of Karl Wittfogel.* Vol. 17, no. 1.
Avineri, S., ed. 1968. *Karl Marx on Colonialism and Modernization.* Garden City, N.Y.: Doubleday.
Baxi, U. 1985. "Understanding the Traffic of Ideas in Law between America and India."

In *Traffic of Ideas between India and America*, ed. R. Crunden, 319–42. Delhi: Chanakya Publications.

Beer, L. W., ed. 1979. *Constitutionalism in Asia: Asian Views of the American Influence.* Berkeley: University of California Press.

Bentham, Jeremy. 1962/1838–43. "Of the Influence of Time and Place in Matters of Legislation." In *The Works of Jeremy Bentham,* vol. 1, ed. John Bowring, 171–94. New York: Russell & Russell.

Bernard, Linda, and K. Reynolds. 1992. "Karl August Wittfogel. A Register of His Papers in the Hoover Institution Archives." Palo Alto: Stanford University, Hoover Institution.

Bernier, Francois. n.d./1670. *Travels in the Mogul Empire AD 1656–1668,* trans. Irving Brock, ed. Archibald Constable. Karachi: Indus Reprint.

Bhutta, M., and E. Vander Velde, Jr. 1992. "Equity of Water Distribution along Secondary Canals in Punjab, Pakistan." *Irrigation and Drainage Systems* 6: 161–77.

Brown, Hanbury. 1905. "Irrigation Under British Engineers." *Transactions of the American Society of Civil Engineers* (54C): 3–31.

Burke, Edmund. 1981. *The Writings and Speeches of Edmund Burke*. ed. Paul Langford. Oxford: Clarendon Press.

Butzer, K. W. 1976. *Early Hydraulic Civilization in Egypt.* Chicago: University of Chicago Press.

———. 1996. "Irrigation, Raised Fields and State Management: Wittfogel Redux?" *Antiquity* 70: 200–04.

Chapman, G. P., and M. Thompson, eds. 1995. *Water and the Quest for Sustainable Development in the Ganges Valley.* London: Mansell.

Chappell, John C. Jr. 1971. "Huntington, Wittfogel, and the Environmental Bases of Ideas and Politics." In *Selected Papers,* vol. 3, *Population and Settlement Geography, Political and Historical Geography* 364–70.21st International Geographical Congress [1968]. Calcutta: National Committee for Geography.

Clapp, Gordon R. 1955. *TVA: An Approach to the Development of a Region.* Chicago: University of Chicago Press.

Colvin, J. 1833. "On the Restoration of the Ancient Canals in the Delhi Territory." *Journal of the Asiatic Society* 15: 105–27.

Commissioners on the Irrigation of the San Joaquin, Tulare and Sacramento Valleys of the State of California. 1874. *Irrigation of the San Joaquin, Tulare and Sacramento Valleys, California.* U.S. Congress House of Representatives, 43rd Cong., 1st sess., Ex. Doc. no. 290. Washington, D.C.: Government Printing Office.

Cosgrove, D., and G. Petts, eds. 1990. *Water, Engineering and Landscape*. London: Bellhaven.

Cosgrove, D., Ulrik Lohm, Kenneth R. Olwig, and G. Zanetto. 1995. "Nature, Environment and Landscape: European Attitudes and Discourses in the Modern Period (1920–1979) with Particular Attention to Water Regulation." EC Environmental Research Program, Consortium Final Report. Egham, England: Geography Department, Royal Holloway, University of London.

Coward, E. Walter, Jr. 1980. *Irrigation and Agricultural Development in Asia.* Ithaca: Cornell University Press.

Damodar Valley Corporation. 1992. *Damodar Valley. Evolution of the Grand Design.* Calcutta: Damodar Valley Corporation.

Davidson, G. 1875. *Irrigation and Reclamation of Land for Agricultural Purposes as now practiced in India, Egypt, Italy, etc.* U.S Senate, 44th Cong., 1st Sess. Ex. Doc. no. 94. Washington, D.C.: Government Printing Office.

Duloy, John H., and Gerald T. O'Mara. 1984. *Issues of Efficiency and Interdependence in Water Resource Investments: Lessons from the Indus Basin of Pakistan.* Washington, DC: World Bank Staff Working Paper No. 665.

Edney, Matthew H. 1997. *Mapping an Empire: The Geographical Construction of British India, 1765–1843.* Chicago: University of Chicago Press.

Emel, Jody, and Rebecca Roberts. 1995. "Institutional Form and Its Effect on Environmental Change: The Case of Groundwater in the Southern High Plains." *Annals of the Association of American Geographers* 85: 664–83.

Farnie, D. A. 1979. *The English Cotton Industry and the World Market, 1815–1896.* Oxford: Clarendon Press.

Flynn, P. J. 1892. *Irrigation Canals and Other Irrigation Works.* San Francisco: G. Spaulding.

Freeman, David, and Vrinda Bhandarkar. 1989. *Local Organizations for Social Development: Concepts and Cases of Irrigation Organization.* Boulder: Westview Press.

Gilmartin, D. 1994. "Scientific Empire and Imperial Science: Colonialism and Irrigation Technology in the Indus Basin." *Journal of Asian Studies* 53: 1127–49.

———. 1995. "Models of the Hydraulic Environment: Colonial Irrigation, State Power and Community in the Indus Basin." In *Nature, Culture and Imperialism: Essays on the Environmental History of South Asia*, ed. David Arnold and R. Guha, 210–36. New Delhi: Oxford University Press.

Gole, S. 1989. *Indian Maps and Plans: From Earliest Times to the Advent of European Surveys.* New Delhi: Manohar Publications.

Gourou, P. 1961. "Letter to the editor." *Annals of the Association of American Geographers* 51: 401–02.

Government of Sind. 1944. *Rejoinder of the Government of Sind to the Representations of the Punjab and Other Units on the Report of the Indus Commission.* Karachi: Government Press.

Gulhati, N. D. 1973. *Indus Waters Treaty: An Exercise in International Mediation.* Bombay: Allied Publishers.

Habib, Irfan. 1961. "An Examination of Wittfogel's Theory of Oriental Despotism." *Enquiry* VI: 54–73.

———. 1983. "Marx's Perceptions of India." In *Marx on Indonesia and India*, 29–66. Trier, Federal Republic of Germany: Karl-Marx-Haus.

Hall, Wm. H. 1886. *Irrigation Development.* Sacramento: State Office.

Hargrove, Erwin C. 1994. *Prisoners of Myth: The Leadership of the Tennesee Valley Authority, 1933–1990.* Princeton: Princeton University Press.

Hilgard, E. W. 1886. *Alkali Lands.* Sacramento: State Office; University of California, College of Agriculture.

Hunt, Robert C., and Eva Hunt. 1976. "Canal Irrigation and Local Social Organization." *Current Anthropology* 17: 389–410.

Jackson, W. T., et al., eds. 1990. *Engineers and Irrigation: Report of the Board of Commissioners on the Irrigation of the San Joaquin, Tulare, and Sacramento Valleys of the State of California, 1873.* Reprinted as Engineer Historical Studies No. 5. Fort Belvoir: Office of History, U.S. Army Corps of Engineers.

Jacobs, J. W., and James L. Wescoat Jr. 1994. "Flood Hazard Problems and Programmes in Asia's Large River Basins." *Asian Journal of Environmental Management* 2: 91–104.

Jensen, J. M. 1988. *Passages from India: Asian Indian Immigrants in North America.* New Haven: Yale University Press.

Jones, S. B. 1958. "Review of Oriental Despotism." *Geographical Review* 48: 306–07.

Kinney, Clesson S. 1912. *A Treatise on the Law of Irrigation.* 2nd ed. San Francisco: Bender-Moss.

La Brack, B. 1988. *The Sikhs of Northern California, 1904–1975.* New York: AMS Press.

Lansing, J. S. 1991. *Priests and Programmers: Technologies of Power in the Engineered Landscape of Bali.* Princeton: Princeton University Press.

Lee, Lawrence B. 1980. *Reclaiming the American West: An Historiography and Guide.* Santa Barbara: ABC-Clio.

Leonard, K. I. 1992. *Making Ethnic Choices: California's Punjabi Mexican Americans.* Philadelphia: Temple University Press.

Lieftinck, Pieter. 1968. *Water and Power Resources of West Pakistan: A Study in Sector Planning.* 3 vols. Baltimore: Johns Hopkins University Press for the World Bank.

Lilienthal, D. 1944. *TVA: Democracy on the March.* New York: Harper & Brothers.

Locke, John. 1965/1690. *Two Treatises on Government*, rev. ed., ed. Peter Laslett. New York: Mentor Library.

———. 1812. "A Catalogue and Character of Most Books of Voyages and Travels." In *The Works of John Locke,* 11th ed., vol. 10, 513–64. London: W. Otridge & Sons.

Maass, A., and R. Anderson 1978. *. . . and the Desert Shall Rejoice: Conflict, Growth and Justice in an Arid Environment.* Cambridge: Massachusetts Institute of Technology Press.

Marsh, George P. 1874. *Irrigation: Its Evils, the Remedies, and the Compensations.* U.S. Congress. Senate. 43rd Cong., 1st Sess., Misc. Doc. no. 55. Washington, D.C.: Government Printing Office.

Masood-ul Hassan, S. ed. 1992. *The Manual of Canal and Drainage Laws.* Lahore: Irfan Law Book House.

Mazumdar, Sucheta. 1984. "Punjabi Agricultural Workers in California, 1905–1945." In *Labor Immigration Under Capitalism: Asian Workers in the United States Before World War II*, ed. Lucie Cheng and Edna Bonacich, 549–78. Berkeley: University of California Press.

Merrey, D. 1979. *Irrigation and Honor: Cultural Impediments to Improvement of Local Level Water Management in Punjab Pakistan.* Fort Collins: Colorado State University, Water Management Technical Report No. 53.

Michel, Aloys A. 1967. *Indus Rivers: A Study of the Effects of Partition.* New Haven: Yale University Press.

Mill, J. S. 1965/1848. *Principles of Political Economy.* Toronto: University of Toronto.

Moncrief, Scott C. C. 1868. *Irrigation in Southern Europe.* London: E. and F. N. Spon.

Montesquieu, Baron de. 1949/1748. *The Spirit of the Laws,* trans. Thomas Nugent. New York: Hafner Press.

National Research Council. Water Science and Technology Board. 1996. *A New Era for Irrigation.* Washington, DC: National Academy Press.

Norton, C. E. 1853. "Irrigation in India." *North American Review* 77: 439.

O'Leary, B. 1989. *The Asiatic Mode of Production: Oriental Despotism, Historical Materialism, and Indian History.* London: Basil Blackwell.

Peet, Richard. 1988. "Political Geographers of the Past VI: 2. Wittfogel on the Nature-Society Dialectic." *Political Geography Quarterly* 7: 81–83.

Radosevich, G., and C. Kirkwood. 1975. *Organizational Alternatives to Improve On-Farm Water Management in Pakistan.* Fort Collins: Colorado State University, Water Management Technical Report No. 36.

Raynal, Abbe. 1776. *A Philosophical and Political History of the Settlements and Trade of the Europeans in the East and West Indies*, 2nd ed., trans. J. Justamond, vol. 2. London: T. Cadill.

Rennell, James. 1785. *Memoir of a Map of Hindoostan, or the Mogul's Empire.* London: M. Brown.

Rogers, Peter, Peter Lydon, and David Seckler. 1989. *Eastern Waters Study.* Arlington, VA: U.S. Agency for International Development.

Said, Edward W. 1978. *Orientalism*. New York: Vintage Books.

Schwartzberg, J. E. 1992. "South Asian Cartography." In *The History of Cartography*, vol. 2, bk. 1, *Cartography in the Traditional Islamic and South Asian Societies,* ed. J. B. Harley and D. Woodward, 293–331. Chicago: University of Chicago Press.

Shabad, T. 1959. "Non-western views of 'The hydraulic society.' " *Annals of the Association of American Geographers* 49: 324- 25.

Sidky, H. 1997. "Irrigation and the Rise of the State in Hunza: A Case for the Hydraulic Hypothesis." *Modern Asian Studies* 31: 995–1017.

Singh, Chhatrapati. 1991. *Water Rights and Principles of Water Resources Management.* Water Project Series, Indian Law Institute. With "Research Programme" by U. Baxi. Bombay: N. M. Tripathi Pvt.

Smith, A. 1976/1776. *An Inquiry into the Nature and Causes of The Wealth of Nations*, 2 vols. in one, ed. E. Cannan. Chicago: University of Chicago Press. [Reprint of 1904 edition published by Methuen, London]

Smith, E. Goodrich. 1861. *U.S. Commissioner of Patents Report, 1860.* U.S. Congress. House of Representatives. 36th Cong., 2nd sess., Ex. Doc. no. 48: 166–224. Washington, D.C.: Government Printing Office.

Smith, R. Baird. 1849. "Canals of Irrigation in the North-Western Provinces." *Calcutta Review* 12: 79–183.

———. 1855. *Italian Irrigation, Report on the Agricultural Canals of Piedmont and Lombardy to the Honourable Court of Directors of the East India Company.* Edinburgh: Blackwood.

Spate, O. H. K. 1959. "Review of Oriental Despotism." *Annals of the Association of American Geographers* 49: 90–95.

Steward, Julian, ed. 1955. *Irrigation Civilizations: A Comparative Study.* Washington, DC: Pan American Union.

Stokes, Eric. 1959. *The English Utilitarians and India.* Oxford: Oxford University Press.

Templer, Otis. 1997. "Water Law and Geography: A Geographic Perspective." In *Geography, Environment and American Law*, ed. Gary L. Thompson, Fred M. Shelley, and Chand Wije, 61–78. Boulder: University of Colorado Press.

TVA (Tennessee Valley Authority). 1961. *TVA—Symbol of Valley Resource Development.* Knoxville: TVA.

Ulmen, G. 1978. *The Science of Society: Toward an Understanding of the Life and Work of Karl August Wittfogel.* The Hague: Mouton.

United Nations. Department of Economic Affairs. 1950. *Proceedings of the United Nations Scientific Conference on the Conservation and Utilization of Resources.* Vol. 1, Plenary Meetings, and Vol. 4, Water Resources. 17 August–6 September 1949. Lake Success, NY: United Nations.

United States. President's Water Resources Policy Commission. 1950. *A Water Policy for the American People.* The Report of the President's Water Resources Policy Commission, Vol. 1. Washington, DC: U.S. Government Printing Office.

Uphoff, Norman. 1992. *Learning from Gal Oya: Possibilities for Participatory Development and Post-Newtonian Social Science.* Ithaca: Cornell University Press.

Vani, M. S. 1992. *Role of Panchayat Institutions in Irrigation Management: Law and Policy.* New Delhi: The Indian Law Institute.

Verghese, B. G. 1990. *Waters of Hope: Integrated Water Resource Development and Regional Cooperation Within the Himalayan-Gangal-Brahmaputra-Barak Basin.* New Delhi: Oxford University Press and IBH Publishing.

Wade, Robert. 1988. *Village Republics: Economic Conditions for Collective Action in South India.* Cambridge: Cambridge University Press.

Wallach, Bret. 1996. *Losing Asia.* Baltimore: Johns Hopkins University Press.

Wescoat, James L. Jr. 1985. "Early Water Systems in Mughal India." *Environmental Design: Journal of the Islamic Environmental Design Research Centre* 2: 50–57.

———. 1991. "Managing the Indus River Basin in Light of Global Climate Change: Four Conceptual Approaches." *Global Environmental Change: Human and Policy Dimensions* 1: 381–95.

———. 1994a. "Water Rights in South Asia and the United States: Comparative Perspectives." Paper for SSRC Comparative Examination of Landed Property Rights Project. Stowe,VT, August 19–22, 1994.

———. 1994b. "Varieties of Geographic Comparison in *The Earth Transformed.*" *Annals of the Association of American Geographers* 84: 721–25.

———. 1996. "Main Currents in Multilateral Water Agreements: A Historical-Geographic Perspective, 1648–1948." *Colorado Journal of International Environmental Law and Policy* 7: 39–74.

Wescoat, James L. Jr., R. Smith, and D. Schaad. 1992. "Visits to the U.S. Bureau of Reclamation from South Asia and the Middle East, 1946–1990: An Indicator of Changing International Programs and Politics." *Irrigation and Drainage Systems* 6: 55–67.

Whitcombe, E. 1982. "Irrigation." In *The Cambridge Economic History of India,* vol. 2, ed. D. Kumar, 677–736. Delhi: Orient Longman.

White, G. F. 1957. "A Perspective of River Basin Development." *Law and Contemporary Problems* 22: 157–87.

———. 1997. "Watersheds and Streams of Thought." *Reviews in Ecology: Desert Conservation and Development: A Festschrift for Prof. M. Kassas on the Occasion of His 75th Birthday*, ed. Hala N. Barakat and Ahmad K. Hegazy, 89–97. Cairo, Egypt: UNESCO; IDRC; South Valley University.

White House. Department of Interior Panel on Waterlogging and Salinity in West Pakistan. 1964. *Report on Land and Water Development in the Indus Plain: The Revelle Report on Land and Water Development in West Pakistan.* Karachi: United States Information Service.

Wilson, Herbert M. 1890–91. *Irrigation in India.* Washington, DC: U.S. Geological Survey 12th Annual Report, Part II, Irrigation.

———. 1894. "American and Indian Irrigation Works." *The Irrigation Age* 10: 107–09.

Wittfogel, Karl A. 1985/1929. "Geopolitics, Geographical Materialism and Marxism," trans. G. L. Ulmen. *Antipode* 17: 21–72.

———. 1954. "Hydraulic Civilizations." In *Man's Role in Changing the Face of the Earth,* ed. W. L. Thomas Jr., 152–64. Chicago: University of Chicago Press.

———. 1955. "Developmental Aspects of Hydraulic Societies." In *Irrigation Civilizations: A Comparative Study*, ed. Julian Steward, 43–52. Washington, DC: Pan-American Union.

———. 1981/1957. *Oriental Despotism: A Comparative Study of Total Power.* 2nd ed. New York: Vintage.

Worster, Donald. 1985. *Rivers of Empire: Water, Aridity and the Growth of the American West.* New York: Pantheon.

Yule, H. 1846. "A Canal Act of the Emperor Akbar, With Some Notes and Remarks on the History of the Western Jumna Canals." *Journal of the Asiatic Society* 15: 171, 213–36.

Zafar, Fareeha. 1985. "The Colonisation Process and Population Changes in the Punjab During British Rule." *Die Erde* 125: 329–40.

6

Wetlands as Conserved Landscapes in the United States

James A. Schmid

Wetlands are those parts of the vegetated landscape where water is present above, at, or near the surface of the ground for long periods during the growing season for higher plants. Many are associated with flowing watercourses, but they also may be found in isolated depressions and broad flats that trap surface runoff or wherever groundwater emerges at the surface in springs and seeps. Popular terms for various wetlands include swamps, marshes, sloughs, bogs, fens, mires, potholes, wet prairies, wet meadows, pocosins, sponges, peatlands, and muskegs. The term "wetland" first appeared in print in 1778 but was popularized during the mid-twentieth century to identify landscapes where land and water meet (Shaw and Fredine 1956). Unlike open-water features, wetlands are not permanently covered by water so deep that emergent plants cannot grow on them; the vegetated, shallow margins of such water bodies, however, may constitute wetlands.

Wetlands have many valuable characteristics. They are part of the natural hydrologic storage system,where floodwaters reside during peak episodes of snowmelt and precipitation without harm to people or property. Wetland plants filter pollutants from, and may contribute oxygen to, surface waters, enhancing their quality. Wetlands typically are occupied by unmanaged communities of higher plants and may provide habitat for many rare species of plants and animals. The biological productivity of many North American wetlands is high, and is comparable to that of the highest-yielding farmlands and tropical rainforests (Lieth 1975). Wetlands are essential habitats during the reproductive cycle for many birds, fish, and other animals. Hence their biological significance far exceeds their relative acreage in the landscape.

Wetlands also are objects of study, of beauty, and of mystical contemplation. In urban and suburban areas wetlands may be virtually the only remaining wild landscapes anchoring residents to their unique geographical place, and current residents may value passing them down to future generations without drastic human alterations. The biota and physical characteristics of many wetlands in North America have been profoundly affected by human activities, especially since the beginning of the sixteenth century, and their ability to perform valued functions has been compromised in many locations.

Wetlands are highly variable. Some support old-growth forest; others, scrub or herbaceous vegetation; still others, a mixture of physiognomic classes of plants. Some are wet only for short periods except during high-rainfall years; others are saturated even during droughts. Some are washed twice daily by the saline waters of the ocean tide. Others are hypersaline as a result of intense evaporation and rare opportunities to have salts washed away. Many inland wetlands are freshwater ecosystems lacking salinity. Some withstand the intensive stress of urban settings, where they continue to provide some measure of ecosystem function.

Constantly wet examples of wetlands are easily recognized. Backbarrier tidal marshes are a prominent element of the coastal landscape along the Atlantic Ocean and Gulf of Mexico. Pond margins are readily identified. Seasonal wetlands may not be obvious to the casual observer during the dry season or unusually dry years, especially if they are farmed during those periods. In mountainous terrain wetlands are confined to seeps and to the limited expanses of flat land or small depressions where water persists for long periods. Where topographic relief is slight, the limits of wetlands are subtle. Wetlands may be as small as a few hundredths of an acre or extend from horizon to horizon, as in the Florida Everglades.

Wetlands are viewed by people in many ways. Some yield poorly when farmed for traditional crops, especially the northern peatlands of the Great Lakes states. In contrast, the once-vast Midwestern wet prairies, now effectively drained, are among the most productive acres that exist today in the Corn Belt. The domain of wild plants and animals, unfarmed wetlands exhibit an ever changing array of colors and biological activities attuned to seasonal rhythms. They may harbor large seasonal populations of mosquitoes and biting flies, creatures universally loathed as sources of annoyance and disease. Wetlands have to be converted with effort and expense through draining and filling in order to become "improved" sites for roads, homes, and commercial enterprises. Prior to such conversion, they are less hospitable for most land uses than well-drained uplands. The current and potential values of a wetland ecosystem are affected not solely by its size, but also by the landscape context in which it is embedded.

HISTORICAL OVERVIEW OF WETLANDS IN THE UNITED STATES

Wetlands never occupied much of the North American landmass. At the time the thirteen British colonies became independent, some 345,000 square miles of wetlands occupied about 11 percent of the lands that now make up the 48 conterminous states (Schmid 1994). These wetlands were unevenly distributed across the continent. Territories that became the states of Florida and Louisiana were more than half wetlands. Ten other states had more than a quarter of their land in wetlands (in descending order of abundance): Delaware, North Carolina, South Carolina, Mississippi, Maine, Michigan, New Jersey, Arkansas, Minnesota, and Wisconsin. All the other states had less, except for Alaska, which remains even today nearly half wetlands. At the bottom end of the scale were the dry, intermountain states of Nevada, New Mexico, Montana, Arizona, Utah, and Idaho. These political entities, along with Hawaii and Kansas, had less than 2 percent of their land in wetlands when the Europeans arrived.

There is a long tradition in Anglo-American thought that associates forested swamp wetlands with moral turpitude, social outcasts, bandits, and renegade Indians (Vileisis 1997). Europeans with this ideological orientation reacted negatively to the wetlands they encountered. When traveling through the eastern United States by steamboat in 1842, the English novelist Charles Dickens (1891: 200) saw several wetlands that inspired negative outbursts. In his reading of these landscapes, swamps were dismal places, teeming with unhealthy vegetation, slimy sites that entrapped the unwary traveler, places without redeeming merit.

For nearly three hundred years following European settlement in North America, wetlands were but one of many kinds of lands open to the taking by industrious individuals as soon as the threat posed by the aboriginal inhabitants was removed. Both tidal and freshwater marshes provided useful hay and pasture, but the pursuit of urban land uses and adoption of more intensive agricultural practices with higher profitability led to uncontrolled marsh destruction from the early seventeenth century onward. Wetlands were not choice lands for settlement by newcomers to the vast American countryside. They were avoided by the earliest settlers; later arrivals had less choice and were more likely to encroach into such undesirable, leftover spaces (McManis 1964).

The pace of wetland conversion in the United States was dictated primarily by the advent of technology for draining, diking, and ditching and by the demand for the agricultural or forest products that could be wrested from them. During the nineteenth century, the federal government provided incentives to states and to private enterprise to convert swamps and other watery wastes into farms, towns, sources of wood products, and mines for gravel, peat, and placer gold. Sixty-four million acres were transferred from the public domain to the states for the specific purpose of conversion into private farmland as a result of the Swamp Land Acts from the 1850s into the 1920s (Prince 1997: 144–46). Railroads opened Midwestern wet prairies to markets during the two decades before 1870; tile drains achieved their conversion to cornfields in the ensuing fifty years, with skyrocketing farmland values. The next generation's attempt to repeat this process and substitute agriculture for lumbering in the peatlands from which the vast pine and other forests of Wisconsin and Minnesota had been cut was an economic and social failure strewn with individual human tragedies, despite subsidies from the federal and state treasuries (Prince 1997: 308–14).

By the mid-twentieth century, the extent of wetlands in the nation had been reduced by half, with more than 182,000 square miles converted into various uses. Wetland elimination had been most assiduous in California and the Midwest (Ohio, Iowa, Indiana, Missouri, Illinois, and Kentucky), where more than 80 percent of the wetlands no longer exist today. Many of these former wetlands still go under water along the Mississippi River valley in great flood years such as 1993, because they are too expensive to protect structurally from infrequent, major floods. The greatest acreage losses of wetlands were in Florida, Texas, and Louisiana, followed by Arkansas and Illinois. As of the 1980s just over 104 million acres (163,000 square miles) of wetlands remained, 5 percent of the conterminous 48 states, with another 170 million acres in Alaska (Table 6.1). By the mid-1990s, the total had dropped to 102 million acres in the conterminous states according to the National Wetland Inventory (NWI), an expanse of wetlands roughly equal to the land area of California (Vileisis 1997: 333). About three-quarters of these remaining wetlands outside Alaska are in private ownership. Only 26 percent are held by governments: 13 percent federal, 11 percent state, and 2 percent municipal (USFWS 1990).

The economic cost of converting wetlands into developed uses anywhere in urban or suburban America today usually is far outweighed by the enhanced value of the land itself in its filled condition for roads, shopping malls, offices, or homesites. In the Hackensack Meadowlands just west of New York City, nearly 850 acres of wetlands are expected to be filled during the next two decades for activities deemed to lack practicable non-wetland alternatives (USEPA and USACE 1995: 5–4). At the current price of $300,000 to $600,000 to compensate locally for each acre of wetland lost, the cost of efforts to replace functional wetland values in the Meadowlands could approach $500 million. Market conditions across the nation still sometimes warrant the clearing and draining of wet bottomlands for crops such as soybeans, rice, or cranberries. Hence landowners both urban and rural have a financial incentive to encroach upon wetlands for many purposes, if they can secure permit approvals within the time frame of a perceived market opportunity.

WETLANDS NEED PROTECTION: A MINORITY VIEW THAT GAINED ACCEPTANCE

The benefits derived from natural wetlands accrue not only to the owners of the land they occupy but also to the public at large (Greeson, Clark and Clark 1979; Mitsch and Gosselink 1993; Tiner 1984). Unless tied to attractive water bodies in suburban or urban areas or sited in rural resorts popular with vacationing urbanites, these benefits are not usually reflected by the marketplace in the monetary value of the property that contains the wetlands. This reflects the hazards that their wetness may present to would-be users prior to draining or filling, and the higher economic costs associated with most potential development there as compared with dry upland sites nearby. The flow of wetland benefits to the public at large, and the fairly recent imposition of some of the regulatory costs on those owners who seek to convert wetlands, set the stage for land-use controversy.

From early colonial times some Americans recognized the value of wetlands as a useful part of the human environment (Vileisis 1997: 29–37). Tidal marshes furnished salt hay to settlers in coastal Massachusetts and New Jersey. Wetlands harbored much of the wild game that bolstered the diet of colonists. They were the repositories for many of the species collected by William Bartram and other pioneer botanists, and their habitat features and birds were captured in paint by John James Audubon. For capitalists, wetlands were storehouses of ancient timber. Slowly, the notion gained adherents that wetlands in their undrained condition had aesthetic as well as economic value.

In his first published essay, Henry David Thoreau expressed a new, introspective appreciation of the wetland landscape quite at odds with the widely held view articulated by Charles Dickens just twenty years before. For Thoreau, hope and the future lay not in the "improved" landscapes of lawns and cultivated fields, but in quaking swamps. He admonished his readers to abolish front yards and site their homes instead on the very edge of the swamp (Thoreau 1862: 666).

Thoreau's vision did not take over the national consciousness overnight, but it gained adherents steadily for the next century, aided by other popular writers such as John Muir (Fox 1981; Turner 1985) and Marjorie Stoneman Douglas (1947; Nash 1967). Wetlands were assigned a spiritual value, as landscapes relatively untouched by human hands, where refuge could be sought from modern society. The painters of the Hudson River School set out to document the beauty of the natural American landscape before it was transformed by industry and agriculture. In Thoreau's New England, the swamps were practically the only unmanipulated landscapes remaining. Thus it was to them that he resorted for solace on his walks.

Frederick Law Olmsted (1968) and George Perkins Marsh (1965/1864) campaigned at mid-century against the wanton destruction of the nation's remaining natural heritage. In Boston's Back Bay Fens, Olmsted provided an early and short-lived example of how to incorporate a wetland into an urban park (Zaitzevsky 1982). Late in the century George Bird Grinnell, the editor of *Forest and Stream*, encouraged sportsmen to protect game and particularly birds. His efforts, along with those of others, led to the establishment of Audubon societies (Vileisis 1997: 151 ff.). In the twentieth century professional naturalists such as Paul Errington (1957), Aldo Leopold (1949), William Niering (1966), and John and Mildred Teal (1969) spread the understanding and appreciation of wetland ecosystems to ever more appreciative audiences.

In the Progressive Era, efforts to protect migratory birds provided some inad-

Table 6.1 Wetlands and Wetland Losses by State in the United States, 1780s–1980s

	1980s		*1780s*	*1780s–1980s*		
State	*Wetland Acres (000s)*	*Wetlands as % of All Land*	*Wetlands as % of All Land*	*% Loss of Wetlands*	*Wetland Acres Lost (000s)*	*Rank by Acres Lost*
AK	170,000	45.3	45.3	0.01	200	45
FL	11,038	29.5	54.2	46	9,287	1
LA	8,784	28.3	52.1	46	7,411	3
MN	8,700	16.2	28.0	42	6,370	6
TX	7,612	4.4	9.4	52	8,388	2
NC	5,690	16.9	33.0	49	5,400	9
MI	5,583	15.0	30.1	50	5,617	8
WI	5,331	14.8	27.3	46	4,468	13
GA	5,298	14.1	18.2	23	1,545	20
ME	5,199	24.5	30.4	20	1,261	23
SC	4,659	23.4	32.3	27	1,755	19
MS	4,067	13.3	32.3	59	5,805	7
AL	3,784	11.5	22.9	50	3,784	15
AR	2,764	8.1	29.0	72	7,085	4
ND	2,490	5.5	10.9	49	2,438	17
NE	1,906	3.9	5.9	35	1,005	26
SD	1,780	3.6	5.5	35	955	28
OR	1,394	2.2	3.6	38	868	29
IL	1,255	3.5	22.8	85	6,957	5
WY	1,250	2.0	3.2	38	750	31
VA	1,075	4.1	7.1	42	774	30
NY	1,025	3.2	8.1	60	1,537	21
CO	1,000	1.5	3.0	50	1,000	27
OK	950	2.1	6.4	67	1,893	18
WA	938	2.1	3.1	31	412	36
NJ	916	18.3	29.9	39	584	33
MT	840	0.9	1.2	27	307	40
TN	789	2.9	7.2	59	1,150	25
IN	751	3.2	24.1	87	4,849	10
MO	643	1.4	10.9	87	4,201	14
AZ	600	0.8	1.3	36	331	38
MA	588	11.1	15.5	28	230	44
UT	558	1.0	1.5	30	244	42
PA	499	1.7	3.9	56	628	32
OH	483	1.8	19.0	90	4,517	12
NM	482	0.6	0.9	33	238	43
CA	454	0.4	4.9	91	4,546	11

	1980s		1780s	1780s–1980s		
State	*Wetland Acres (000s)*	*Wetlands as % of All Land*	*Wetlands as % of All Land*	*% Loss of Wetlands*	*Wetland Acres Lost (000s)*	*Rank by Acres Lost*
MD	440	6.5	24.4	73	1,210	24
KS	435	0.8	1.6	48	406	37
IA	422	1.2	11.1	89	3,578	16
ID	386	0.7	1.6	56	491	35
KY	300	1.2	6.1	81	1,266	22
NV	236	0.3	0.7	52	251	41
DE	223	16.9	36.4	54	257	39
VT	220	3.6	5.5	35	121	46
NH	200	3.4	3.7	9	20	49
CT	173	5.4	20.9	74	497	34
WV	102	0.7	0.9	24	32	48
RI	65	8.4	13.2	37	38	47
HI	52	1.3	1.4	12	7	50
48 Conterminous States:						
	104,374	5.0	11.0	53	116,576	
50 States:						
	274,426	11.9	17.3	30	116,962	
48 Conterminous States 1780s Wetland Total: 221,130,000 acres						
50 States 1780s Wetland Total: 391,388,000 acres						

Sources: National Wetland Inventory data from T. E. Dahl, *Wetlands Losses in the United States, 1780's to 1980's* (Washington, D.C.: U.S. Department of the Interior, Fish and Wildlife Service, 1990). Table is reprinted from Rutherford Platt et al., eds., *The Ecological City: Preserving and Restoring Urban Biodiversity* (Amherst: University of Massachusetts Press, 1994). Copyright © 1994 by the University of Massachusetts Press.

Note: States are listed in order of 1980s remaining wetland acreage. The extent of regulated wetlands could differ substantially from the reported values, depending on the definition of wetlands used.

vertent protection to wetlands, yet the federal government continued to foster the drainage of wetlands on a massive scale in order to promote farming during the 1930s and 1940s, to build dams, and to pile dredged spoils into marshes. More than a quarter century of effort by many was required to establish the Everglades National Park in 1947, a wetland ecosystem seriously impacted by government hydraulic interference that has proved much more costly to restore than it was to channelize initially.

The Bureau of Biological Survey (now the Fish and Wildlife Service) was established in the Department of the Interior, and its biologists quickly recognized the necessity of preserving wetlands as waterfowl habitat outside the National

Wildlife Refuges they administer. In the 1930s, efforts were begun to restore water to submarginal farmlands in order to foster breeding habitat for ducks. One of the agency's key accomplishments was the mapping and analysis of the major wetland resources of the nation, most recently in the National Wetland Inventory (NWI; <<http://www.nwi.fws.gov>). Efforts to limit the drastic alteration of wetlands are but one manifestation of twentieth-century Americans' growing concern for environmental quality, which by the 1970s had garnered sufficient political consensus to enact environmentally protective laws at all levels of government. As the nation's population became increasingly suburban and urban, homeowners financially able to choose their environments came to regard the remaining open spaces, including wetlands, as key amenities in their everyday surroundings. Proposals to alter such lands now typically receive intense scrutiny, and recourse often is made to the regulatory agencies and to the courts to limit proposals for what is perceived as excessive environmental destruction.

Much larger segments of the American landscape today are occupied by farms and forests than by cities, suburbs, or towns. Most farmers who forgo profits by not cultivating prairie potholes are not directly reimbursed by duck hunters from the city seeking outdoor recreation. Rural residents typically are less concerned about protecting a few acres of wetlands here and there in the wild landscape than are urbanites who see precious few such acres remaining in their immediate surroundings. Implementation of laws designed to protect wetlands has lagged in rural areas, as the current litigation against Appalachian coal-mining activities in West Virginia wetlands and streams demonstrates. Political turmoil seethes around wetland protection, especially among the rural constituencies who still enjoy political power in excess of their actual numbers of voters. Wetland regulation has been a perennial target of those who embrace the rhetoric of limited government. Federal wetland regulation narrowly survived the environmentally hostile administration of Ronald Reagan and the 103rd and 104th Congresses.

WETLAND REGULATION

Political support for wetland protection reached a peak during the last quarter of the twentieth century. Some wetlands were purchased outright by government agencies or conservation groups for permanent protection. Primary reliance was placed on a regulatory approach that aims to slow, but not necessarily stop, wetland losses. Future alterations of wetlands in general were not prohibited outright; instead, government agencies were assigned to review proposals for wetland alteration so as to minimize unnecessary impacts and were authorized to prohibit those activities deemed too damaging or for which non-wetland sites were available. Shortly after mid-century several states initiated wetland protection through laws aimed at regulating the use of tidal marshes along the Atlantic coast: Massachusetts (1963), Rhode Island (1965), Connecticut (1969), New Jersey (1970),

Maryland (1970), Georgia (1970), and New York (1972). All of the Atlantic and Gulf Coast states later followed suit in seeking to protect their coastal wetlands under state law (except Texas), as did the states bordering the Pacific Ocean (except Alaska and Hawaii). These laws put a sudden halt to the development of tidal marshes for residential and other purposes that had burgeoned after World War II and thereby boosted the price of already filled seaside land.

As long ago as 1899, Section 3 of the Rivers and Harbors Act made it unlawful to dump refuse into the navigable waters of the United States (33 *U.S. Code* 401 et seq.). This prohibition was not interpreted to protect wetlands against filling until the 1970s, after other federal statutes were enacted or strengthened to protect water quality, air quality, and the environment in general nationwide during the administration of Richard Nixon.

In 1972, Section 404 was added to the Federal Water Pollution Control Act (42 *U.S. Code* 7401 et seq.). This provision required a permit from the Army Corps of Engineers to authorize the placement of dredged or fill material into the waters of the United States, over which the federal government asserted regulatory control pursuant to the Commerce Clause of the Constitution and the international treaties to protect migratory birds. Those waters were redefined explicitly to include wetlands. Ongoing farming and forestry were exempted from Clean Water Act regulation, but the further expansion of farming into wetlands (except in Alaska) and the conversion of still wet, but cropped, wetlands to urban and suburban uses were not exempted. The Clean Water Act was amended again in 1977, but political consensus has not been reached since that time to allow reauthorization or changes of the Act through further amendment.

Direct federal subsidies for converting wetlands to agricultural uses were officially withdrawn by the "Swampbuster" provisions of the National Food Security Act of 1985 (Title XII of Public Law 99-198), which was amended in 1990 (Public Law 101-624) and 1996 (104-127). The Tax Reform Act of 1986 (Public Law 99-5114) reduced tax benefits of agricultural modification of wetlands. The Natural Resources Conservation Service (NRCS) was assigned responsibility for measures to protect wetlands in farms, and the Agricultural Stabilization and Conservation Service was directed to withhold payment of subsidies for commodity crops produced in lands that were wetlands as of December 1985. The actual removal of government subsidy for agricultural activities in wetlands proceeded very slowly, in as much as locally administered agencies had traditionally promoted drainage rather than wetland preservation (Vileisis 1997: 293–316). Such subsidies were regarded by the recipients of drainage benefits as virtual entitlements.

The Army Corps of Engineers, another bureaucracy long assigned to alter wetlands, also was reluctant to implement its regulatory authority over new construction in wetlands during the 1970s, but subsequently increased its efforts after successful litigation by conservation groups (such as *Natural Resources Defense Council v. Calloway,* 392 *F. Suppl.* 685, D.D.C. 1975). Substantial discretionary authority resides in Corps district offices, and local interpretations by the Corps and other federal agencies have led to regional differences in the Section 404 fill-permit program. Only two states (Michigan and New Jersey) have assumed administration of the federal program for inland, nonnavigable waters since such delegation was authorized in 1977.

About 15 states now regulate inland, freshwater wetlands in parallel programs that more or less overlap the federal program. Some of the state wetland statutes are more protective of wetlands than Section 404, which does not regulate the logging or farming or draining of wetlands or any activities in upland buffers surrounding them. Other states have minimum acreage thresholds for regulation that exempt projects affecting small wetlands. Some municipalities have attempted wetland regulation, either on their own initiative pursuant to state enabling legislation or through delegation of state permitting authority. It is ironic that wetland protection at present is not being implemented primarily at the level of local government across most of the nation, inasmuch as most of the human benefits derived from wetlands, whether in their natural or in their developed condition, are experienced locally. The political and economic forces that seek to convert wetlands, of course, also are most powerful locally.

At present wetland regulation is primarily the responsibility of the federal and state levels of government, and most technical expertise is found among federal and state regulatory personnel rather than at the municipal level. The interior western states and some states in the Midwest lack wetland protection programs at the state level; these nonregulating states encompass about one quarter of the remaining wetlands in the conterminous United States. Municipalities in some states can supplement other governmental protection of wetlands where the locally scarce wetlands are too small to warrant state or federal concern, and several have done so. In a few states of the Northeast, wetland delineation and regulation are primarily municipal responsibilities (Aurelia 1988).

Proposed fills or other drastic alteration of wetlands to accommodate new uses of the land are regulated in order to protect the public against harm. Most wetland regulatory procedures call for some measure of public disclosure and opportunity for public comment on applications for proposed construction. Corps regulations presuppose that requested federal permits will be granted, unless substantial objections are raised during a review. Regulators seek to balance technical, economic, and political-legal concerns when reaching decisions on specific proposals for wetland alteration. Such regulation exposes agencies to the risk of depriving landowners of some or all new economic uses they might want to make of their property, but outright permit denials are few. At some point regulation to prevent harm can become so onerous that courts may consider a specific property as "taken" from its owner for public benefit.

"Takings" claims were raised in more than half of some 400 wetland cases reported over a thirty-year period, but in very few cases were the takings claims sustained by the courts (Kusler 1990). Most governments and conservation groups lack the funds to pay full (development) market value for wetlands and seldom purchase wetlands outright from private owners to ensure the preservation of long-standing public benefits. Hence the threshold of regulation deemed by the courts to require takings compensation in accordance with constitutional guarantees is closely watched by developers and conservationists alike. Various development interests constantly lobby Congress and state legislatures, seeking to limit government activity in general and to dismantle wetland protection in particular (Echeverria and Eby 1995; Vileisis 1997).

Such concerns occasionally have inhibited municipal ordinances protecting wetlands, even though municipalities are primarily responsible for land-use regulation and are closest to the actual wetland benefits and losses on the ground. Few municipalities have on staff or choose to retain experts in wetland analysis. An encouraging trend among the more sophisticated municipalities is the imposition of review fees for development applications sufficient for the municipality to retain outside professionals to review the work of project sponsors. Even small municipalities can thereby obtain wetland expertise and other professional assistance when reviewing proposed projects without incurring costs to existing taxpayers or increasing their permanent staff.

When examined closely, the present regulatory system is not notably efficient in protecting wetlands or arriving at decisions. Wetland losses exceed 100,000 acres (156 square miles) annually nationwide, losses only partially offset by efforts to create new wetlands (Lewis 1997; Tolman 1997). Regulation has slowed wetland losses substantially in many parts of the nation, particularly in urban and suburban areas, but also in the rural landscape where it often is cheaper to purchase idle dry farmland than to drain additional wetlands. Wetland regulation provides many opportunities for interagency rivalry as well as for evasion of the responsibility for enforcement and for escaping the financial risk of permit denials that might be judged to be takings. Apparent inequities abound but are seldom publicly identified or recorded systematically.

American culture generally rewards efforts to maximize individual gain and encourages the alteration of land in support of that goal. Yet many Americans are motivated by a strong conservative tendency to preserve environmental quality in general and wetlands in particular. Caught in the middle is government. Fragmented, inconsistent, alternately heavy-handed and timid, government efforts since mid-century have been only moderately successful in protecting wetlands from destruction.

WETLAND DEFINITIONS

The definition of wetlands across the United States for regulatory purposes is technically complex, given the array of dissimilar wetland ecosystems found in this vast nation.[1] Wetlands need not be inundated or saturated continuously to incur regulation, but there must be a reasonable likelihood that any regulated wetland is at least saturated to the surface for prolonged periods in most years. Inferences concerning the presence of probable wetness can be made from soil and plants observed during seasonally dry periods, provided there is not substantial evidence for past hydrologic alteration. In a wood or marsh where water can be observed above, at, or just below the ground surface for long periods, there is little disagreement that a wetland exists. Many commonly acknowledged wetlands, however, typically are not inundated or saturated at all times. Moreover, the ecosystem of even a ponded wetland may not stop abruptly at the edge of the ponded water. A short-term field inspection during a dry season or rainless period may not document the actual presence of water on or near the surface of a wetland. The precise length of time that water must be present in or above the surface soil to warrant regulation is a subject of enduring controversy that spills from the scientific realm into the political arena.

The delineation of wetlands on specific parcels of land has proven a complex undertaking, especially as regulation has been extended inland beyond the tidal marshes easily identified on aerial photographs. Precise boundaries are necessary for regulation, however inexactly they may capture ever changing natural systems. To this end the Army Corps of Engineers issued a manual for wetland delineation in 1987 (EL 1987). Information is required regarding the presence of hydrophytic vegetation, hydric soils, and hydrology, all of which must be present in excess of minimum threshold amounts for a parcel of land to be regulated as wetland.

THE NITTY GRITTY: APPLYING DEFINITIONS IN THE FIELD

The detailed rules for recognizing wetlands in the field, as applied by regulatory personnel, establish the actual geographical scope of protected wetlands. Because no one has determined how many acres of wetland should be protected in the United States, wetland regulation is carried out incrementally, site by site, on lands discovered to meet the individual regulator's definition of wetlands. However the rules are crafted, landowners can be expected to say "too much" land is being regulated; conservationists, "too little." Overall, the administrative agencies have slowly been adopting the findings of scientists into the process of recognizing wetlands, but that process remains unfinished. What is a wetland and what field data are necessary to support wetland delineations are far from being certified truths of "cold" science devoid of controversy (Killingsworth and Palmer 1992: 107).[2]

Unlike the resource-based mapping of forests, crops, soils, floodplains, and wet ecosystems, regulatory wetland maps showing the actual extent of confirmed regulatory boundaries (with a few state and local exceptions) do not exist, despite the ready availability of computerized geographical information systems. In consequence, the actual extent of regulatory control across the American landscape remains shrouded in mystery and political obfuscation.

Vegetation

Plant communities are the most readily apparent wetland indicators. They often are the most significant determinants of major wetland functional values such as wildlife habitat and the ability to enhance water quality. Plant communities can be recorded on aerial photographs and observed readily throughout much of the year. Based on the analysis of aerial photographs, the more than 50,000 NWI maps prepared as overlays to U.S. Geological Survey topographic quadrangles provide a general guide to the location of prominent wetlands. NWI maps are exceptionally well suited to the wildlife management purpose for which they were originally designed (Stolt and Baker 1995; Klemow 1998). For regulatory purposes, however, the NWI maps are not sufficiently accurate to define jurisdiction on specific tracts of land. Many plants (obligate hydrophytes) that virtually require saturated soil conditions to compete successfully in the wild and that almost always are found growing in wetlands do not require those saturated conditions throughout the year. Most can tolerate occasional prolonged periods of soil dryness. Robust stands of such plants typically suggest prolonged wetness and provide strong evidence for the minimum extent of any wetland boundary. Those state and local regulatory definitions of wetlands based on plants alone generally focus on the species now characterized as obligate hydrophytes.

Many more plants not only have roots capable of growing in the oxygen-poor conditions of saturated soils but also can compete successfully in well-drained uplands. Thus the mere presence of stands of these species (facultative hydrophytes) may not be conclusive for wetland determination. Many of these broadly tolerant plants are among the most prominent components of the vegetation, particularly along the upslope margins of easily recognized wetlands (Schmid and Kartesz 1994).

The most rigorous method for identifying communities of plants as hydrophytic was developed by the U.S. Environmental Protection Agency (USEPA) and was most fully set forth in a 1989 interagency federal Manual, which for a time replaced the manuals of four federal agencies (FICWD 1989). Several alternative ways are provided for analyzing descriptive field data to reach the conclusion that the vegetation is hydrophytic. That manual as a whole subsequently was disallowed by statute for use in Clean Water Act Section 404 regulation. The 1989 vegetation assessment methodology, however, is authorized by the Corps, along with the less complex methodology of the 1987 Manual, at the user's discretion when making regulatory determinations (Williams 1992). In some plant communities the methods from these two manuals lead to opposite conclusions regarding the presence of hydrophytic vegetation.

Soils

Soil morphology responds to wetness conditions that range from ever-saturated or ponded, to infrequently saturated, to wet only during and immediately following precipitation events. The colors and textures of soil layers provide clues to the likely presence and duration of water in the plant root zone, reflecting the chemistry of oxygen, iron, and other elements in the soil solution under the local moisture regime. Nevertheless, some very wet habitats that support typical obligate hydrophytes may never experience anaerobic conditions during the growing season, because the water moves sufficiently to maintain free oxygen or the soil is too cold when wet or too poor in nutrients to support oxygen-sequestering bacteria. Where there is too little iron or manganese to provide telltale stains on sand grains near the soil surface, evidence for protracted wetness may not be noticeable during dry periods. Even some organic soils are not typically associated with regulated wetlands.

Most soil morphologic features have the advantage of being available for examination all year long, and NRCS recently adopted field indicators by land resource region for recognizing soils that meet the National Technical Committee on Hydric Soils (NTCHS) definition of hydric as "a soil that formed under conditions of saturation, flooding, or ponding long enough during the growing season to develop anaerobic conditions in the upper part" (SCS 1994). More than 3,100 named soil series have been deemed hydric nationwide (NTCHS 1991). After widespread testing nationwide, 36 direct field indicators were adopted by NRCS for use; 17 additional indicators currently are under review for potential adoption (NRCS 1998).

The adopted NRCS indicators represent the best currently available guidance for recognizing hydric soils. To date they have not been adopted by the Corps, which requires that the 1987 Manual field indicators be used instead for regulatory determinations (Studt 1997). The 1987 Manual and NRCS indicators produce dramatically different configurations of hydric soils on some sites, typically more dramatic than the differences among hydrophytic and nonhydrophytic plants occasioned by the 1988 and 1996 NWI plant indicator status designations. Five of the eleven 1987 Manual indicators for hydric soils are not capable of application in the field (NRCS 1995). The current Corps directive not to use the NRCS field indicators (or the 1996 NWI plant classifications) appears to contradict the general mandate of its 1987 Manual that delineators should always utilize best professional judgment when identifying wetlands (EL 1987: 12; Kunz 1997a, 1997b, 1998).

Hydrology

Wetland hydrology is the most important of the three defining parameters, but it can be the most difficult to document during short-term fieldwork. Just because a tract of land is ponded or saturated at the time it is examined does not mean that it experiences the prolonged and recurrent wetness required for wetland regulation. All uplands are wet during major precipitation events, and there are millions of acres of non-wetlands within floodplains subject to occasional inundation that may last several days. Historic data on inundation may be available for some areas, but data on saturation are lacking for most sites nationwide.

Water must be present longer in a wetland to satisfy the hydrology criterion than to make vegetation hydrophytic or soils hydric. The 1987 Corps Manual definitively classifies lands never or intermittently (less than 5 percent of the local growing season, 50 years out of 100) inundated or saturated as non-wetlands (EL 1987: 36). In contrast, lands saturated or inundated at least seasonally (more than 12.5 percent of the growing season) exhibit wetland hydrology sufficient for regulation. Many irregularly inundated or saturated lands (5 to 12.5 percent of the growing season) with hydric soils and hydrophytes are considered not to have wetland hydrology sufficient for universal regulation, but some may be regulated on a case-by-case basis. The threshold duration of saturation or inundation must be met by continuous wetness, not cumulative days of wetness during the growing season after discrete rainy periods.

Most of the field indicators of hydrology identified by the 1987 Manual bear little relationship to the duration of wetness and saturation. NRCS subsequently developed various analytical procedures to assist in the analysis of wetland hydrology (NRCS 1997), although no comprehensive technical guide has yet been published. Without an accepted methodology for monitoring soil saturation, no current technical procedure exists for resolving controversy over hydrology through direct field monitoring in disputed wetland boundary determinations.

Resulting Controversies

When applied to specific parcels, state and local wetland definitions may yield wetland boundaries very different from those of federal agencies. In some instances the Corps has agreed to adopt the wetland boundaries imposed by state or local agencies. In other cases, the state usually adopts boundaries established by the Corps. It is not helpful for agencies to adopt dissimilar wetland regulatory boundaries on the same parcel of ground, although this situation often arises. The public understandably becomes skeptical when regulators are unable to agree as to what lands are to be regulated as wetlands on a given property.

Differences of a few feet in the placement of a wetland boundary may have little significance when determining the approximate extent of wetland resources nationwide, but they strongly influence what landowners are or are not allowed to build within a specific property. Wetland boundaries also dictate the extent of regulated buffers in adjacent uplands. The extent of regulated wetlands to be disturbed generally determines the costs of compliance with permits. Thus there is considerable incentive for differences of opinion to arise in cases involving all but the most obvious of wetland boundary determinations.

These differences of opinion arise from various sources. Agencies and individuals within agencies differ by applying wetland definitions broadly or narrowly. One Corps office may throw out from jurisdiction all areas except those that never could be construed as uplands. In a different Corps office, field representatives may try to protect every inch of hydric soil. A Corps district can refuse to assert jurisdiction over dozens of acres with hydrophytes, hydric soils, and documented wetness on one site, while regulating a nearby site that lacks documentation of one or more of the three parameters. Courts generally allow broad discretion to regulatory agency representatives when making such determinations.

Such differences in outlook are not confined to regulators. Competing consultants interpret wetland definitions differently at the margin on specific tracts of land. It is common practice for landowners to hire multiple consultants to perform independent wetland boundary analyses. After reviewing the results of several efforts to apply the same rules, the landowner then selects the most favorable boundary for submission to the regulatory agencies.

Who Decides?

Clean Water Act jurisdictional determinations are made formally by the Corps based on information supplied by applicants for each property, except in unusual cases where the determination is made by USEPA (33 *CFR* 325.9). Except in those states that have published detailed, large-scale maps of state-regulated (usually tidal) wetlands, landowners are responsible for identifying the extent of wetlands on their properties and for avoiding such areas during new construction. That means establishing lines on the ground where wetlands give way to uplands, and then transferring those lines to property maps and engineering drawings. Governments consider it too expensive to map wetlands in detail, because activities requiring regulation are highly localized and most of the costs of such mapping logically can be imposed upon the parties expecting to benefit from the new construction in or adjacent to the wetlands.

Agencies either may confirm a boundary as proposed or may revise it on the basis of field inspection or reinterpretation of definitional guidance. The Supreme Court encouraged agencies to define wetlands broadly, inasmuch as development can be allowed through the permit process (*United States v. Riverside Bayview Homes, Inc.*, 474 *U.S.* 121, 1985). Conservation groups often experience great frustration when agencies fail to assert jurisdiction over obvious wetlands. The Corps regards the establishment of a wetland boundary as a private matter between the agency and the landowner; other agencies, such as the New Jersey Department of Environmental Protection, require public notice and invite comments when establishing regulatory boundaries. Reliable boundaries are essential to the orderly process of land sales and development. The regulatory flux in defining and identifying wetlands during the past twenty years has been far from optimal for either wetland protection or cost-effective land development.

PERMITS FOR ALTERING WETLAND LANDSCAPES

Just because a tract of land contains a regulated wetland does not preclude it from alteration or development. Some activities in wetlands, such as forestry and ongoing farming, are exempted from federal regulation under the Clean Water Act. Project sponsors have a universal incentive to avoid new construction in regulated areas. Regulated activities require permits. Few permits are denied, but acquisition of agency approval is time consuming and may entail costly modification of a construction project. The costs of compensating for lost wetland values may be significant where wetland use is allowed.

Where it is not possible to avoid working in regulated wetlands, project sponsors are required to secure permit approval prior to undertaking regulated activities, and projects are vulnerable to agency and public criticism during the review process. Certain classes of minor activities in wetlands and waterways have been authorized by the Corps through nationwide permits that allow work without a more complex individual permit application. Some of these nationwide permits require agency notification before the work can proceed, and all are premised on the permittee's compliance with ostensibly enforceable conditions designed to minimize environmental damage (33 *CFR* 330 et seq.). The terms of nationwide permits, like those of general permits issued by some states, are revised every few years, and are closely watched as de facto standards that denote impacts below thresholds requiring detailed scrutiny and justification for using the wetland project site. Regional general permits may lend routine federal sanction to activities that have been authorized by permits from state agencies.

Wetland fill projects that do not fit within general permits require an individual Corps permit application triggering public notice and opportunity for review by various agencies. For activities deemed to be not water dependent, the applicant must demonstrate that no practicable alternative exists that would avoid or reduce the proposed fill in wetlands. The question of alternatives often generates protracted argument, and there are no clear standards by which to demonstrate that no practicable upland alternative site exists. States can block Federal Clean Water Act fill permits by denying the requisite Section 401 water quality certification. The Corps performs a public interest review encompassing many aspects of the proposed construction activity and may prepare a formal environmental impact statement on projects with major impacts. On the other hand, the Corps and state wetland agencies sometimes dismiss objections from the public and from other agencies and may issue permits after cursory review. Generally the review process for federal or state individual permits is slow, and it may impose substantial front-end costs on applicants, costs that are not necessarily proportionate to the wetland values at risk. Permit approvals based on grossly inadequate permit reviews can be challenged in court, as can permit denials, but in either case the process of judicial appeal may be slow, costly, and of uncertain outcome.

Approvals of those projects deemed to warrant the filling of wetlands generally are conditioned upon the permittee's attempted replacement of wetland values lost, as partial mitigation for the unavoidable harm to public resources. On paper, significant compensation for wetland losses is required. Such conditions can be expensive for permittees to implement, especially in urban areas with high land values. Many promised replacement wetlands simply have not been built, even though the wetlands were filled. Agency efforts to monitor wetland mitigation long have been minimal, but recently issued permits may include requirements that the permittee report annually on the success of mitigation efforts over a period of several years. Performance bonds can be demanded by the Corps to guarantee the achievement of required mitigation, but seldom are, and litigation to compel permittees to remedy noncompliance or unsuccessful mitigation is nonexistent. Some state agencies may be more consistent in requiring performance bonds. Agency failures to require permit compliance generally cannot be challenged legally, and the historic record of compliance with permit conditions is dismal. Like regulated wetlands themselves, the alterations allowed by permits and any required or successful mitigation wetlands are virtually never shown on regional maps.

CREATING AND PRESERVING WETLAND LANDSCAPES

Governmental requirements are leading to wetland and stream corridor preservation across the United States. Where wetlands exist in suburban areas, they can develop a significant constituency willing and able to question proposals for future development. Suburbs built in the closing years of the twentieth century in areas where there are many wetlands look different from those built previously nearby, in part because their wetland ecosystems remain more or less intact. The public scrutiny of wetland development in rural areas typically is less rigorous than in the urban areas where most Americans now live, but even in farmlands some marginal fields are being protected as wetlands.

When unavoidable wetland encroachment exceeding the allowable threshold is authorized by a permit, compensation for the lost wetlands typically is required, and there now exist examples of successful implementation. If sufficient new wetlands are created to offset the wetlands lost, the future extent of wetlands presumably would remain stable or increase. Plans for wetland creation or restoration too often undergo only cursory review, and proposals to plant nonnative trees, shrubs, or herbs may fail to garner objections.

Outside the regulatory process, wetland preservation and enhancement efforts can be anticipated chiefly along rivers and estuaries. In many metropolitan areas, the functions of remaining public and private wetlands have been impaired by diking, ditching, or other historic alterations, and such wetlands offer the opportunity for restoration to a full array of functions at relatively low cost. Whatever public funds become available for urban wetland improvement are likely to be spent to restore degraded wetlands along waterways. In rural areas, the limited federal funds authorized for wetland restoration have been woefully insufficient to accommodate the acreage proffered by interested landowners pursuant to the Wetland Reserve Program and Conservation Reserve Program of the Department of Agriculture and the Partners for Wildlife Program of the Fish and Wildlife Service. The Nature Conservancy nationwide and many smaller local conservation groups continue to acquire wetlands through purchase or donation and have placed thousands of acres under permanent protection. Ducks Unlimited for many years has supported the return of prairie potholes to use by waterfowl.

Wetlands are scarce and are valuable to the public far in excess of the proportion of the landscape that they occupy. Combined with water bodies, they can dramatically enhance the value of adjacent lands in urban and suburban areas when incorporated into aesthetic landscape designs (Tourbier and Westmacott 1992). Other wetlands, just as valuable for many functions, are not appreciated, and most are discounted by the marketplace for real property.

At present few new wetlands are being constructed intentionally except as required compensation for permits to fill other wetlands, or as regional mitigation banks where compensation for several projects can be aggregated efficiently (Dennison 1997). Many regulators remain hostile to mitigation banking, despite encouragement from the upper echelons of the federal hierarchy (DACE et al. 1995). Landowners understandably are reluctant to cause or allow new wetlands to arise because of the potential for incurring regulation. Some jurisdictions regulate wetlands that form in stormwater management facilities; others do not. The present regulatory framework thus serves both directly and indirectly to discourage wetland creation as part of stormwater runoff controls, where new wetlands could accomplish substantial benefits for water quality, wildlife habitat, and landscape aesthetics.

CONCLUSION

The American experience over the past three hundred years proves that public policies can transform landscapes through wetland destruction on a grand scale. The emergence of public opinion that now generally supports wetland conservation was a dramatic development of the late twentieth century that portends a much diminished loss of our remaining wetlands in the years ahead. Yet federal and state wetland protection measures are routinely attacked in the political arena (Helvarg 1997). Wetland protection depends on broad public understanding of the reasons why wild, muddy places should be kept that way and on a widespread recognition of the kinds of benefits that flow from wetlands to the public at large as well as to wetland owners. Wetlands have utility in guarding people from floods and protecting water quality, but they also exhibit beauty and provide places for plants and animals to survive through future human generations. Laws are indispensable, along with effective funding and administration of regulatory and land-preservation programs. Compliance with and enforcement of existing laws against wetland destruction to date have been variable across the nation.

The widespread occurrence of wetlands on private landholdings, technical difficulties in defining the extent of wetlands regulated by various political entities, the uneven geographic distribution of wetlands across properties and regions, the diffuse nature of wetland benefits flowing to people at large rather than to individual landowners, the reluctance of government agencies to regulate in the absence of consensus, the slowness that typically characterizes government action in contrast to private-sector eagerness for rapid construction in response to fleeting opportunities for profitable land development, and the erratic enforcement of wetland laws and regulations have spawned controversies unlikely to disappear soon. Legal and regulatory systems struggle to balance public interests against landowner property rights. Wetlands are a prime example of a natural resource, once ignored or heedlessly exploited but now scarce in many locales, that is increasingly valued by an ever more populous society for reasons other than its potential for transformation into dry land for the sake of human convenience or monetary gain.

Efforts are being made to create replacement wetlands where road improvements and other construction cannot avoid wetland encroachment. Wetland functions are being restored in idle farmlands in response to private efforts and to minor funding by the federal government and some states. New construction in upland buffers is being regulated in some places to protect adjacent wetlands. Real estate values and property tax assessments are being adjusted downward to acknowledge the regulatory difficulty of converting wetlands into other uses, despite readily available and generally affordable technology for transforming them into valuable building sites. Consequently most of the nation's remaining wetlands seem likely to survive through the twenty-first century, provided that current regulatory requirements remain in place and gain compliance.

Whether future Americans will conserve wetlands effectively for the long term—not just by slowing the losses but by actually restoring degraded wetlands and bringing forth new wetlands that achieve multiple functions across the landscape in significant quantity and quality—is a major question for the new millennium. How well the nation conserves its wetlands will be a key indicator of how well it preserves a sustainable human environment for all its citizens.

NOTES

1. Since 1975, Clean Water Act wetlands have been defined as "those areas that are inundated or saturated by surface or groundwater at a frequency and duration sufficient to support, and that under normal circumstances do support, a prevalence of vegetation typi-

cally adapted for life in saturated soil conditions" (33 *CFR* 328.3; 40 *CFR* 230.3). The definition of wetlands on farms is similar (NRCS 1996: 525–15).

2. Congress obtained a review of the scientific underpinning of wetland identification procedures from the National Academy of Sciences (Committee on Characterization of Wetlands 1995).

REFERENCES

Aurelia, Michael. 1988. "Protecting Freshwater Wetlands and Riparian Habitats in Southwestern Connecticut." In *Proceedings of the National Wetland Symposium on Urban Wetlands at Oakland CA, 26–29 June 1988*, ed. J. A. Kusler, S. Daly, and G. Brooks, 253–61. Berne, NY: Association of State Wetland Managers.

Committee on Characterization of Wetlands. Commission on Geosciences, Environment, and Resources. National Academy of Sciences. 1995. *Wetlands: Characteristics and Boundaries*. Washington, DC: National Research Council, National Academy Press.

DACE (Department of the Army Corps of Engineers), Environmental Protection Agency, Department of the Interior, Fish and Wildlife Service, and Department of Commerce, National Oceanic and Atmospheric Administration. 1995. "Federal Guidance for the Establishment, Use, and Operation of Mitigation Banks." 60 *Federal Register* 228: 58605–58614 (28 November 1995).

Dahl, T. E. 1990. *Wetlands Losses in the United States, 1780s to 1980s*. Washington, DC: U.S. Department of the Interior, Fish and Wildlife Service.

Dennison, Mark S. 1997. *Wetland Mitigation: Mitigation Banking and Other Strategies for Development and Compliance*. Rockville, MD: Government Institutes.

Dickens, Charles. 1891. *American Notes for General Circulation*. Philadelphia: B. Lippincott.

Douglas, Marjorie Stoneman. 1947. *Everglades, River of Grass*. Marietta, GA: Mockingbird Books.

Echeverria, John, and R. B. Eby. 1995. *Let the People Judge: Wise Use and the Private Property Rights Movement*. Washington, DC: Island Press.

EL (Environmental Laboratory). 1987. *Corps of Engineers Wetland Delineation Manual*, Technical Report Y-87–1. Vicksburg, MS: U.S. Army Waterways Experiment Station. (Available at: <<http://www.wetlands.com/regs/tlpge02h.htm>)

Errington, Paul L. 1957. *Of Men and Marshes*. New York: Macmillan.

FICWD (Federal Interagency Committee for Wetland Delineation). 1989. *Federal Manual for Identifying and Delineating Jurisdictional Wetlands*. Washington, DC: Army Corps of Engineers; Environmental Protection Agency; Department of Agriculture, Soil Conservation Service; Department of the Interior, Fish and Wildlife Service.

Fox, Stephen. 1981. *John Muir and His Legacy*. Boston: Little, Brown.

Greeson, P. E., J. R. Clark, and J. E. Clark, eds. 1979. *Wetland Functions and Values, the State of Our Understanding*. Minneapolis: American Water Works Association.

Helvarg, David. 1997. *The War Against the Greens: The 'Wise Use' Movement, the New Right, and Anti-Environmental Violence*. San Francisco: Sierra Club Books.

Killingsworth, M. J., and J. S. Palmer. 1992. *Ecospeak: Rhetoric and Environmental Politics in America*. Carbondale: Southern Illinois University Press.

Klemow, K. L. 1998. “Wetland Mapping.” In *Ecology of Wetlands and Associated Systems*, ed. S. K. Majumdar, E. W. Miller, and F. J. Brenner, 1–26. Easton: Pennsylvania Academy of Science.

Kunz, Stephen P. 1997a. “Soil Color Charts: The Rest of the Story.” *Wetland Journal* 9 (4): 8–12.

———. 1997b. “What Color Is Your Hydric Soil?” *National Wetlands Newsletter* 19 (5): 5–6.

———. 1998. “More ‘Dirt’ on Hydric Soil Indicators.” *Wetland Journal* 10 (4): 16–20.

Kusler, John. 1990. “Takings: Is the Claims Court All Wet?” *National Wetlands Newsletter* 12 (6): 6–7.

Leopold, Aldo. 1949. *A Sand County Almanac*. New York: Oxford University Press.

Lewis, Roy R. 1997. “Paper Numbers.” *National Wetlands Newsletter* 19 (4): 22.

Lieth, H. 1975. “Primary Productivity of the Major Vegetative Units of the World.” In *Productivity of the Biosphere*, ed. H. Lieth and R. H. Whittaker, 203–316. New York: Springer-Verlag.

Marsh, George Perkins. 1965/1864. *Man and Nature*, ed. David Lowenthal. Cambridge, MA: Belknap, Harvard University Press.

McManis, Douglas R. 1964. *The Initial Evaluation and Utilization of the Illinois Prairies, 1815–1840*. Chicago: University of Chicago, Department of Geography, Research Paper No. 94.

Mitsch, William J., and James G. Gosselink. 1993. *Wetlands*, 2nd ed. New York: Van Nostrand Reinhold.

Nash, Roderick F. 1967. *Wilderness and the American Mind*. New Haven: Yale University Press.

Niering, William. 1966. *The Life of the Marsh: The North American Wetlands*. New York: McGraw-Hill.

NRCS (Natural Resources Conservation Service). 1995. *Field Indicators of Hydric Soils in the United States*, Draft Version 1.3. Lincoln, NE: Natural Resources Conservation Service.

———. 1996. *National Food Security Act Manual* (NFSAM). 3rd ed., Amendment 2. Washington, DC: U.S. Department of Agriculture.

———. 1997. “Hydrology Tools for Wetland Determination.” In *National Engineering Field Handbook*, Part 650. Washington, DC: U.S. Department of Agriculture.

———. 1998. *Field Indicators of Hydric Soils in the United States, A Guide for Identifying and Delineating Hydric Soils*, Version 4.0, ed. G. W. Hurt, P. M. Whited, and R. F. Pringle. Fort Worth, TX: U.S. Department of Agriculture. (Available at: <<http://www.statlab.iastate.edu/soils/hydric>)

NTCHS (National Technical Committee on Hydric Soils). 1991. *Hydric Soils of the United States*. 3rd ed. Miscellaneous Publication 1491. Washington, DC: U.S. Department of Agriculture.

Olmsted, Frederick Law. 1968. *Landscape into Cityscape*, ed. Albert Fein. Ithaca: Cornell University Press.

Prince, Hugh. 1997. *Wetlands of the American Midwest: A Historical Geography of Changing Attitudes*. Chicago: University of Chicago Press.

Schmid, James A. 1994. “Wetlands in the Urban Landscape.” In *The Ecological City, Preserving and Restoring Urban Biodiversity*, ed. R. H. Platt, R. A. Rountree, and P. C. Muick, 106–33. Amherst: University of Massachusetts Press.

Schmid, James A., and John T. Kartesz. 1994. *Checklist and Synonymy of Higher Plants in New Jersey and Pennsylvania, with Special Reference to Their Rarity and Wetland Indicator Status*. Vol. 2, Desk Manual, rev. ed. Media, PA: Schmid & Company, Consulting Ecologists.

SCS (Soil Conservation Service), U.S. Department of Agriculture. 1994. "Changes in Hydric Soils of the United States." 59 *Federal Register* 133: 35680–35695 (13 July 1994).

Shaw, Samuel P., and C. Gordon Fredine. 1956. *Wetlands of the United States: Their Extent and Value to Waterfowl and Other Wildlife*. Washington, DC: Department of the Interior, Fish and Wildlife Service, Office of River Basin Studies, Circular 39.

Stolt, Mark H., and James C. Baker. 1995. "Evaluation of National Wetland Inventory Maps to Inventory Wetlands in the Southern Blue Ridge of Virginia." *Wetlands* 15 (4): 346–53.

Studt, John F. 1997. "Memorandum for All Major Subordinate Commands, District Commands (Attn: Regulatory Chiefs), Subject: NRCS Field Indicators of Hydric Soils." Washington, DC: U.S. Department of the Army, Corps of Engineers, 21 March 1997.

Teal, John, and Mildred Teal. 1969. *The Life and Death of a Salt Marsh*. Boston: Little, Brown.

Thoreau, Henry David. 1862. "Walking." *Atlantic Monthly* 9 (56): 666.

Tiner, Ralph W., Jr. 1984. *Wetlands of the United States: Current Status and Recent Trends*. Department of the Interior, Fish and Wildlife Service, National Wetlands Inventory. Washington, DC: Government Printing Office.

Tolman, Jonathan. 1997. "How We Achieved No Net Loss." *National Wetlands Newsletter* 19 (4): 1, 19–22.

Tourbier, J., and R. Westmacott. 1992. *Lakes and Ponds*, 2nd ed. Washington, DC: Urban Land Institute.

Turner, Frederick. 1985. *Rediscovering America: John Muir in His Time and Ours*. New York: Viking Penguin.

USEPA (United States Environmental Protection Agency). 1998. *Clean Water Action Plan: Restoring and Protecting America's Waters*. Available at: <http://www.epa.gov/cleanwater/action/toc.html>.

USEPA (U.S. Environmental Protection Agency, Region II) and USACE (United States Army Corps of Engineers, New York District). 1995. *Draft Environmental Impact Statement on the Special Area Management Plan for the Hackensack Meadowlands District, New Jersey*. New York: Army Corps of Engineers.

USFWS (United States Fish and Wildlife Service). 1990. *Wetlands: Meeting the President's Challenge*. Washington, DC: Department of the Interior.

Vileisis, Ann. 1997. *Discovering the Unknown Landscape: A History of America's Wetlands*. Washington, DC: Island Press.

Williams, Arthur E. 1992. "Memorandum on Clarification and Interpretation of the 1987 Manual." Washington, D.C.: U.S. Department of the Army, Corps of Engineers, 6 March 1992.

Zaitzevsky, Cynthia. 1982. *Frederick Law Olmsted and the Boston Park System*. Cambridge, MA: Belknap, Harvard University Press.

7

Navigability of American Waters: Resolving Conflict through Applied Historical Geography

John A. Kirchner

Conflict over the allocation or use of water may be due to cultural, economic, or political variables, or a combination of these factors. The rules of law which apply in any given case are defined by the nature of the dispute and the ways that a particular culture or society chooses to resolve them. Like many cultural elements, defining navigability requires an appreciation of process—the interpretation may be stable, based on years of tradition, or dynamic, changing significantly at the moment a new court decision is rendered.

To navigate, or, more simply put, to move upon or travel over a body of water, implies a basic function. Indeed, international disputes, such as the right of passage through the Straits of Hormuz or the Gulf of Aqaba, are not so much a matter of the ability to navigate as a matter of the desire by various parties to prevent it. A blockading gunboat may seem to have little in common with a dam, but in fact they both constitute a possible impediment to free movement.

Because navigation manifests a variety of conditions and concerns, there is no simple definition of navigability that will suffice. There are many different and even disparate reasons to be concerned with navigation and navigability. The applicable reason for studying a particular topic or case may be tied to the traditional spatial or temporal concerns of transportation, may involve a major environmental issue, or may be expressed in defining or redefining the public interest. Thus the power to regulate dam safety may rest in the power to regulate commerce, with navigability, in its many different contexts, forming the critical link between the two.

Within the United States the definition of navigability is quite complex. What the public perceives as navigation or navigability may be interpreted in quite different ways through the judicial process. In cases involving both states' rights and interstate commerce law, disputes have arisen over who has the power to control a river: for example, the right to erect a dam or diversion, or build and operate a hydropower plant. For the federal government, through an agency like the Federal Energy Regulatory Commission (FERC), to claim jurisdiction, the watercourse must be "navigable and a part of Interstate Commerce." But what about a river or lake that was altered in the past, where physical navigability is today impossible? Over the years the definition of *navigability* has broadened, so that *navigation* can now be argued in past, present, and even future contexts. Thus a finding that Native American canoes, fur traders in buffalo-skin boats, or log drivers at one time in the past transited a given watercourse may be sufficient, under the right conditions, to argue federal jurisdiction.

NAVIGATION AS AN EVOLVING CONCEPT

English common law drew many of its concepts from Roman law. Under Emperor Justinian, the Roman *Corpus Juris Civilis* of A.D. 528 differentiated between private and public waters. All free-flowing rivers and streams were considered public, unless the watercourse was of intermittent or ephemeral character. The state held title to the bed of public rivers, while the bed of private watercourses remained under control of adjacent riparian landowners. The water itself could not be owned by individuals. Likewise, the Romans also protected navigable waterways to ensure free passage for vessels of various kinds (MacGrady 1975: 530; Mathews 1984: 23–24; Bayle 1994: 9–10).

The English initially adopted the Roman rules, but as early as A.D. 1216 they redefined the basics of authority, giving control of most riverbeds to private parties. By the seventeenth century English law had established two basic interpretations to define the public character of streams and watercourses. To be *public* the river had to be affected by the ebb and flow of the tide, or had to be navigable in fact (Mathews 1984: 25; Bayle 1994: 10). Tidelands as well as the sea bottom belonged to the Crown, and anybody had the right to navigate upon or fish on those waters. Upstream, beyond the influence of the tide, if a waterway was navigable in fact, the public was guaranteed the right of passage, but was denied other uses, such as fishing (Morreale 1963: 26–28; Bayle 1994: 11). *Navigability in fact* implied that tidal or freshwater rivers could carry products of the soil in commercial quantities (Johnson 1949: 268; Bayle 1994: 11).

THE UNITED STATES

For the United States, the short rivers and streams linking the Atlantic tidewater with the interior seemed well suited to the rules of English law.[1] But with sizable inland seas, such as the Great Lakes, and major rivers, such as the Mississippi and its tributaries, the issues were different. The emergence of commerce on these waters created a need for new definitions.

Admiralty law and commerce were intertwined, and were dealt with in cases like *Gibbons v. Ogden* (1824) and *The Steamboat Thomas Jefferson* (1825), but without clear resolution of the major jurisdictional issues.[2] A catalyst for change came in 1851 with *The Propeller Genesee Chief v. Henry Fitzhugh*. Brought about by an 1847 collision on Lake Ontario, the finding clearly expanded federal control. Invoking the Commerce Clause (Article I, Section 8) of the U.S. Constitution, Justice Taney argued:

> There is certainly nothing in the ebb and flow of the tide that makes the waters peculiarly suitable for admiralty jurisdiction, nor anything in the absence of a tide that renders it unfit. If it is a public navigable water, on which commerce is carried on between different states or nations, the reason for jurisdiction is precisely the same. (*The Propeller Genesee Chief v. Henry Fitzhugh* 1851: 454–55)

In 1870, the Supreme Court handed down one of the most important navigation decisions of the nineteenth century (Guinn 1966: 560–61; Bayle 1994: 31). The case of *The Steamer Daniel Ball v. United States* (1870) involved a small unlicensed vessel operating on the Grand River wholly within the state of Michigan. Because the steamer carried interchange cargo destined for points outside the state of Michigan, the court determined that:

> Those rivers must be regarded as public navigable rivers in law which are navigable in fact when they are used, or are susceptible of being used, in their ordinary condition, as highways for commerce, over which trade and travel are or may be conducted in the customary modes of trade and travel on water. And they constitute navigable waters of the United States within the meaning of the Acts of Congress, in contradistinction from the navigable waters of the States, when they form in their ordinary condition by themselves, or by uniting with other waters, a continued highway over which commerce is or may be carried on with other States in which such commerce is conducted by water. (*The Steamer Daniel Ball v. United States* 1870: 563)

In a similar case four years later (*United States v. The Steamer Montello* 1874), the court added a new dimension. The Fox River flowed northeasterly into Green Bay and Lake Michigan. In its natural state it had several rapids and falls which precluded continual movement along the route. These were removed to create artificial navigation, and by building a short canal it was connected to the Wisconsin River, allowing access to the Mississippi system.

The river was held jurisdictional, the court noting that just because a river was exceedingly difficult to navigate, and needed artificial improvements, did not preclude its consideration as an interstate waterway. The ruling also made mention of lumber rafts, pointing out that vessels "of any kind" that can float, whether propelled by animal power, by wind, or by steam, are or may become the mode by which commerce can be carried out (*United States v. The Steamer Montello* 1874: 441; Guinn 1966: 561, 576; Bayle 1994: 31). Further clarification came in *Ex Parte Boyer* (1883). In this case, which involved the Illinois and Michigan Canal, the Court expanded the definition of navigable waters clearly to include artificial waterways (Guinn 1966: 562–63; Bayle 1994: 26).

Up to this point the major issues of jurisdiction related primarily to the regulation and licensing of vessels. While interstate commerce was a critical test of federal authority, the issues decided on constitutional grounds were commonly ones dealing with the actual operation of vessels, and the government's power to control those operations. However, in 1898, a major case was decided which dealt with the impact of dam building on what appeared to be a non-navigable waterway. In *United States v. Rio Grande Dam & Irrigation Company* (1898), the Court ruled that the federal government could control even nonnavigable waters if their use presented a threat to or impacted the navigable portion of jurisdictional waters. A large dam on the upper Rio Grande might reduce flow into the navigable lower Rio Grande; thus, the potential to *affect* navigability became the determining factor (Mathews 1984: 25; Bayle 1994: 37–38).

Congress, through the river and harbor acts of 1890 and 1899, better defined the government's authority to control the construction and operation of dams. The first made it unlawful to build structures that might interfere with navigation, commerce, or anchorage without the permission of the Secretary of War, while the second made these obstructions on navigable waters subject to the consent of Congress, requiring planning approval by both the Chief of Engineers (Army Corps of Engineers) and the Secretary of War. Two general dam acts of 1906 and 1910 added detailed conditions governing the building, maintenance, and operation of projects on navigable waters (Ecton, n.d.: 2).

With the passage of the Federal Water Power Act (FPA) of 1920, the Federal Power Commission (FPC) was created. At this point regulation became specifically focused on controlling and licensing the hydropower industry; navigation became an element of jurisdiction, rather than a focus in itself. The FPC included the Secretaries of Agriculture, Interior, and War, but it was initially without its own staff and had a very limited role (Ecton, nd.: 2).

In 1921, a very interesting case came before the Supreme Court. In *Economy Light & Power v. United States,* the high court declared that the Des Plaines River in Illinois was navigable, reversing the contrary rulings of two lower courts. At the time, because of certain changes in the river course, the waterway was not commercially viable. Delineating the concept of *past use,* the court found the Des Plaines to be jurisdictional because during the Fur Trade period (1675–1832) it formed part of one of the main trade routes between the Great Lakes and the Mississippi. Justice Pitney spelled this out:

> A river having actual navigable capacity in its natural state, and capable of carrying commerce among the states, is within the power of Congress to preserve for pur-

> poses of future transportation, even though it be not at present used for such commerce, and be incapable of such use according to present methods, either by reason of changed conditions or because of artificial obstructions. (*Economy Light & Power Company v. United States* 1921: 123)

Thus the concept of *indelible* navigability was established: if in its past a river or stream had been transited in interstate commerce, the existence of obstructions, seasonal fluctuations in water level, or other impediments did not affect the navigable character of the route. Furthermore, occasional natural obstructions or portages, which technically broke the continuity of water travel, did not destroy the legal definition of navigability (*Economy Light & Power Company v. United States* 1921: 122).

In *United States v. Appalachian Electric Power Company* (1940), the meaning of the Commerce Clause was expanded to justify even future uses of a river or stream. If by reasonable improvements a watercourse could be made navigable, it might be jurisdictional. Reinforcing earlier decisions, the Court ruled that navigability did not depend on a river's ordinary condition, and that it could occur despite the existence of falls, rapids, sandbars, or shifting currents. Nor could disuse over time change the constitutional character of navigation (Ecton, nd.: 5).

In the *Tomahawk* case of 1945, the Court of Appeals agreed with the FPC that the floating of logs on the Wisconsin River, in the course of a continuous movement from one state to another, clearly constituted interstate commerce. And much later, in 1965, the *Taum Sauk* decision extended FPC power to include hydropower projects on non-navigable waters if those plants were tied into the interstate power grid (Ecton, nd.: 6).

Over the years the FPC, and its 1977 successor, the Federal Energy Regulatory Commission (FERC) followed a variable path that reflected both administrative attitudes and funding constraints. The FPC became an independent agency in 1930, but was hardly an activist organization. During most of its early years it declined jurisdiction in the majority of cases brought before it. This was the result in 22 cases in which rivers had been used for the transport of logs, notwithstanding the fact that many of these rivers, like the Androscoggin, Saco, Connecticut, Menominee, Chippewa, and Flambeau, would later be reclassified as navigable (Ecton, nd.: 4).

NAVIGATION STUDIES

In 1937, the FPC initiated a series of comprehensive water basin studies to determine the status of the nation's dams and rivers. These became the basis for much of the investigative work which would follow, and are still an invaluable aid in geographical and historical research. In the 1980s, FERC's water basin research intensified, with particular emphasis on high hazard unlicensed projects (Ecton, nd.: 4–6).

Since historical navigability is the critical factor in many navigation cases, the work is clearly within the scope of the historian or historical geographer whose purpose is not so much to judge the merits of a case as to develop evidence of navigability or its absence. Using the tools of historical investigation, the researcher studies the water basin in its entirety, focusing on settlement, industrial growth, transportation networks, and the specific uses of the river itself. The investigation is often on site, working in the field, and usually involves local archival and library research, using both primary and secondary sources. The critical evidence is often fleeting and ephemeral in character—a few lines in a diary, the mention of an accident involving log drivers, an old photograph or postcard showing river uses, or even an offhand remark from a contemporary resident.

A number of the studies carried out through FERC's Office of Hydropower Licensing have been selected to illustrate the investigative process and the critical tests of navigability. The cases have been chosen to represent a variety of historical and geographical contexts, including several examples where, for quite different reasons, the research failed to produce results that meet the tests of navigability.

It should be noted that minor streams and rivers may be just as important from a regulatory point of view as major watercourses; indeed, from a dam-failure standpoint the little known and the obscure have contributed a disproportionate share of dam-related disasters. Federal, state, and local standards of dam safety may differ, but generally the federal rules and regulations represent the highest order of public safety. The structural integrity or condition of a dam, as well as the potential hazard to downstream peoples and communities, may determine the priorities in seeking to prove it jurisdictional.

SACO RIVER

The Saco River has its headwaters in Saco Lake at the northern end of Crawford Notch in the White Mountains of New Hampshire (Figure 7.1). Along its upper course to Conway, New Hampshire, the river falls steeply, with an average slope of 43.8 feet per mile. Rapids, rips, and gorges are common. Below Conway, through Fryeburg, Maine, and down to mile 50 near Hiram, the river becomes relatively smooth, before dropping through a series of intermittent falls prior to reaching the coast just below the urban, industrial towns of Biddeford and Saco. A number of tributary streams feed the Saco and these were also, in several cases, the subject of FERC investigations (FERC 1989: 2).

The Swan Falls Hydropower Project is located near Fryeburg, Maine, at river-mile 76.9. The 240-foot long, rock-filled, timber crib dam was built in 1923, and

Figure 7.1 Saco River Drainage System.

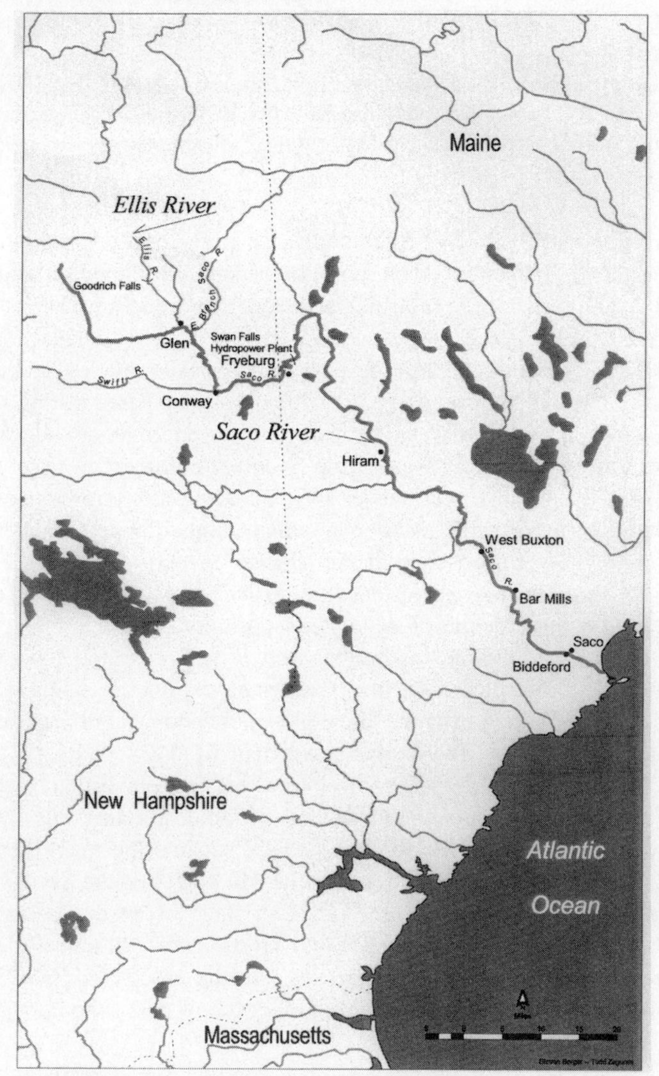

at the time of the study (1989) supplied power to Public Service of New Hampshire through a substation adjacent to the site (FERC 1989: 3).

The Saco had been used for a variety of purposes since the colonial era, including early canoe and *batteau* navigation, log drives and log rafting, as well as recreational boating. Each of these activities involved use of the river for the transportation of persons or property in interstate commerce (FERC 1989: 3–4).

Several writers allude to Indian navigation of the upper Saco in the precolonial era, and early European explorations clearly made use of the river.

> They went up the Saco River in birch canoes, and that way, they found it 90 miles to Pegwagget, an Indian town, but by land it is but 60. Upon Saco River, they found many thousands of acres of rich meadow, but there are ten falls, which hinder boats. (Spencer 1930: 180)

Having canoed past the site of Swan Falls, the party beached their boat about three miles above, and continued on foot into the White Mountains, before returning down river. There are similar narratives of other early journeys (Spencer 1930: 180–81; FERC 1989: 3–4).

By the mid-eighteenth century, the upper Saco was being settled by Englishmen from Maine and Massachusetts. Fryeburg was populated during the 1760s, while Conway and Barlett in New Hampshire followed in the next decade. There is evidence that some settlers boated the Saco for commercial purposes in the vicinity of the Swan Falls hydropower site, including many references to *"battowing"* between the Fryeburg settlements and points downriver. These vessels carried grain and other produce to market (FERC 1989: 6–7).

The region supported an important lumber industry, and from an early date the Saco was the primary means of log transportation. There is much historical evidence to show that logs and cut lumber were driven or floated down the watercourse, and this traffic clearly involved interstate commerce. Mills were established at Saco and Biddeford in the late eighteenth century, and the river became the main means of supply from points upriver (FERC 1989: 7–8).

A legal case in 1831 provides some interesting insight. Logger John Spring filed suit against the owners of the Fryeburg Canal, a man-made bypass that linked two sections of the river (and eventually became the main course of the river), charging that $10,000 worth of logs he had placed in the Saco were stranded in the artificial diversion. At the ordinary stage of water, timber could be floated "by the use of competent skill and proper care, without difficulty, and with little or no loss." Spring lost his case, but the associated investigation of river conditions provides an abundance of evidence of past river uses and conditions (FERC 1989: 9–11).

One of the most critical tests of interstate commerce was realized when logging activity moved into neighboring New Hampshire. Trees felled along the Pequawket River, a tributary of the Saco near Conway, New Hampshire, were floated to the Biddeford and Saco mills, and in time many other tributaries, including the Swift, contributed to the Saco lumber trade. In the mid-1850s logs were still being brought down from the upper waters around Bartlett and Conway, and the practice would continue well into the next century (FERC 1989: 13–14).

While the multiplicity of accounts makes the Saco case for navigability fairly easy to prove, several narratives are worth focusing upon. A river man named

Kilgore told of how he and his father put logs into the upper Saco in New Hampshire. These logs were bound for the Usher mills at Bar Mills, Maine, and in transit had to have passed the Swan Falls site. In 1879 another source reported that logs, representing about a million feet of lumber, had been placed along the banks of the river near Conway, destined to float down in the spring to the mills below (FERC 1989: 13–14).

> During the time of the big spring freshets, when the Saco River was very tempestuous, logs which the woodsmen had felled during the winter months would be driven nearly the whole length of the Saco to the sea. (Horne 1963: 55–56; FERC 1989: 15)

There were other tests of navigability. Log drives continued until 1948 when the large Skelton Dam was erected near Buxton, Maine. But as far back as 1905 and 1915 smaller dams on the route were required to have log sluices, and there is evidence that the Swan Falls project was itself once smaller and did not interfere with the log traffic. In the absence of log driving, another factor might also have made the Saco jurisdictional—recreational boating. As early as the 1890s, boating began at points in New Hampshire and traversed the river to points in Maine below the waterpower site. The practice has continued and expanded in recent years, with the modern Saco being recognized as one of America's great canoe streams, attracting hundreds of canoeists on summer weekends (FERC 1989: 18–26).

ELLIS RIVER

The abundance of evidence supporting interstate commerce on the Saco seemed to make its tributary stream, the Ellis, a good candidate for investigation. With other parts of the Saco system, most particularly the Pigwacket and Swift, contributing their share of log traffic to the main river, the research team fully expected to add another navigable waterway to the list (FERC 1992b).

The eighteen-mile Ellis River flows out of the White Mountains from near Mount Washington and joins the Saco south of Glen, New Hampshire (Figure 7.1). Along its course are several cataracts or falls, including Goodrich Falls, which is the site of the Goodrich Falls Dam and Power Plant. Located one mile above the confluence with the Saco, the plant was classified as low hazard by the New Hampshire Department of Water Resources (FERC 1992b: 21).

By the middle of the nineteenth century, lumbering in the area had established itself as a major industry. Sawmills and gristmills appeared in nearby Conway and Bartlett soon after their settlement, and one of the earliest local mills was located on the Ellis itself—it was later swept away in a flood. In 1827 a sawmill and gristmill was built by Jeremiah Goodrich at the Ellis River site that bears his name—the first of a number of uses based upon the natural drop in the river at that point (FERC 1992b: 12–14).

Notwithstanding the local importance of logging, little evidence was encountered to prove historical navigability. By the time intensive logging moved into the area above the Goodrich Falls site, rail transport rather than water had become the major means of exploiting the resource (FERC 1992b: 15–20).

The only evidence of log flotation on the Ellis came from old photographs, which show what appears to be cut timber in the area below the falls. While there was nothing to preclude log driving from the Ellis into the Saco, and it may have occurred, this and other evidence was at best circumstantial, leaving the Ellis as an unlikely candidate for federal jurisdiction.

NIOBRARA RIVER

The Niobrara is a tributary of the Missouri River. Following a course just south of the South Dakota/Nebraska border, the 431-mile Niobrara begins in eastern Wyoming and converges with the Missouri at a point 941 miles northwest of the junction of the Missouri and the Mississippi (Figure 7.2). Varying in character from deeply entrenched canyons in the west to a broader, fairly open channel in the east, the river has often been noted for its abundance of sand and silt and for the frequent shifting of its channels in the lower section of the river (FERC 1988: 2–3). The hydroplant in question was the Spencer Hydropower Project on the lower section of the river, some five miles south of the town of Spencer, Nebraska (FERC 1988: 3).

The Niobrara is an interesting case because so many observers wrote it off as a non-navigable river. While often referred to by early explorers as *L'eau qui court*, or the Water That Flows, it was viewed as too fast and too shallow to be useful. French explorer Jean Baptiste Truteau, who ascended the Missouri in 1794 and 1795, described the Niobrara as "the most abundant one in the entire continent of beaver and otter, but the waters are so swift that one is not able, so to speak, to either ascend or descend" (Truteau 1952 vol. 1: 266; FERC 1988: 4). Lewis and Clark, in their famous expedition to the Northwest, considered the Niobrara unsatisfactory because "the current is verry rapid, not navagable for even canoos without Great difficulty owing to its Sands" (Moulton 1987, vol. 3: 47).

Nevertheless, despite the difficulties attributed to the river, its importance in the fur trade stimulated efforts to exploit it for transportation. Fur trading forts were established at various points along the Missouri, including a small post at Fort Mitchell near the confluence of the Niobrara and Mississippi. Steamboats

Figure 7.2 Missouri River Drainage System.

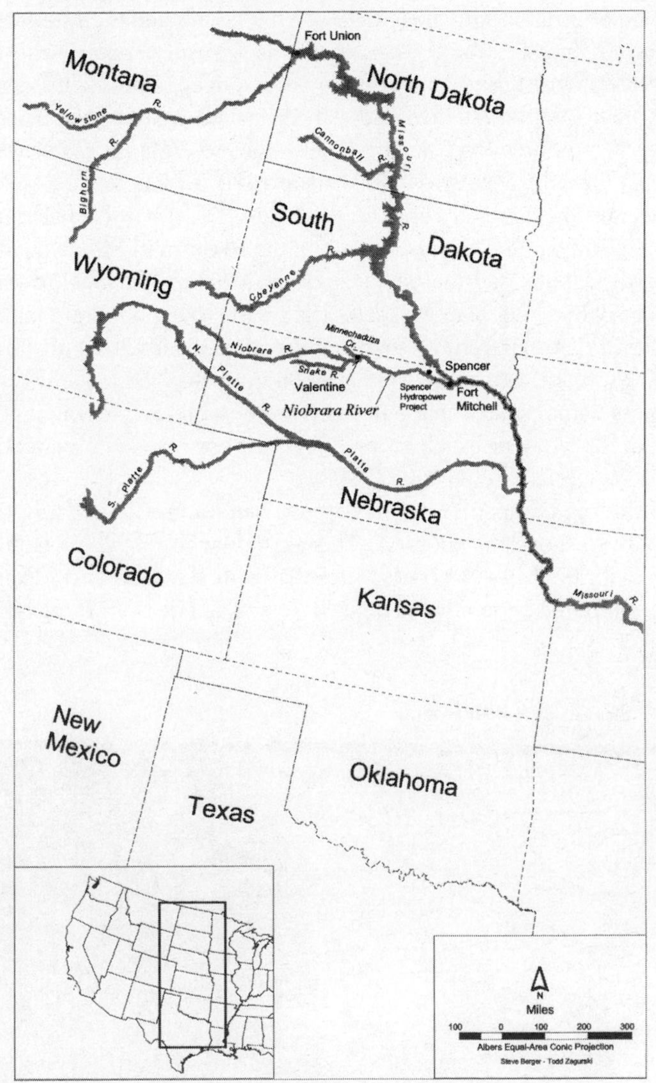

worked the Missouri to this point and north to Fort Union, at the junction of the Missouri and Yellowstone rivers (Chittenden 1903; FERC 1988: 5).

The key to using the Niobrara was having the right technology, and the bull boat or skin canoe was ideally suited to this end (Figure 7.3). Developed by local Indians, the small tubular boats were constructed by wrapping or stretching a single buffalo hide around a circular frame. The fur traders, who needed additional capacity, utilized the skin-covered frame in creating their own version of the bull boat. Several buffalo hides were sewn together and then stretched around a willow or cottonwood frame. The vessel was typically about twenty feet long, four to five feet across, and about twenty inches deep. The seams were caulked and the skins treated with tallow and ash. Propelled by two men using poles, it could carry two to three tons, and drew only four to eight inches of water (Chittenden 1903: 96–102; Metcalf 1972; Hanson 1980; FERC 1988: 5–6).

Bull boats were extensively used on the Platte River, but were also found on a variety of shallow, difficult rivers, including the Bighorn in Montana, the Cannon Ball in North Dakota, and the Yellowstone in Montana. Critical to the issue of the Spencer Hydropower plant was their use on the lower Niobrara in the vicinity of the modern dam. Historian Hiram Chittenden placed the bull boats on the Cheyenne, Niobrara, and other tributaries, noting that their use was extensive. Other authors demonstrated that bull boats were commonly used on the shallow tributaries of the Missouri, such as the Platte, Bighorn, and Niobrara (Chittenden 1903: 100; Winnart 1979: 85; FERC 1988).

Several trading companies, including the American Fur Trading Company, maintained posts along the Niobrara. This particular firm had an operation at the confluence with the Missouri (Fort Mitchell), with a second site 125 miles west at the junction of the Niobrara and Snake rivers. The Niobrara furnished the most

Figure 7.3 Sketch of a Bull Boat.

practical route between these two points, and was the likely choice when water conditions permitted such movements (Warren 1855; FERC 1988: 6, 8).

The actual evidence of early river navigation is limited, and perhaps circumstantial, but there is enough of a record to argue historical navigability. What is interesting is that there is no further evidence of navigability after the end of the fur trade.

The river played no transportation role in the region's later settlement, and while it might have been suitable for driving logs, the absence of timber along or near its banks precluded such uses. When the author carried out a 1991 navigation study of the Minnechaduza, a Niobrara tributary near Valentine, Nebraska, he was surprised to find extensive stands of conifers in the Sand Hills region, including the banks of the area's rivers and streams. This was in part the result of government forestry efforts, for historical photographs from the nineteenth century showed this to be a land of sod houses and buffalo chips that was virtually devoid of trees (FERC 1988: 9; 1991).

In 1934, the Army Corps of Engineers concluded that the Niobrara was nonnavigable and could not be made navigable (War Department 1934; FERC 1988: 10). While canoeing as a recreational use was (and is) possible on parts of the river, this was not a direct concern of the Corps. Nevertheless, because at one time in the past the river was used in the performance of interstate commerce, that historical navigability in fact carries forward to the present through the indelibility of its onetime navigable character.

LOGAN RIVER

The Logan River, a tributary of the Little Bear and Bear Rivers, has its origins in southeastern Idaho, flowing in a southerly direction into northern Utah (Figure 7.4). It then turns southwestward, passing through Logan Canyon in the Wasatch Mountains, and once out of the range flows west through the Cache Valley to its link with the Little Bear and Bear rivers. A dam and power plant known as Logan #1, owned and operated by Utah State University, exists near the mouth of the canyon. Directly below the dam, which was opened in 1913, are a public park and a number of university facilities, but the flood plain also includes a considerable area of urbanized settlement (FERC 1992a: 1).

The larger Bear River, while not navigable for many years, has a history that suggests historical navigability. When fur trapper Jim Bridger made his much heralded discovery of the Great Salt Lake in 1824, he had arrived via the Bear in a buffalo-hide bull boat. And for a time in the 1870s the shallow draft steamer *City of Corinne* operated on the lower course of the river. The Bear was also used to float logs and lumber, including railroad ties, an activity that in some cases may have involved interstate commerce (FERC 1992a: 4).

Figure 7.4 Logan River Drainage System.

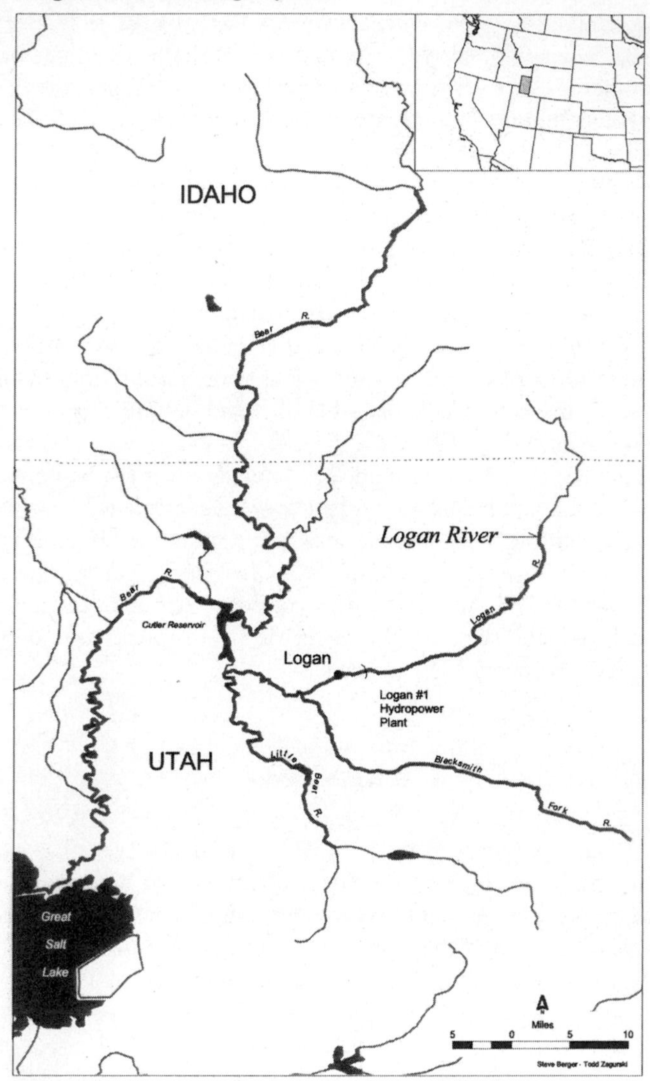

The literature of settlement in the Cache Valley contains many references to the Logan River and to the early fur trade, but it was with the coming of the Mormon settlers that the river began to play a role in local commerce. With considerable variability in annual precipitation, diversion of irrigation water from local streams was critical to the success of Mormon farming.

Trees cut in the Wasatch had three primary uses: (1) cordwood or firewood for local heating and industrial needs; (2) timbers or cut lumber for construction purposes; and (3) railroad ties. Early efforts, in the 1860s, to drive logs down the Logan failed (Wright 1924; FERC 1992a). But during the 1870s, a period of abundant moisture, the river ran full and was at times successfully utilized for log driving (Bird 1964: 21–26; FERC 1992a: 28–30).

> Logan Canyon had the necessary wood supply but hauling it out of the canyon proved a difficult task. The river was used during high water in the late spring to float the logs down the canyon but the river was too low for this purpose during the summer and fall. (Moses 1971: 6)

There are many references to the log drives, including a number of commentaries related to the building of the Logan Tabernacle, a structure which required a million feet of lumber and 256,000 cubic feet of rock (Hovey 1925: 43; Bird 1964: 27; FERC 1992a: 28–31).

> A thick grove of Douglas-fir was opened up in Logan Canyon, at Tabernacle Hallow, near what is now called Woods Camp. The trees were cut at this point and slid down the mountain slope to the Logan River. The logs were kept in the river until high water, and then were floated down the river to the City of Logan. A log boom was set up on the river adjacent to the Card Sawmill where the logs were hauled out of the river and taken to the mill. Here they were sawed into proper dimensions for the Tabernacle. (Bird 1964: 1–22; FERC 1992a: 30)

While these movements were not unlike those reported on the Saco River and elsewhere, they did have one major distinction—they were local, entirely within Utah, and therefore could not be considered interstate commerce. However, the traffic in railroad ties suggested another possibility.

> Charged with supplying the railroads with large numbers of ties, lumbermen scoured the region in search of adequate timber. As in previous cases, hand hewn ties were stockpiled along the Logan and its tributaries during the winter and at full water in the Spring were floated down into the Little Bear and Bear Rivers where they were snaked out and trans-loaded to railroad cars. Shipments went to a variety of destinations, among them rail construction sites well beyond the state of Utah, including the Oregon Short Line (Union Pacific) project from Granger, Idaho, to Puget Sound. (Bird 1964: 30–36; FERC 1992a: 31–33)

While it had its origin in an adjacent state, the Logan was used primarily for local transport within Utah, without real continuity of water travel in interstate commerce. The waters, in fact, eventually flow, via the Little Bear and Bear, into the Great Salt Lake, which is itself located wholly within Utah. When shipments across state lines did occur, they involved the use of the railroad, and thus interstate commerce occurred only through a combination of several transport modes.

The researchers argued for jurisdiction on this basis, but as with any legal interpretation, this became the task of FERC officials and lawyers to interpret, rather than the researchers who carried out the navigation study.

CONCLUSION

This discussion has cited a limited number of case studies that dealt with navigation of the waters of the United States, with particular focus on the issue of historical navigability. To the researcher the true test of jurisdiction is clearly the issue of commerce between the states, based on a long history of legal cases and an evolving set of definitions as to what constitutes the movement of people and goods in interstate commerce.

Often at considerable odds with the public's perception of navigability, the federal application of the term implies a spatial continuity of movement, past, present, or even future, across recognized boundaries or on clearly jurisdictional waters, such as the Great Lakes or the oceans. The kind of vessel is immaterial, so long as it floats, or floated, or possibly swims. Thus even migratory fish may now be interpreted as a form of interstate movement.

For a cultural geographer trained in the 1960s, this kind of research represented a distinct refocusing of traditional interests and goals. These are tasks assigned to the researcher, to meet the needs of a government agency, and are intended as evidence in a court of law.

While cultural and historical geographers have long contributed to environmental and social theory, their ties to argument and implementation of public policy have been limited. The environmental movement of the 1970s provided some opportunities, particularly for the biogeographer and cultural ecologist, whose training was (and is) well suited to the writing of environmental impact reports and related documents.

Application, which has become a major focus of modern geography, is hardly exclusive, nor is it new. The University of Chicago graduate student of the 1960s was witness to nascent movements in behavioral geography, economic geography, urban geography, and quantitative methods. Natural hazards research, with its policy implications, often dealt with the perception of environment, and thus contemplated cultural issues related to decision making.

In that sense, navigation studies are adversarial, and there may be an opposing side seeking to disprove the findings. The test is obviously the accuracy of the reports and the power of their documentation. That they study culture is not an end in itself, but reflects the need to determine how man has used the environment in a variety of ways that might impact federal authority. However clothed in legal jargon, the studies are inherently geographical, reflecting both the importance of place and the spatial processes of navigation itself. They are also part of an increasingly interdisciplinary world which in this case ties together the social sciences, engineering, and the law.[3]

Interpretations of navigation law and related spatial processes will continue to evolve, and new legal issues will undoubtedly arise. Navigation studies will continue to provide a unique opportunity to link the past with the future.

NOTES

The author wishes to acknowledge the support of the Office of Hydropower Licensing, Federal Energy Regulatory Commission (FERC). Particular thanks go to Dr. Henry "Hank" Ecton, historian at FERC's Project Evaluation Branch. Credit is also due to former graduate students Karen Myioshi, Maria Dolores Bayle, and Viva Nordberg, and to Dr. Larry McGlinn of Valdosta State University. Dolores Bayle deserves particular note because her participation in the FERC projects led to a master's thesis (Bayle 1994). Thanks also goes to Steve Burger and Todd Zagurski for their cartography and to Gigi Ying for her artwork.

1. Following British practice, admiralty courts were established in the United States and charged with dealing with a variety of issues, including piracy, smuggling, and buccaneering (see Dana 1871; Conover 1958).

2. While the first case, based on the Commerce Clause, reaffirmed federal control over interstate navigation, the second case, only a year later, limited admiralty jurisdiction, restricting it, as was traditionally the case, to areas affected by the ebb and flow of the tide.

3. For the author, the navigation studies provided an opportunity to combine a variety of interests. But more important, the graduate students, who were trained in statistics, computer cartography, and geographic information systems, had an opportunity to exploit more traditional research skills and thus enrich their value as scholars.

REFERENCES

Bayle, Maria Dolores. 1994. "Navigable Waters of the United States: Geography and the Law." Master's thesis, California State University, Los Angeles.

Bird, Douglas H. 1964. "A History of Timber Resource Use in the Development of Cache Valley, Utah." Master's thesis, Utah State University, Logan.

Chittenden, Hiram M. 1903. *History of Early Steamboat Navigation on the Upper Missouri River*. New York: Francis P. Harper.

Conover, Milton. 1958. "The Abandonment of the 'Tidewater' Concept of Admiralty Jurisdiction in the United States." *Oregon Law Review* 38: 34–53.

Dana, Richard Henry. 1871. "History of Admiralty Jurisdiction in the Supreme Court of the United States." *American Law Review* 5: 581–621.

Economy Light & Power Company v. United States, 256 U.S. 113 (1921).

Ecton, Henry G. nd. In Search of Navigable Waters. Washington, DC: Federal Energy Regulatory Commission. [Photocopy]

Ex Parte Boyer, 109 U.S. 629 (1883).

FERC (Federal Energy Regulatory Commission). 1988. Prepared by Crosspaths, Inc.

"Navigability of the Niobrara River (Spencer Hydropower Project), Holt and Boyd Counties, Nebraska." Washington, DC: FERC.

———.1989. Prepared by Michael Reis. "Navigability Report, Saco River (Swan Falls Hydropower Project), Oxford County, Maine." Washington, DC: FERC.

———.1991. Prepared by John A. Kirchner. "Preliminary Navigability Report, Minnechaduza Creek (Minnechaduza Power Plant), Cherry County, Nebraska." Washington, DC: FERC.

———.1992a. Prepared by John A. Kirchner and Maria Dolores Bayle. "Navigability Report, Logan River, Utah." Washington, DC: FERC.

———.1992b. Prepared by John A. Kirchner and Maria Dolores Bayle. "Navigability Report, Ellis River, New Hampshire." Washington, DC: FERC.

Gibbons v. Ogden, (9 Wheaton) 1 (1824).

Guinn, David M. 1966. "An Analysis of Navigable Waters of the United States." *Baylor Law Review* 18: 559–80.

Hanson, James A. 1980. "James Bridger et al., Boat Builders: The Skin Canoes of the Great Plains and Rockies." *Museum of the Fur Trade Quarterly* 16 (Spring): 1–7.

Horne, Ruth B. D. 1963. *Conway Through the Years and Whither.* Conway, NH: Conway Historical Society.

Hovey, M. R. 1925. "An Early History of Cache County." Compiled January 1, 1923 to January 1, 1925. In Special Collections, Merrill Library, Utah State University, Logan, Utah.

Johnson, Ralph W. 1949. "Comment—Navigable Rivers and Streams of Oregon." *Oregon Law Review* 28: 267–81.

MacGrady, G. 1975. "The Navigable Concept in the Civil and Common Law: Historical Development, Current Importance, and Some Doctrines That Don't Hold Water." *Florida State University Law Review* 3: 513–615.

Mathews, Paul Olen. 1984. *Water Resources—Geography & Law*. Washington, DC: Association of American Geographers.

Metcalf, George. 1972. "The Bull Boats of the Plains Indians and the Fur Trade." *Museum of the Fur Trade Quarterly* 8 (Summer): 1–10.

Morreale, Eva H. 1963. "Federal Power in Western Waters: The Navigation Power and the Rule of No Compensation." *Natural Resources Journal* 33: 1–77.

Moses, Dennis J. 1971. "Transportation and Road Development in Logan Canyon." Research paper, History 193, Utah State University. In Special Collections, Merrill Library, Utah State University, Logan, Utah.

Moulton, Gary E., ed. 1987. *The Journals of the Lewis and Clark Expedition,* August 25, 1804–April 6, 1805, vol. 3. Lincoln: University of Nebraska Press.

Spencer, William D. 1930. *Pioneers on Maine Rivers, with Lists to 1651.* Portland, ME: Lakeside Printing.

The Propeller Genesee Chief v. Henry Fitzhugh, 53 U.S. (12 Howard) 443 (1851).

The Steamer Daniel Ball v. United States, 77 U.S. (10 Wallace) 557 (1870).

The Steamboat Thomas Jefferson, 23 U.S. (10 Wheaton) 428 (1825).

Truteau, Jean Baptiste. 1952. "Journal of Truteau on the Missouri." In *Before Lewis and Clark: Documents Illustrating the History of the Missouri 1785–1804,* ed. Abraham P. Nasatir, 259–311. St. Louis: St. Louis Historical Documents Foundation.

United States v. Appalachian Electric Power Company, 311 U.S. 377 (1940).

United States v. Rio Grande Dam & Irrigation Company, 174 U.S. 690 (1898).

United States v. The Steamer Montello, 87 U.S. (20 Wallace) 430 (1874).

War Department. 1934. *Letter from the Secretary of War . . . Containing a General Plan for the Improvement of Niobrara River Nrbr. And Wyo., for the Purpose of Navigation and Efficient Development of its Water Power, the Control of Floods and the Needs of Irrigation*. Washington, DC: U.S. Government Printing Office.

Warren, Gouvernor [sic] K. 1855. *Explorations of the Dacota Country in the Year 1855*. Washington, DC: Senate Doc. No. 76, lst Session 34th Congress, Appendix.

Winnart, David J. 1979. *The Fur Trade of the American West 1807–1840: A Geographical Synthesis*. Lincoln: University of Nebraska Press.

Wright, John F. 1924. *Logan Journal*. February 2, 1924. Transcribed by Joel Ricks. Special series. In Special Collections, Merrill Library, Utah State University, Logan, Utah.

8

Environmental History: From the Conquest to the Rescue of Nature

David Lowenthal

The word "environment" generally refers to our material surroundings, above all to those we regard as natural. Its derivative "environmentalism" connotes the current crusade to amend our relations with nature. The reformist gloss on environment is quite new, dating back little more than fifty years. The shift in meaning has profound implications, both positive and negative, for environmental understanding and behavior.

This chapter is about changing environmental attitudes over the past two centuries.[1] Widespread faith in progress, based first on scripture and then on science, long stressed the benefits of subduing nature and exploiting the environment. Cautionary warnings of the harmful side effects of technological impact began to be raised by the mid-nineteenth century, but confidence in the utility of most environmental change abated but little for another hundred years. Over the last half century, the potentially dire consequences of a wide range of newly recognized dangers—nuclear radiation, global warming, antibiotic attrition, ecosystem degradation—have seriously eroded faith in progress, rousing widespread concern for an environment now seen as fragile and at risk. I conclude that present alarmism, no less than past hubris, rests on outmoded ecology and unrealistic views of human agency.

Classically, environment meant "surroundings" of every sort. It embraced all aspects of existence, ideas and feelings along with nature, artifacts, and other people—John Gower wrote of "Lucrece all environed with women." Early usage implied that it is we ourselves who environ, circumnavigating our own beings; Thomas Traherne wrote of "the holy soul of a quiet man invironed with its own repose."[2] Only in the nineteenth century did "environment" come to stress *exter-*

nal determinants. And not till the twentieth century did these specifically denote the circumstances of nature, as distinct from, or even opposed to, those of culture.

"Environmentalism" underwent a more drastic shift in meaning. Echoing Hippocrates, nineteenth- and early twentieth-century savants supposed that environment—landforms, climate, soils, vegetation—formed national culture and personal traits. Friedrich Ratzel, Ellen Churchill Semple, Ellsworth Huntington, Griffith Taylor, and their "environmentalist" devotees held physical geography the key determinant of history and human destiny.

"The question of questions for mankind—the problem which underlies all others," held Darwin's supporter T. H. Huxley, was to "ascertain . . . the place which Man occupies in nature . . . What are the limits of our power over nature, and of nature's power over us?" (Huxley 1959/1863: 71). Most then believed that nature called the tune. Human acts, intended or accidental, could but dent the geological and biological fundament. Mankind might improve nature but could not substantially transform it (Spate 1968).

Almighty nature is now dethroned. "Environmentalism" today means the reverse: it expresses fears for a fragile nature degraded by human derangement of ecological order. Ecology became a scholarly discipline only in the 1920s, but ecological awareness of human impact goes back to eighteenth- and nineteenth-century observers. Almost coincident with Huxley's question came another and unexpected answer: George P. Marsh's *Man and Nature; or, Physical Geography as Modified by Human Action* (1965/1864).

Environmental despoliation in imperial and ex-imperial realms—America, Australia, India—catalyzed radical reassessment of nature, now seen as easily damaged by technology. Marsh's insights were historically imperial: he likened ongoing change in North America and alpine Europe to Mediterranean lands devastated by tyrranical misuse in the Roman Empire. *Man and Nature* was persuasive to imperial stewards because it drew on current forestry and engineering to assess environmental damage in colonial and ex-colonial realms, where the pace of deforestation, erosion, and species extirpation was fearsome. Colonial rulers faced agonizing choices between market-driven extractive enterprise and precautionary stewardship. More often than is usually realized, they aimed, though usually in vain, to stem present-minded gutting of resources that made parents rich but children poor.

We inherit wasteful habits along with degraded habitats. Again and again environmental stewards seek to curtail entrepreneurial practices harmful to soils, vegetation, wildlife, even climate. Again and again such efforts prove nugatory in the face of public ignorance, greed, or unconcern. Again and again catastrophe generates public demands for protection and renovation, followed by new cycles of oblivious exploitation. Jeremiahs now outshout Pollyannas, however. Previously nature's hapless pawns or ordained improvers, we now see ourselves as prime agents in its historic destruction and, we hope, its future regeneration.

ENVIRONMENT AS HERITAGE

Legacies of environment, like those of culture, are keenly cherished—and bitterly contested by rival claimants. Our own environments, individual, local, and national, are seen as uniquely precious, as *unlike* all others.

Each people treasures physical features felt to be distinctively their own. Landscapes are compelling symbols of national identity. Patriotic feeling builds on talismans of space and place. Every national anthem praises special scenic splendors and natural bounties. Rugged mountains, dense forests, deep lakes, storm-scarred coasts and cliffs equip myriad chauvinisms. Thus the Swiss ascribe their sturdy freedom to frugal, communitarian mountain life and pure alpine air. In contrast, the English vaunt humanized rural scenes as schools of decorum (Lowenthal 1994, 1996).

How much the English landscape matters emerged during the 1990s mad-cow debacle, when a letter in the *Times*, signed by the heads of a dozen environmental groups ranging from the Council for the Protection of Rural England to the Royal Society for the Protection of Birds, exhorted the public to attend to an even graver loss than that of the meat industry and of the millions of cattle to be slaughtered: the pastoral landscape on which those cows had grazed and had kept beautiful. Not only were these pastoral scenes important for tourism, but they lay at the heart of the English psyche, and their loss meant a loss of national soul.

Environmental identity today is often expressed in terms of the "purity" of native flora and fauna. The English oak is held threatened by the importation of cheap foreign seedlings (Nuttall 1996). Rowan trees, anciently English, are said to suffer culture shock after being raised from Russian seedlings or in Dutch nurseries, then transplanted into British car parks (Vines 1997). Exotic pets—frogs, toads, turtles, snakes—are held to threaten genetic pollution through hybridization in "Sites of Special Scientific Significance" (Nuttall 1997). Zeal for plant purity is rife in America, where wildflower-seed customers have to be assured that none of the seeds sold them comes from any other part of the world (Jenkins 1994: 176; Begley 1998).[3] "I'm a conservationist of African species," says a South African reptile park guide, urging any finder of a North American turtle to "bash in its skull. This is an alien" (McNeil 1998).

To be sure, introduced pests do devastate fragile ecosystems, notably on islands. But the long history of plant and animal dispersal means that few if any forms of life are truly "native"; the idea that things belong in their original place is a human construct. And the notion of "native good, alien bad" has become a widespread analogue for racist exclusion among humans, a kind of nascent eco-fascism (Kendle 1996; Langston 1995: 240; Lowenthal 1997: 194–99).

Other environmental concerns are also intensely parochial. In enhancing surroundings and in warding off risks, the general good normally takes a backseat to the local, to what is most deeply our own. The NIMBY syndrome—put that obnoxious waste somewhere else, "not in my back yard"—is a universal and largely rational response to threats of impact. "Pollution will occur," in Edelstein's (1988: 194–95) summary, "victims will suffer. The challenge is to avoid being one of those victims." Hence global and even national concerns ever founder on parochial rocks. Yet the global reach of environmental heritage is today beyond dispute. Unlike most of our precursors, we now view the living globe as a common legacy requiring communal custody. Fresh water and fossil fuels, rain forests and gene pools are recognized as legacies common to all and needing all our care.

Environmental legacies overspill all bounds; they are the heritage of the whole world. International treaties prohibit development in Antarctica; international skills and resources are enlisted to combat marine pollution and space debris. Such awareness of global interdependence is newly potent. Only since Rachel Carson's *Silent Spring* (1982/1962) has realization of environmental hazard become pervasive. As never before, millions express alarm about nuclear decay, global warming, ozone depletion, species loss, ecodiversity—the end of nature as we know and cherish it.

ENVIRONMENTALISM AS REACTION TO THREATS OF CRISIS

So great is contemporary concern about environmental impact that many suppose it unique to our era. But acceleration blinds us to the antiquity of environmental transformations. Humankind has been altering the earth since before the dawn of history, albeit more slowly and less visibly in earlier times. Yet even long ago, appalled by some adverse effects, many blamed themselves or others for fancied environmental degradation. Those who think modern awareness of impact unprecedented wrongly suppose our predecessors as little aware of their environmental powers as of their environmental perils. "The idea that humans can change nature" is quite recent, supposed Ehrenfeld (1986: 168), but many premoderns attributed environmental change more readily to human agency than to nature, god, or chance.[4]

The roots of modern concern are mainly traceable to eighteenth-century observations and nineteenth-century ecological insights. Yet analogous environmental fears had engaged previous epochs, when men were likewise seen as major agents of terrestrial change—notably for the worse.[5] Were those who found man responsible for nature's decay preternaturally perceptive about human impact? Not at all: they fantasized change, used spurious data to forecast degradation, and "explained" natural events as divinely ordained. Wanting to believe that sin caused decay, they massively misread the environmental evidence to limn a reprehensible and admonitory chronicle. Many today remain prescientific catastrophists. Because the notion of a nature unresponsive to man's desires seems to them impious, intolerable, or incomprehensible, modern doomsters echo antiscientific millenarian apprehensions. Along with worn-out environments we inherit outworn environmental attitudes.

Fears about the decay of nature have been voiced throughout history. Even some of our scientific precursors thought they faced environmental threats no less dire than ours. Consider the portent in George Perkins Marsh's pioneering *Man and Nature* (1965/1864: 42–43):

> [In] parts of Asia Minor, of Northern Africa, of Greece, and even of Alpine Europe, . . . the operation of causes set in action by man has brought the face of the earth to a desolation almost as complete as that of the moon. . . . The earth is fast becoming an unfit home for its noblest inhabitant, and another era of equal human crime and human improvidence . . . would reduce it to such a condition of impoverished productiveness, of shattered surface, of climatic excess, as to threaten the depravation, barbarism, and perhaps even extinction of the species.

That tirade was penned 135 years ago. What environmental impacts then menaced? Deforestation, overgrazing, erosion, flooding, and desiccation. These still haunt us, but they are not now our prime concerns. New environmental threats recurrently overshadow old ones. In the 1960s they were pollution and chemical poisons and the Bomb. Today they are acid rain, stratospheric ozone depletion, global warming, nuclear waste. Except for the last, none of the key issues raised at the environmental congress in Rio in 1992 had seemed worth stressing at Stockholm just two decades before (White 1996).

Today's perils differ neither in imminence nor in apocalyptic immediacy; past like present Jeremiahs enjoined instant reform against impending doom. What is new is that the modern menaces often cannot be seen: they are invisible to everyday view. The effects of soil erosion, even of DDT, were patent to any observant eye. But today's risks are clear only to arcane experts themselves deeply at odds over how much they see: the 1995 Madrid working group on climatic change barely agreed to call human influence on global climate "discernable" instead of "appreciable," "notable," "measurable," or "detectable" (Sawyer 1995; Houghton et al. 1996).[6]

Those afflicted by toxins that cannot be tasted, touched, smelled, or seen, like the victims of Bhopal or Three Mile Island, feel petrified by deadly poisons that "slink in without warning," in Erikson's phrase, "and then begin their deadly work from within—the very embodiment of stealth and treachery" (1994: 150–51). Hence the drive to prohibit chemical warfare. In the First World War shrapnel was far more lethal than gas, but it did a straightforward job of killing. Asbestos seems to us more fearsome than fire. Toxic poisons provoke special dread as contaminants that deceive the body's alarm systems, crouch for years, even generations, ticking away like a time bomb in our tissues; those afflicted feel utterly helpless, out of control. Deprived of confidence in a fruitful and manageable environment, they lose faith in the goodwill as well as the good sense of public officials and scientific experts who also turned out to be baffled and impotent.

Lengthening time lapses between cause and effect make it ever more difficult to ensure precautions, assign liability, or provide compensation. In Ulrich Beck's telling example, "the injured of Chernobyl, years after the catastrophe, are not even all *born* yet" (1996: 31). At Yucca Mountain, Nevada, the American government plans to bury nuclear waste in containers designed to be leakproof for ten thousand years. Maybe the containers will be safe, but how will humans ten thousand years from now interact with these sites? Social change cannot be forecast a generation on, let alone three hundred generations. But even given civil continuity and social stability without precedence, this span of time would be far too brief. For radioactive carbon-14 remains lethal in air or groundwater up to a million years.[7] The Yucca Mountain nuclear waste solution, enacted out of a felt need to appear reassuring, mocks common sense. It would be mature to admit, as Kai Erikson (1994: 224) puts it, that "we don't know, we can't know, and we dare not act as though we know," but what government is likely to do that? Instead, officials behave like the late lamented Brooklyn Dodgers' manager, who when his team was, as usual, well behind late in the game, would say, "OK, guys, go out there and hold 'em while I think of something."

Science is feared and resented as both remote and authoritarian and because its unintended consequences seem ever more ominous. Once radiant innovations now cast the darkest shadows. Nuclear power was but yesterday a glittering technological panacea. Today, economic, health, and safety fears all but throttle the nuclear industry in many lands (Pasqualetti and Pijawka 1996: 57). We lose faith more and more in both the goodwill and good sense of those in charge of a dangerous universe.

Popular confidence that science can or government will mount effective controls erodes year by year for other reasons as well. Scientific enterprise is seen to be ever more costly, and the public, accustomed to progress, now largely discounts its miracles in advance. The social effects of this double disillusionment—that technological progress brings happiness; that saving miracles will continue to unfold—are as depressing as the physical anxieties they foreshadow. The failure of inflated expectations and of utopian reform and the loss of faith in progress induce despondency, impotence, and *après nous le déluge* escapism.[8] Whether he dared disturb the universe was T. S. Eliot's famous query in "The Love Song of J. Alfred Prufrock" (1915), an ironical overstatement capped by wondering if he even dared to eat a peach. Such queries are not jokes anymore; who knows what genes might be in that peach. Given the decay of trust and faith, it is hardly surprising that environmental activism of the 1970s—holistic, small-scale, participatory—has lost ground in recent years.

FROM THE CONQUEST OF NATURE TO THE COLLAPSE OF NATURE

The shift in public sentiment can be simply stated. In the past, the forces of nature seemed vastly to exceed those of man. Whatever humans did could not seriously harm the natural fundament. Technology would remain of minor global consequence; the gravest man-made disasters were only small and temporary setbacks in progressive mastery of an infinitely resourceful earth. Eighteenth- and nineteenth-century observers detailed impacts of forest loss, intensive farming, and wildlife extermination. But these untoward effects were noted in scattered and abstruse sources and were little heeded. Meanwhile the faith of environmental determinism left ultimate power safely sheltered in nature's might. And cornucopian progressivists went on assuming that science could safely enlarge its power over malign nature.

For nature *was* malign; ruthless, cruel, savage, wasteful, and selfish, as Huxley and Spencer termed it, nature must be subdued to ensure civilized virtue. "Visible nature is all plasticity and indifference," wrote William James (1896: 43); "to such a harlot we owe no allegiance." Marsh (1860: 34, 56, 60–61) inveighed against *mis*use, not *any* use; do not stop conquering, do it *better*, for "wherever man fails to master nature, he can but be her slave."

As late as 1929 Freud extolled the conquest of nature in terms of imperious hubris:

> A country has attained a high state of civilization when . . . everything in it [is used] in exploiting the earth for man's benefit and in protecting him against nature . . . the course of rivers . . . regulated . . . the soil industriously cultivated . . . mineral wealth brought up assiduously from the depths . . . wild and dangerous animals . . . exterminated (Freud 1946: 53–54).

Many, maybe most, still think technology's impacts largely benign. But how much less sanguine we now are! Two generations on, Freud's accolade to progress seems naively anthropocentric. Such blithe neglect of technology's malign effects flouts today's conventional piety, if not wisdom. Almost two out of three Americans identify themselves as environmentalists, valuing environment over economic growth.[9] That few behave in accordance with such intent is another tale.

UNDERLYING MYSTIQUES

Environmentalists characteristically embrace certain creeds about nature, people, and culture. Like other heritage givens, these creeds defy evidence or reason; some are demonstrably false. But they are taken as true because deeply desired. I look at three such faiths: that supreme virtue lies in nature unspoiled by culture; that environmental wisdom is inherent among primitive indigenes; that environmental impacts must be reversible.

The Legacy of Natural Purity and Stability

Environmentalists polarize nature and culture as moral opposites. In their eyes, human manipulation spoils natural purity and degrades the environment. The notion of nature as sacred has classical, Christian, and romantic roots. Its modern version owes more, perhaps, to nineteenth-century Americans who deified New World scenery to make up for absent history. They set primordial nature above degenerate human annals. They decried "temples built by Roman robbers" and "towers of feudal oppression" so as to extol their own "deep forests which the eye of God has alone pervaded. What is the echo of roofs that a few centuries since rung with barbaric revels . . . to the silence which has reigned in these dim groves since the first Creation?"[10]

Americans early developed a "paradoxical ability to devastate the natural world," as one historian put it, "and at the same time to mourn its passing" (Ekirch 1963: 189). Finding God's promise in "the sublimity of nature," an 1835 panegyrist intoned that "the august TEMPLE in which we dwell was built for lofty purposes."[11] In perpetual touch with Nature, Americans need not fear the debauchery of the artificial, the urban, the civilized. America's sublime natural backdrop not only compensated for the lack of picturesque ruins and hoary legends but guaranteed Americans would never be contaminated by artificiality. Guided by mountains and chastened by cataracts, they could progress indefinitely without sinful delusions of grandeur. They vaunted proximity to nature while simultaneously despoiling nature (Miller 1956: 211–12).

American forest worship was long more literary than literal. The actual wilderness was a mortal menace—a howling waste to be extirpated for farms and towns and factories, scenes of civilized progress. Only with the closing of the frontier did more than a handful of Thoreauvians find comfort in the wild—and even Thoreau famously balked at the bleak rawness of truly savage Mount Katahdin in Maine.

Wilderness worship today owes much to remorse for its felt disappearance. The cult of the wild ranges from rural nostalgia to radical rejection of all trace of culture, viewing humanity itself as a terrestrial cancer. Some invest nature with a plan, a purpose, an order it achieves when undisturbed. Others posit innate human affinity with nature, a "biophilia" that requires ecodiversity for our own well-being.[12] Still others damn only *previous* impacts, arguing that when we repair environmental damage we are working *with* nature.

Modern environmentalism comprises several related credos, often unconsciously held. In brief: existence is profoundly dualistic; nature and culture are separate and incompatible essences; untouched nature is materially and morally superior to any human impress; wilderness is health, civilization a disease. Imagining nature as an escape, a place to be born again, a paradise where work is play, environmentalists reserve wild lands for sport or spiritual contemplation.[13]

The ideal of untouched nature gained a scientific imprimatur in the 1920s from

Frederic Clements's equilibrium model of ecology. This model held nature most fruitful when least altered. If left undisturbed, flora and fauna in time attained maximum diversity and stability. Extractive despoliation thwarted or abridged this beneficent climax; technology did not improve nature, but degraded it.[14] This ecological mystique, which seemed consonant with mounting environmental disasters, became sacrosanct.

Enthroning stability and passive noninterference mirrored determinist views about human nature. Ecological utopia became the supreme moral order. To "replace the chaos of a world torn by human greed and voraciousness with a well-ordered moral universe," reformers called for curtailing human numbers, technology, and consumption (Taylor 1986: 258).[15] This reversed eighteenth-century views of nature as an unfinished fabric to be perfected by human conquest and ingenuity. Enlightenment savants and Social Darwinists had envisioned artifice ever improving environment. Twentieth-century reformers thought to restore environmental well-being by curtailing human impact. By the 1950s "ecology" had become a token of right thinking even in government agencies.[16]

To be sure, ecologists had by then disowned the Clementsian paradigm. It was mainly *non*-ecologists who extolled equilibrium, ecodiversity, and noninterference. And today's environmentalists still deploy these views as "ecological." Aldo Leopold's famous Land Ethic (1949: 224–26)—"A thing is right when it tends to preserve the integrity, stability, and beauty of the biotic community. It is wrong when it tends otherwise"—remains gospel. Nature is cast as normative good, technology as aberrant evil.[17]

Manipulative greed is held to subvert most environments and to threaten the few remaining pristine locales. Though divine, nature can no longer mend itself; voiceless nature is "an oppressed and silent class, in need of spokespersons."[18] Such spokespersons, animal rightists more caring of other creatures than of people, are now legion. Images of wild nature as exemplary, culture as toxic, suffuse environmental discourse. Even those whose raison d'être is to exploit resources feel forced to don Thoreauvian masks as "greenscammers" with environmentally friendly names. As people become "more and more environmentally sensitive, it's more difficult to be burdened with a name that speaks the truth about your intentions." On getting the 1997 World Environment Center Gold Medal for Corporate Environmental Achievement, Compaq took a half-page ad to show a misty sun in a forest glade: "Computers make the world a better place. We kind of like the world the way it is . . . At Compaq, we believe in changing the world but we don't believe in changing the Earth . . . Success shouldn't be at the expense of the environment."[19] Lauding their own recycling of plastic as reducing landfill waste, Phillips Petroleum boasts of thereby having "left another little corner of the world all alone."[20]

Leaving nature alone is a dogma that shapes international land reform in defiance of local realities, notably in the Third World. "Wherever there are some trees, a presumption is made that there was [and hence] ought to be continuous forest cover" (Fairhead and Leach 1996). Even a 1995 UNESCO volume explicitly aimed at promoting cultural landscapes reiterates the conventional wisdom that "every natural region of the world loses much of its intrinsic value under human influence" (Plachter and Rössler 1995: 16). One lone contributor attests that "in many parts of the world human intervention has created and maintained environments . . . arguably richer and more diverse . . . than the natural forest and other ecosystems they have replaced" (Green 1995: 405). The other essayists mainly demur, one citing "harmony with nature" as a prime criterion for placing cultural landscapes on the World Heritage list (Phillips 1995: 390).

National parks' policy in America long reified imaginary wildness. To restore nature by erasing traces of occupance was gospel from Yellowstone on. Anything suggestive of progress or advancement was taboo. These sacred Edens were emptied of people in order to be preserved as pure wilderness. Parks could contain no permanent settlers; those already there were uprooted.

But the illusion of supposedly untouched nature required massive intervention. Park managers exterminated wolves, cougars, and coyotes lest they eat the deer. They introduced exotic fish and European grasses. They suppressed fire, which they considered an "unnatural" evil. As a result, elk proliferated and ate up the more palatable trees; and undergrowth accumulated over a half century of fire prevention fueled huge conflagrations.[21]

Now wolves are brought back, and Smokey the Bear has shed his incendiary warning. But it is hard to persuade a public brought up to loathe fire that keeping nature "natural" means letting fires burn unchecked, even deliberately setting them. Only now are people becoming aware, as parks' spokesmen put it, that "the human presence has always been part of the wilderness experience"; even Yellowstone is celebrated as a "cultural landscape."[22]

In the U.S. Virgin Islands, Laurance Rockefeller envisioned St. Johns National Park as tropical wilderness. To "restore" a pre-Columbian paradise the island's slave descendants were brought out and expelled and traces of two and half centuries of shifting cultivation expunged. As Karen Fog Olwig (1980; 1985: 162–73) relates, this zeal for wilderness mystified the native islanders, and no wonder. For what emerged was no primeval forest but a thorny tangle of alien plants once imported to keep slaves far away from plantation mansions. In this bush roamed not jungle beasts but feral jackasses, mongooses originally brought in to control snakes in canefields, and voracious mosquitoes.

Environmentalists often find that goals they strive for fly in the face of values they espouse. Thus human greed is driving more and more species to extinction; yet to save some species, it may be necessary to put them on the menu. We have to learn to eat rare breeds of livestock in order to make their breeding economically desirable enough to save them (Hoge 1997). Similarly, nature reserves sometimes thrive best in places made dreadful by humanity, as in Korea's demilitarized zone; if peace were to come, the pristine green would go (Jordan 1997). Sites poisoned by mining and smelting at Røros, Norway, and Queenstown, Tasmania, are now felt to deserve protection as scenically and botanically unique landscapes (Anderson 1993). The bizarre topography of Lausitz, the sandy, acidic legacy of decades of intensive coal mining in East Germany, is now admired for its own strange beauty—and as a prime locale for waterskiing (Charles 1998).

Yet the legendary virtues of nature undefiled continue to dominate environmental discourse. "We continue to think and speak . . . in terms of climaxes, optima, balance, harmony, equilibria, stability," chides a biogeographer. This "overarching language . . . is deeply flawed, yet it remains so powerful that it continues to override what 'real' ecology is actually telling us," namely that we live in "a non-equilibrium world, in which change takes place all the time, in all sorts of directions and at all sorts of scales, catastrophically, gradually, and unpredictably" (Stott 1998: 1).

The Legacy of Indigenous Virtue

Many today see environmental evil emanating only from Western culture. Blameless are indigenous peoples whose "nurturant tribal ways, integrative communitarian values, and rich interplay with nature" respect its balance (Sale 1990: 368–69). Heeding tribal wisdom might enable technological humanity to repair the damage and restore environments fit to live in and to hand down.

For our own sake as well as nature's, we are told, we ourselves must revert to the wild. "Before agriculture was midwifed," holds an Earth First! guru, "humans were in the wilderness and *we were a part of it.* [Unhappily], irrigation ditches, crop surpluses, and permanent villages [set us] *apart from* the natural world . . . Between the wilderness that created us and the civilization that we created grew an ever-widening rift."[23] But ecological nous, lost by technocrats estranged from nature, still survives among indigenes. Backwardness has become a bona fide of stewardship. American Indians once loathed as uncouth savages arise as ecological saviors, their innate tribal wisdom undimmed over the millennia. Respect for all nature is held to be a "particularly Indian" legacy (Laxson 1991: 374–75; see also Maybury-Lewis 1992).

Current pieties get read back as indigenous tribal heritage. The nonpareil anachronism is Chief Seattle, whose 1854 plea to President Franklin Pierce, "Brother Eagle, Sister Sky," is ritually intoned at Earth Day pow-wows:

> The earth is our mother. I have seen a thousand rotting buffaloes on the prairies left by the white man who shot them from a passing train. What will happen when the buffalo are all slaughtered? The wild horses tamed? . . . when the secret corners of the forest are heavy with the scent of many men and the view of the ripe hills is blotted by talking wires?

Yet no buffalo had roamed within six hundred miles of Chief Seattle's home in Puget Sound; the railroad crossed the Plains only in 1869, three years after his death; the horrific buffalo slaughter came a decade later. In fact, the letter that made the chief an ecological guru was penned only in 1971 by Texan scriptwriter Ted Perry. "The environmental awareness was based on my own feelings," Perry later confessed; he had not "the slightest knowledge of Indian views on the environment."

But for Susan Jeffers, whose children's book of the speech sold 250,000 copies, Chief Seattle still incarnates a creed that made sacred "every creature and part of the earth . . . Basically, I don't know what he said—but I do know that the Native American people lived this philosophy, and that's what is important."[24] She "knows" this because modern rhetoric has made it a heritage virtue. In the same fashion, the ecological maxims of Amazon rainforest defenders gain an Amerindian imprimatur.

Tribal others soon learn Western ecospeak. Maoris and Ghanaians, Inuits and Aborigines refer in identical terms to respect for nature, to ancestral instincts for conserving, to tribal taboos against degradation, and so forth. Other minorities emulate these tribal virtues. "Reaffirm[ing] traditional . . . support for the natural world," spokesmen at the First National People of Color Environmental Leadership Summit conference of 1991 boasted of being uniquely one with nature.[25] But the rhetoric, like the sentiment, is that of the Western mainstream.

Some who damn high-tech farming as environmental sin would replace it with low-tech virtue. The nature writer Wendell Berry relies on animal power on his land and urges others to do likewise. His advice is best taken by literary farmers. Only his writing enables Berry to farm with horses. Such work resembles gardening, a favored model these days for getting reconciled with nature. It yields lovely florid insights, but it does not yield a living.

Modern romantics hold that those who live in and depend on a place will not harm it. Yet Berry "restores" land that others, just as fully of that place, had "destroyed" through their use. Both destruction and construction bring a knowledge of nature. Sometimes work is destructive and constructive at the same time, as when a meadow is cut or burned to prevent forest encroachment (White 1995).

The notion of indigenous rapport with nature is not wholly invented; it stems in part from two evident contrasts. Primitive economies past and present impact environments more slowly and less intensively than do modern agriculture and engineering. And tribal folk live closer to "nature," as usually defined, than do urbanized and industrialized folk. These differences matter; in Ray Dasmann's (1975) comparison, "ecosphere people" confined to their local milieu impact resources more lightly than do "biosphere people" who draw indiscriminately on the resources of the whole world.[26]

But this does not make tribal indigenes necessarily wiser, more caring, or in the long run less destructive. Indigenous ignorance or shortsightedness induced the extinction of moas in New Zealand, soil exhaustion in central Mexico, salinization in the Tigris-Euphrates valley, and wholesale ecocide among Arizona Anazasi and Guatemala Mayans. Tribal Papuans, to whose rituals anthropologists assigned ecological functions, knew nothing about "carrying capacity" and expressed no moral concern for nature.[27] But like the unicorn, the ecologically noble savage is a mythical creature too useful to disavow as a chimera. And tribal indigenes, from the sacred groves of Ghana to the Pequot gambling casino of Connecticut, are now clad as environmentalists in Western garb.

The Myth of Reversible Harmony

Reversibility is a third shibboleth. Environmentalists posit two sorts of change: irrevocable events that, like extinct species, can never be retrieved; and lesser changes that might be halted before some point of no return, allowing reversion to an original or previous state. Untouched nature is presumed stable, undergoing only seasonal and other cyclic oscillations—Mircea Eliade's (1959) myth of the eternal return applied to old-fashioned ecology. In contrast, most human alterations are deemed irreversible, subverting the natural order, at length fatal to all life. In this view, only restoring and husbanding a stable equilibrium can avert catastrophe.[28]

Such fears are not confined to the realm of nature. Since John Ruskin and William Morris, would-be restorers of art and architecture have been accused of irreversible damage to fabric and quality—as in the National Gallery cleaning controversy forty years ago, and in current alarm over the radical restoration of the Sistine Chapel. Connoisseurs exhort conservators that "every method must be reversible; do nothing which cannot be undone."[29] But more than any artistic canvas, it is the irrevocable fabric of nature that arouses the gravest anxiety, as in Marsh's portent cited above.

That environmental impacts are subtle, multiple, and long-delayed has become more than ever apparent. Predictive doubts augment fears of technological damage. Faced by myriad unknowns, environmentalists incline to embargo any action that cannot surely be reversed. But such a stance is wholly quixotic. Art historians rightly term irreversibility a myth used by conservators to justify their own interventions (Talley 1996: 169). Aging and accretions of memory and history implacably alter every object, as they do each sentient being. All acts, individual or collective, are fundamentally irreversible (Cramer 1994). Whether heroic or horrific, no deeds can be undone. W. W. Jacobs's "The Monkey's Paw" (1994/1902) limns the futility of yearning, like Shakespeare's Richard II, to "call back yesterday, bid time return." Nothing in natural history, any more than in human history, is truly reversible.

We are mostly resigned to seeing life as a stream that flows in one direction only. But some reformers persuade themselves that the environment, like incorruptible relics, should be exempt from time's arrow. Environmentalists idolize a fictive nature unchanged by history, in eternal sacred equilibrium.

CONCLUSION

Such mystiques suffuse environmental concern. And media hype elevates that concern to catastrophe. Urgency is assumed. "We must make the rescue of the environment the central organizing principle for civilization," said Vice President Albert Gore (1992: 269). A Harvard litterateur finds Gore's mission so self-evident that "I hardly need," he begins his own book, "to spend many pages defending the reasonableness of the claim" (Buell 1995: 2). Ardent environmentalists need no proof to feel their cause supreme. Any trumpeted warning at once persuades them that the sky is falling (Langston 1995: 285). So we have the scare of DHMO (dihydrogen monoxide), held implicated in the deaths of thousands annually through accidental ingestion. In gaseous form, DHMO

> can cause severe burns . . . The chemical is so caustic that it accelerates the corrosion and rusting of many metals . . . is a major component of acid rain and . . . has been found in excised tumors of terminal cancer patients. . . . For those who have developed a dependency on DHMO, complete withdrawal means certain death.

Told these facts, most queried agree dihydrogen monoxide should be banned (Glassman 1997: 9).[30]

To be sure, cornucopians who find the environment in fine fettle likewise deploy invective unbacked by evidence or logic. Indeed, few environmental texts eschew partisan extremism. The environmental historian William Cronon is execrated for writing of nature as a *cultural* construct. Such "ecofascist relativism" is held to give comfort to the enemy—no entrepreneur or politico need take reform seriously when a noted scholar says environment is all in the mind![31]

Like nature, like wilderness, environment *is* a construct, not an essence; its existence predicates our own. The idea of preserving "virgin" nature is "entirely a creation of the culture that holds it dear, a product of the very history it seeks to deny." The home of a god who remains unchanged by time's arrow, the deified wilderness leaves humanity outside nature. If nature must be wild to be truly nature, then our very presence implies its fall (Cronon 1995b: 79–81). Save as a museum of relics or a retreat for contemplation, nature so conceived has no place for humans. To actual environmental problems, such a vision offers only passive despair, not practical solutions.

Our environmental heritage is embedded in history and culture. How can we learn to cheer rather than to scorn these inescapable links, to view our own environmental habits not just as burdens but as benefits, not as pathologies but as opportunities? Here are a few suggestions:

1. *Community*. Let us stress the universal need for community. Communities are organisms that transcend the life spans of individuals and attach us to the history and heritage of forebears and the inheritance of our descendants.

They represent a compact between the dead, the living, and the still unborn. Faith in the community, as Durkheim (1995/1912: 213–14, 351–52, 372, 379) put it, is a necessary religion. With it individual lives are enriched by myriad linkages; without it life is shorn of purpose and meaning.

2. *Intergenerational equity*. Faith in community requires taking future generations into greater account than is done nowadays by most societies, rich or poor. Among the rich, the naked market compels an economy that favors the present and the immediate over any more distant future. The poor confront agonizing choices between extraction and stewardship. "We have to cut down trees to feed our families," a Mexican farmer puts it; "we're living in the present so that our children can have enough to eat and go to school"—and perhaps become environmentally concerned. "The tragedy," comment his interlocutors, "is that to feed his children today, he has to destroy that which would give them sustenance tomorrow" (Arizpe, Paz, and Velásquez 1996: 66). Envisioning more efficient and less destructive future modes of retrieval, a few farsighted folk recommend leaving some fossil fuels untapped, some prehistoric sites unexcavated. But such selfless stewardship is uncommon.
3. *Inheritors and stewards*. Even if we have no responsibility to nature itself, we are responsible for the world our descendants will inherit. The bedrock of environmental reform, it requires more clear-cut and strenuous restrictions on private property than our epoch has yet understood. "Man has too long forgotten," as Marsh (1965/1864: 36) put it, "that the world was given to him for usufruct alone." The environment we inherit from myriad forebears includes all their transformations, unwitting and otherwise. As temporary denizens we make the best of that environment according to our own lights. As stewards we pass it on to future generations, hoping that our heirs may also wish to become stewards.

Yet for all its obvious benefits, stewardship is not normal or natural in most societies. Immediate pressing needs, increasing mobility, responses to urgent crises, the faceless unaccountability of enterprise, the fraying of community ties, the democratic process itself impose a tyranny of the present that throttles stewardship. Care for the future requires education, training, and active involvement in public affairs. All these derive, in turn, from present concern for the environment we have inherited.

To be valuable enough to be cared for, the environment must feel truly our own as lived experience, not simply as commodity. Like our forebears and our heirs, we make it our own by adding to it our own stamp, now creative, now corrosive. The environment is never merely conserved or protected; it is modified—both enhanced and degraded—by each new generation. We should strive not to lament but to laud our own creative contributions. Learning to laud rather than deplore, we become more apt to make changes that we and our successors feel worthy of praise.

4. *Overcoming narrow specialization.* In no realm is the tyranny of the expert so socially obnoxious as in environmental management. It is not merely the boundaries of academic disciplines we need to breach, but the walls between academe and active life. The inculcation of stewardship demands active engagement in everyday affairs on the part of those who would reform our environmental behavior. In these days of invisible forces scarcely perceptible even to experts, it is all the more urgent that ordinary citizens strive to become familiar enough with all aspects of the forces that make and shape us to play an intelligent role in accepting or rejecting, using, controlling and disposing of them and of their waste products. The public must become more confident about its own ability to assess even the arcane and the specialized and to realize that the experts are no less irrational, defensive, and culture-bound than everyone else (Wynne 1996; Lidskog 2000). Finally, decision makers should realize that they cannot be effective without public respect and trust. The mad-cow scandal and responses to nuclear radiation leakages in Britain typify the backlash that occurs when the public is ignored.

In bridging the gulf between professional expertise and amateur commitment, we benefit from many new studies that present valuable environmental insights in accessible fashion. The work of William Cronon, noted above, is exemplary; so is that of Stephen Jay Gould, justly the Association of American Geographers' first honorary geographer. For me, three recent books highlight progress in linking historical with ecological cognizance.

Tom Wessels' *Reading the Forested Landscape* (1997), illustrated by Brian D. Cohen, shows how humans have reshaped the central New England scene from precolonial times to the present. Chronicles of trees and streams, granite ledges, pastured slopes, and alluvial meadows come alive through vivid portrayals of pine and maple and beech, stumps and logs, relic features and abandoned enterprises, stone walls and potash and merino sheep, fence posts and fungus and insect pests. Not since the time of Marsh and Thoreau has there been so exciting a synthesis of this region's natural and human history.

Emergent fascination with antipodean nature and human nature is the focus of Tom Griffiths's *Hunters and Collectors* (1996), which tells how Australian antiquarians, genealogists, memoirists, scientists, litterateurs, and painters set about domesticating their little-known—and often bizarrely misunderstood—environment. Anglo-centered fears of a degenerate continent of fossils and relics (not least of Aborigines), then gave way to zest for exotic discovery and finally to admiring pride in a distinctively interlinked land and people. Showing how "the past forces itself shockingly upon the present, like an intrusion across a geological fault line," Griffiths's (1996: 278) congeries of antipodean worlds brilliantly illumine our own.

Nancy Langston's *Forest Dreams, Forest Nightmares* (1995) recounts the fate of the Blue Mountains of eastern Oregon and Washington, whose lovely open ponderosa pine forests were paradise to nineteenth-century settlers. How that paradise was lost—through massive logging, misguided Forest Service fire control efforts, and later mismanagement—is a cautionary ecological tragedy. "Forest problems did not come about because of greed, incompetence, or poor science," Langston concludes; they stemmed rather from "American visions of the proper human relationship to nature." Scriptural devotees initially saw land and people alike in a state of sin requiring conversion. Early conservers aimed to transform inefficient and messy old-growth woodland into regulated sustained-yield forest. Later environmentalists sought to retrieve an ideal prehuman nature. Notions of "what the perfect forest ought to look like" and how people should help shape it all failed, because of built-in assumptions that natural processes and human impacts were simple, predictable, and morally definable—such as views that dead wood was either good or bad for the soil, or that fire was ipso facto evil (Langston 1995: 5, 62–63, 261–62).

Wise cautions permeate all these texts: stability is not necessarily a virtue; ecodiversity is no uniform condition but varies with epoch, type, scale, and place; there is no natural balance to which ecosystems can be returned, nor any way to reverse past impacts. We should beware of clear guidelines, for they promote delusions of being in control of processes whose cascading effects ever outpace our understanding; more information cannot save us from error, for we can never learn "enough knowledge to manipulate natural systems without unanticipated changes" (Langston 1995: 177).

Yet recognizing environmental complexity does not mean leaving nature alone: unavoidably implicated in environmental change, we must try to manage it despite knowing we will always make mistakes.

NOTES

1. I deal only with some aspects of environmental history; as Williams (1994: 3) notes, the field is vast.

2. For these usages, the Oxford English Dictionary cites John Gower, *Confessio Amantis* (1390), and Thomas Traherne, *Christian Ethics* (1675). Rolston (1996: 10–15) posits a similarly inclusive use of "environment."

3. On America's ecological independence crusade against such pests as the Hessian fly, Russian thistle, European corn borer, English sparrow, and gypsy moth, see Pauly (1996); Devine (1998).

4. These paragraphs are condensed from Lowenthal (1990: 121).

5. For seventeenth-century environmental views, see Davies (1969); Harris (1949); Lowenthal (1985: 87–88, 136–37); Nicolson (1959: 100–04).

6. Intergovernmental Panel on Climate Change, Working Group I, "Summary for Policy Makers," reported in Sawyer (1995).

7. On Yucca Mountain, see Erikson (1994: 203–25) and Fri (1995) (summarizing U.S.

National Research Council, *Technical Bases for Yucca Mountain Standards*, Washington, D.C. 1995.

8. Declining faith in science is explained in Rescher (1980) and Cohen (1996).

9. 1995 Gallup Poll figures in Fritsch (1996).

10. Charles Fenno Hoffman and other 1830s observers quoted in Lowenthal (1976: 101–04).

11. James Brooks, "Our Country," *The Knickerbocker*, 1835, quoted in Miller (1956: 210).

12. Exemplified in Kellert (1995), these views of nature are savaged by Budiansky (1995).

13. On the range of environmentalist postures, see Ellis (1995).

14. For such views see Sears (1935); Murray (1954); Worster (1977: 209–42); Rolston (1979); Barbour (1995: 252).

15. See also Birch and Cobb (1981: 273–74, 282–83); Evernden (1985: 15–16).

16. On the canonization of ecology, see Hays (1987: 27, 347); Barbour (1995).

17. On these environmentalist trends, see F. F. Darling in Thomas (1956: 407–08); Frank Egler in Thomas (1956: 447, 940–41); Cowan (1966: 56); Flader (1978: 34–35, 270–71); Oldfield (1983). Typical environmentalist credos are McKibben (1990) and Sagoff (1990).

18. John Tallmadge (*Orion* 9 (3) [1990]: 64), quoted approvingly by Buell (1995: 20–21).

19. This advertisement appeared in the *International Herald Tribune* (22 October 1997: 18), as well as in other publications.

20. Phillips Petroleum advertisement, *New York Review of Books* (19 September 1996: 7).

21. On U.S. National Parks policy, see Chase (1986); Sellars (1997); Young in press; on unintended consequences, Rackham (1996: 55).

22. U.S. Department of the Interior, *Archeology in the Wilderness* (1993), quoted in Layton and Titchen (1995: 179–80). On public perceptions of fire, see Kleiner (1996).

23. Earth First! founder Dave Foreman, *Confessions of an Eco-Warrior* (New York: Harmony Books, 1991), 69, quoted in Cronon (1995b: 83).

24. Chief Seattle's "letter" is in Jeffers (1991, not paged); Ted Perry is quoted by John Lichfield in *Independent on Sunday* (London, 26 April 1992: 13); Susan Jeffers is quoted in Egan (1992).

25. People of Color delegates quoted in Di Chiro (1995: 305–06).

26. On this distinction, see also McNeely and Keeton (1995: 31–32).

27. See also Denevan (1992); Krech (1999); Roy Rappaport on the Tsembaga of Papua-New Guinea, cited in Lash, Szerszynski, and Wynne (1996: 128).

28. On the economics and psychology of reversibility, see Arrow and Fisher (1974: 318–19); Bishop (1978); Meadows et al. (1972: 72); Randall (1986); SCEP (1970: 125–26).

29. Caroline Keck, letter, *New York Review of Books* (24 June 1983: 4); on the National Gallery controversy, see Lowenthal (1985: 161).

30. This lethal substance is, of course, water.

31. Review of Cronon (1995a) by David Rothenberg, *Amicus Journal* (Summer 1996: 41–44). In the same genre is Alexander Cockburn, "Roush xed," *Nation* (8 April 1996:

10). See also Ehrlich and Ehrlich (1996); Rolston (1996). Far from terming wilderness *just* a state of mind, Cronon (1994: 40) critiques environmentalist pieties from a belief that nature is shaped by human intervention but is "not *entirely* our own invention."

REFERENCES

Anderson, Ian. 1993. "Environmental Cleanup? No Thanks." *New Scientist*, 13 November: 6.

Arizpe, Lourdes, Fernanda Paz, and Margarita Velásquez. 1996. *Culture and Global Change: Social Perceptions of Deforestation in the Lacandona Rain Forest in Mexico*. Ann Arbor: University of Michigan Press.

Arrow, K. J., and A. C. Fisher. 1974. "Environmental Preservation, Uncertainty and Irreversibility." *Quarterly Journal of Economics* 88: 312–19.

Barbour, Michael G. 1995. "Ecological Fragmentation in the Fifties." In *Uncommon Ground: Toward Reinventing Nature*, ed. William Cronon, 233–55. New York: W. W. Norton.

Beck, Ulrich. 1996. "Risk Society and the Provident State." In *Risk, Environment and Modernity: Towards a New Ecology*, ed. Scott Lash, Bronislaw Szerszynski, and Brian Wynne, 27–43. London: Sage.

Begley, Sharon. 1998. "Invasion of the Aliens," *Newsweek*, 10 August: 58–59.

Birch, Charles, and John E. Cobb, Jr. 1981. *The Liberation of Life*. Cambridge: Cambridge University Press.

Bishop, R. C. 1978. "Endangered Species and Uncertainty: The Economics of a Safe Minimum Standard." *American Journal of Agricultural Economics* 60: 10–18.

Budiansky, Stephen. 1995. *Nature's Keepers: The New Science of Nature Management*. London: Weidenfeld & Nicolson.

Buell, Lawrence. 1995. *The Environmental Imagination: Thoreau, Nature Writing, and the Formation of American Culture*. Cambridge, MA: Harvard University Press.

Carson, Rachel. 1982/1962. *Silent Spring*. Harmondsworth, Middlesex, England: Penguin.

Charles, Dan. 1998. "Wasteworld." *New Scientist*, 31 January: 32–35.

Chase, Alston. 1986. *Playing God in Yellowstone: The Destruction of America's First National Park*. Boston: Atlantic Monthly Press.

Cohen, Maurie J. 1996. *Risk Society, Modernization, and Declining Public Confidence in Science*, Environmental Risk Management Working Paper ERC, 96–97. Edmonton: University of Alberta.

Cowan, Ian McTaggert. 1966. "Management, Response, and Variety." In *Future Environments of North America: Transformation of a Continent*, ed. F. Fraser Darling and John P. Milton, 55–65. Garden City, NY: Natural History Press.

Cramer, Friedrich. 1994. "Durability and Change: A Biochemist's View." In *Durability and Change: The Science, Responsibility, and Cost of Sustaining Cultural Heritage*, ed. Wolfgang E. Krumbein et al., 19–25. Chichester: John Wiley.

Cronon, William. 1994. "Cutting Loose or Running Aground?" *Journal of Historical Geography* 20: 38–43.

———. 1995a. "Introduction: In Search of Nature." In *Uncommon Ground: Toward Reinventing Nature*, ed. William Cronon, 23–56. New York: W. W. Norton.

———. 1995b. "The Trouble with Wilderness; or, Getting Back to the Wrong Nature." In *Uncommon Ground: Toward Reinventing Nature*, ed. William Cronon, 69–90. New York: W. W. Norton.

Dasmann, Ray. 1975. "National Parks, Nature Conservation, and 'Future Primitives.' " *Ecologist* 65: 164–67.

Davies, G. L. 1969. *The Earth in Decay: A History of British Geomorphology 1578–1878*. London: Macdonald.

Denevan, William M. 1992. "The Pristine Myth: The Landscape of the Americas in 1492." *Annals of the Association of American Geographers* 82: 369–85.

Devine, Robert S. 1998. *Alien Invasion: America's Battle with Non-Native Animals and Plants*. Washington, D.C.: National Geographic Society.

Di Chiro, Giovanna. 1995. "Nature as Community: The Convergence of Environmental and Social Justice." In *Uncommon Ground: Toward Reinventing Nature*, ed. William Cronon, 298–320. New York: W. W. Norton.

Durkheim, Emile. 1995/1912. *The Elementary Forms of Religious Life*, trans. Karen E. Fields. New York: Free Press.

Edelstein, Michael R. 1988. *Contaminated Communities: The Social and Psychological Impacts of Residential Toxic Exposure*. Boulder: Westview Press.

Egan, Timothy. 1992. "Mother Earth? From the Film, Not the Indian." *International Herald Tribune*, 22 April: 2.

Ehrenfeld, D. 1986. "Life in the Next Millennium: Who Will Be Left in the Earth's Community?" In *The Last Extinction*, ed. L. Kaufman and K. Mallory, 167–86. Cambridge: Massachusetts Institute of Technology Press.

Ehrlich, Paul R., and Anne H. Ehrlich. 1996. *Betrayal of Science and Reason: How Anti-Environmental Rhetoric Threatens Our Future*. Covelo, CA: Island Press.

Ekirch, A. A. 1963. *Man and Nature in America*. New York: Columbia University Press.

Eliade, Mircea. 1959. *Cosmos and History: The Myth of the Eternal Return*, trans. W. R. Trask. New York: Harper.

Ellis, Jeffrey C. 1995. "On the Search for a Root Cause: Essentialist Tendencies in Environmental Discourse." In *Uncommon Ground: Toward Reinventing Nature*, ed. William Cronon, 256–97. New York: W. W. Norton.

Erikson, Kai. 1994. *A New Species of Trouble: Explorations in Disaster, Trauma, and Community*. New York: W. W. Norton.

Evernden, Neil. 1985. "Constructing the Natural: The Darker Side of the Environmental Movement." *North American Review* 270: 15–19.

Fairhead, James, and Melissa Leach. 1996. *Misreading the African Landscape: Society and Ecology in Forest-Savanna Mosaic*. Cambridge: Cambridge University Press.

Flader, Susan L. 1978. *Thinking Like a Mountain: Aldo Leopold and the Evolution of an Ecological Attitude toward Deer, Wolves and Forests*. Lincoln: University of Nebraska Press, Bison Book.

Freud, Sigmund. 1946. *Civilization and Its Discontents* [1929], trans. J. Rivière. London: Hogarth Press and the Institute of Psycho-Analysis.

Fri, Robert W. 1995. "Using Science Soundly: The Yucca Mountain Standard." *RFF [Resources for the Future] Review* (Summer): 15–18.

Fritsch, Jane. 1996. "Ecologists Spy Facade: Do Pillagers Hide Behind Cute Names?" *International Herald Tribune*, 26 March.

Glassman, James K. 1997. "It's in Your Own Home and It's a Real Killer!" *International Herald Tribune*, 22 October: 9.

Gore, Albert. 1992. *Earth in Balance: Ecology and the Human Spirit*. Boston: Houghton Mifflin.

Green, Bryn H. 1995. "Principles for Protecting Endangered Landscapes: The Work of the IUCN-CESP Working Group on Landscape Conservation." In *Cultural Landscapes of Universal Value*, ed. Bernd von Droste, Harald Plachter, and Mechthild Rössler, 405–11. Jena: Gustav Fischer for UNESCO.

Griffiths, Tom. 1996. *Hunters and Collectors: The Antiquarian Imagination in Australia*. Melbourne: Cambridge University Press.

Harris, Victor. 1949. *All Coherence Gone*. Chicago: University of Chicago Press.

Hays, Samuel P. 1987. *Beauty, Health, and Permanence: Environmental Politics in the United States, 1955–1985*. Cambridge, MA: Harvard University Press.

Hoge, Warren. 1997. "To Save Rare Livestock, Put Them On the Menu." *International Herald Tribune*, 30 January.

Houghton, J. T., L. G. Meiro Filho, B. A. Callender, A. Kaltenburg, and K. Maskell, eds. 1996. *Climate Change 1995: The Science of Climate Change*. Cambridge: Cambridge University Press.

Huxley, T. H. 1959/1863. *Man's Place in Nature*. Ann Arbor: University of Michigan Press.

Jacobs, W. W. 1994/1902. "The Monkey's Paw." In W. W. Jacobs, *The Monkey's Paw and Other Stories*, 139–53. London: Robin Clarke.

James, William. 1896. *The Will to Believe*. New York: Longmans, Green.

Jeffers, Susan. 1991. *Brother Eagle, Sister Sky: A Message from Chief Seattle*. New York: Dial Books.

Jenkins, Virginia Scott. 1994. *The Lawn: A History of an American Obsession*. Washington, DC: Smithsonian Institution Press.

Jordan, Mary. 1997. "Where Wildlife Thrives in No-Man's-Land." *International Herald Tribune*, 9 October.

Kellert, Stephen. 1995. *The Value of Life: Biological Diversity and Human Society*. Covelo, CA: Island Press.

Kendle, Tony. 1996. "Front Line." *National Trust Magazine* [UK] 77 (Spring): 23.

Kleiner, Kurt. 1996. "Fanning the Wildfires." *New Scientist*, 19 October: 14–15.

Krech, Shepard, III. 1999. *The Ecological Indian: Myth and History*. New York: W. W. Norton.

Langston, Nancy. 1995. *Forest Dreams, Forest Nightmares: The Paradox of Old Growth in the Inland West*. Seattle: University of Washington Press.

Lash, Scott, Bronislaw Szerszynski, and Brian Wynne, eds. 1996. *Risk, Environment and Modernity: Towards a New Ecology*. London: Sage.

Laxson, Joan D. 1991. "How 'We' See 'Them': Tourism and Native Americans." *Annals of Tourism Research* 18: 365–91.

Layton, Robert, and Sarah Titchen. 1995. "Uluru: An Outstanding Australian Aboriginal Cultural Landscape." In *Cultural Landscapes of Universal Value*, ed. Bernd von Droste, Harald Plachter, and Mechthild Rössler, 174–81. Jena: Gustav Fischer for UNESCO.

Leopold, Aldo. 1949. "The Land Ethic." In Aldo Leopold, *A Sand County Almanac and Sketches Here and There*, 201–26. New York: Oxford University Press.

Lidskog, Rolf. 2000. "Scientific Evidence or Lay People's Experience? On Risk and Trust with Regard to Modern Environmental Threats." In *Risk in the Modern Age*, ed. Maurie J. Cohen, 196–224. London: Macmillan.

Lowenthal, David. 1976. "The Past in the American Landscape." In *Geographies of the Mind*, ed. David Lowenthal and Martyn J. Bowden, 89–117. New York: Oxford University Press.

———. 1985. *The Past Is a Foreign Country*. Cambridge: Cambridge University Press.

———. 1990. "Awareness of Human Impacts: Changing Attitudes and Emphases." In *The Earth as Transformed by Human Action*, ed. B. L. Turner II et al., 121–35. New York: Cambridge University Press.

———. 1994. "European and English Landscapes as National Symbols." In *Geography and National Identity*, ed. David Hooson, 15–38. Oxford: Blackwell.

———. 1996. "Paysages et identités nationales." In *L'Europe et ses campagnes*, ed. Marcel Jollivet and Nicole Eisner, 245–71. Paris: Presses de la Fondation Nationale des Sciences Politiques.

———. 1997. *The Heritage Crusade and the Spoils of History*. London: Viking.

Marsh, George Perkins. 1860. "The Study of Nature." *Christian Examiner* 58: 33–62.

———. 1965/1864. *Man and Nature; or, Physical Geography as Modified by Human Action*, ed. David Lowenthal. Cambridge, MA: Harvard University Press.

Maybury-Lewis, David. 1992. *Millennium: Tribal Wisdom and the Modern World*. New York: Penguin Viking.

McKibben, Bill. 1990. *The End of Nature*. Harmondsworth, Middlesex: Penguin.

McNeely, Jeffrey A., and William S. Keeton. 1995. "The Interaction between Biological and Cultural Diversity." In *Cultural Landscapes of Universal Value*, ed. Bernd von Droste, Harald Plachter, and Mechthild Rössler, 25–37. Jena: Gustav Fischer for UNESCO.

McNeil, Donald G., Jr. 1998. "In South Africa, Environmentalists Pick Up Axes to Chop 'Alien' Species." *International Herald Tribune*, 16 June: 4.

Meadows, Donella H., D. L. Meadows, J. Randers, and W. W. Behrens. 1972. *The Limits to Growth: A Report for the Club of Rome's Project on the Predicament of Mankind*. New York: Universal Books.

Miller, Perry. 1956. *Errand into the Wilderness*. Cambridge, MA: Harvard University Press.

Murray, E. G. D. 1954. "The Place of Nature in Man's World." *American Scientist* 42: 130–35, 142.

Nicolson, Marjorie Hope. 1959. *Mountain Gloom and Mountain Glory: The Development of the Aesthetics of the Infinite*. Ithaca: Cornell University Press.

Nuttall, Nick. 1996. "English Oak Threatened by Cheap Foreign Seedlings." *The Times* (London), 25 November: 12.

———. 1997. "Alien Invaders Threaten Survival of Native Species." *The Times* (London), 21 July.

Oldfield, Frank. 1983. "Man's Impact on the Environment: Some Recent Perspectives." *Geography* 68: 245–56.

Olwig, Karen Fog. 1980. "National Parks, Tourism, and Local Development: A West Indian Case." *Human Organization* 39: 22–31.

———. 1985. *Cultural Adaptation and Resistance on St. John: Three Centuries of Afro-Caribbean Life*. Gainesville: University of Florida Press.

Pasqualetti, Martin J., and K. David Pijawka. 1996. "*Un*siting Nuclear Power Plants: Decommissioning Risks and Their Land Use Context." *Professional Geographer* 48: 57–69.

Pauly, Philip J. 1996. "The Beauty and Menace of the Japanese Cherry Trees." *Isis* 87: 51–73.

Phillips, Adrian. 1995. "Cultural Landscapes: An IUCN Perspective." In *Cultural Landscapes of Universal Value*, ed. Bernd von Droste, Harald Plachter, and Mechthild Rössler, 380–92. Jena: Gustav Fischer for UNESCO.

Plachter, Harald, and Mechtild Rössler. 1995. "Cultural Landscapes: Reconnecting Culture and Nature." In *Cultural Landscapes of Universal Value*, ed. Bernd von Droste, Harald Plachter, and Mechthild Rössler, 15–18. Jena: Gustav Fischer for UNESCO.

Rackham, Oliver. 1996. "Hatfield Forest." In *The Remains of Distant Times: Archaeology and the National Trust*. Society of Antiquaries of London, Occasional Paper no. 19, ed. David Morgan Evans, Peter Salway, and David Thackery, 47–58. London: Boydell Press.

Randall, Alan. 1986. "Human Preferences, Economics, and the Preservation of Species." In *The Preservation of Species: The Value of Biological Diversity*, ed. Bryan G. Norton, 79–109. Princeton: Princeton University Press.

Rescher, Nicholas. 1980. *Unpopular Essays on Scientific Progress*. Pittsburgh: University of Pittsburgh Press.

Rolston, Holmes, III. 1979. "Can and Ought We to Follow Nature?" *Environmental Ethics* 1: 7–30.

———. 1996. *Nature for Real: Is Nature a Social Construct?* Odense, Denmark: Odense University, Humanities Research Center, Man & Nature Working Paper 78.

Sagoff, Mark. 1990. *The Economy of the Earth: Philosophy, Law, and the Environment*. Cambridge: Cambridge University Press.

Sale, Kirkpatrick. 1990. *The Conquest of Paradise: Christopher Columbus and the Columbian Legacy*. New York: Knopf.

Sawyer, Kathy. 1995. "Panel: Humans Affect Climate." *International Herald Tribune*, 2–3 December.

SCEP (Study of Critical Environmental Problems). 1970. *Man's Impact on the Global Environment*. Cambridge: Massachusetts Institute of Technology Press.

Sears, Paul B. 1935. *Deserts on the March*. Norman: University of Oklahoma Press.

Sellars, Richard West. 1997. *Preserving Nature in the National Parks: A History*. New Haven: Yale University Press.

Spate, O. H. K. 1968. "Environmentalism." *International Encyclopedia of the Social Sciences* 5: 93–97. New York: Macmillan.

Stott, Philip. 1998. "Biogeography and Ecology in Crisis: The Urgent Need for a New Metalanguage." *Journal of Biogeography* 25: 1–2.

Talley, M. Kirby, Jr. 1996. "The Original Intent of the Artist." In *Historical and Philosophical Issues in the Conservation of Cultural Heritage: Readings in Conservation*, ed. Nicholas Stanley Price, M. Kirby Talley Jr., and Alessandra Melucco Vaccaro, 162–75. Los Angeles: Getty Conservation Institute.

Taylor, Paul W. 1986. *Respect for Nature: A Theory of Environmental Ethics*. Princeton: Princeton University Press.

Thomas, William L., Jr., ed. 1956. *Man's Role in Changing the Face of the Earth*. Wenner-Gren International Symposium, 1955. Chicago: University of Chicago Press.

Vines, Gail. 1997. "Culture Shock in the Car Park." *New Scientist*, 14 June: 48.

Wessels, Tom. 1997. *Reading the Forested Landscape: A Natural History of New England.* Woodstock, VT: Countryman Press.

White, Gilbert F. 1996. "Emerging Issues in Global Environmental Policy." *Ambio* 25: 58–60.

White, Richard. 1995. " 'Are You an Environmentalist or Do You Work for a Living?': Work and Nature." In *Uncommon Ground: Toward Reinventing Nature*, ed. William Cronon, 171–85. New York: W. W. Norton.

Williams, Michael. 1994. "The Relations of Environmental History and Historical Geography." *Journal of Historical Geography* 20: 3–21.

Worster, Donald. 1977. *Nature's Economy: A History of Ecological Ideas*. Cambridge: Cambridge University Press.

Wynne, Brian. 1996. "May the Sheep Safely Graze? A Reflexive View of the Expert-Lay Knowledge Divide." In *Risk, Environment and Modernity: Towards a New Ecology*, ed. Scott Lash, Bronislaw Szersynski, and Brian Wynne, 44–83. London: Sage.

Young, Terence. In press. "Virtue and Irony at Cades Cove." In *The Landscapes of Theme Parks: Antecedents and Variations*, ed. Terence Young and Robert B. Riley. Washington, DC: Dumbarton Oaks.

III

CLAIMING PLACES

9

Place Metaphor and Milieu in Hemingway's Fiction

Anne Buttimer

"Geography is a science of places rather than of people."

Vidal de la Blache (1913: 299)

Marvin Mikesell introduced me to *la géographie humaine* during the lovely summer of 1963 when, as visiting professor at University of Washington, he taught a course in cultural geography. *Readings in Cultural Geography*, which he had jointly edited with Philip Wagner (1962), was a treasure trove of intellectual alternatives to reigning orthodoxies: essays on nature and culture, language and religion, social diversity and agrarian folkways. To my novice queries about social geography—a field unheard of in Seattle during those days—Marvin described the work of Bobek in Vienna, of Hartke in Munich, of Hägerstrand in Lund, of van Paassen in the Netherlands, and of Fleure, Evans, and Jones from the British Isles. *Crême de la crême*, however, would be the French school, *la géographie humaine*.

Three decades later, it was heartwarming to receive an invitation to contribute to a session on "sense of place" at the Eighth International Hemingway Conference of the Hemingway Society and Foundation, held in Les-Saintes-Maries-de-la-Mer in May 1998. It is with even greater delight that I offer this essay for inclusion in a volume honoring Marvin Mikesell.

La géographie humaine, in its classical form, was neither a natural nor a social science; it was a new field of study on relationships between people and places, with a particular analytical focus on the interactions of lifeways (*genres de vie*) and environments (*milieux*). How elegant it sounded, even in translation! And how totally out of step it was in that place and time in which the mysteries of spatial analysis and prospects of computer-based linear programming were beacons for so many. In Seattle, at any rate, the term "cultural" might then have evoked expressions of disdain rather than discovery—something it did in Berkeley perhaps. The Young Turks in Seattle were leading a reformation of the discipline; henceforward geographers would be known as regional scientists. I remained out of step.

Among the fascinating qualities of *la géographie humaine* were ways in which geography and literature could complement each other in elucidating realities of place. How much were our images of Provence shaped by Mistral, for example, and how could anyone understand the poetic works of Mistral without visiting the landscapes of Provence? Other examples abound—those of Shelley, Hardy, Lawrence, and Sibelius, to mention but a few. Nineteenth-century regional novels indeed epitomized the potentially fruitful dialogue between geography and literature. They unmasked the deep-seated loyalties and cultural identities of regional communities whose solidarity rested largely on taken-for-granted social practices vis-à-vis milieu (Buttimer 1971: 17–19). Hippolyte Taine (1828–93) and Ernest Rénan (1823–92) called attention to racial and political dimensions of place identity. Taine suggested that literary styles reflected particular racial *mentalités*, and Rénan's *Souvenirs d'enfance et de jeunesse* (1893) illustrated well the regional character of French literature. Twentieth-century regional novels are among the finest examples: Pierre Loti's and Chateaubriand's graphic descriptions of Brittany, René Bazin's descriptions of local life in Nivernais, the Tharaud brothers' dramatic account of peasant life in Limousin, and Barrès' poignant pieces from Alsace-Lorraine around the time of World War I (Buttimer 1971: 17–19). These place-based stories also evoked attention to wider issues of society and milieu. Loti's *Pêcheur d'Islande* drew attention to general challenges facing fishing societies, and Eugène Leroy's *Le moulin du Frau* (1895) and *Jacquou le Croquant* (1913) dramatized the challenges facing peasant societies in the early decades of the twentieth century (Monod 1894).

In recent years, geographers again witness a growing interest in cultural geography. Issues of culture, place, identity and meaning fill the pages of leading international journals such as *Ecumene*, *Géographie et Cultures*, and *Worldviews* (see also Keith and Pile 1993; Kearns and Philo 1993). Literary works, films, paintings and posters now replace the census and other statistical sources as bases for interpreting landscapes and lifeways (Cosgrove and Daniels 1998; Tall 1993). An important part of this turn is the attempt to forge bridges between geography and literature, and to reintroduce notions of "place" in a discipline that had been so seriously warped by the reformation of spatial analysts. Already in the late 1970s, there was evidence of progress along these lines (Relph 1976; Tuan 1977; Salter and Lloyd 1977; Ley and Samuels 1978; Seamon 1979; Buttimer and Seamon 1980). "The geographer's engagement with literature," Douglas Pocock wrote, "varies along a continuum between landscape depiction and human condition" (Pocock 1981: 12).

Literary works afford fresh insight into regional cultures, life experiences, everyday practices, symbolic and emotional meanings of "home" (Yaeger 1996). In literary circles, too, it is acknowledged that geography could afford insight into the contexts and sites where literary works were composed, unmasking the salience of physiography, landscape character, folkways and traditions which have inspired or shaped them (Schama 1995; Linde-Laursen and Nilsson 1995). Shared interests over the years have been questions of creativity and context—the whereness of literary inspiration—and in recent years, interactions of mindscapes and landscapes and the "situatedness" of all knowledges (Paasi 1996; Buttimer 1998). And in all of this, the concept of place plays a cardinal role. "We got landscape from you," a colleague in English literature remarked in 1998, "and you got metaphor from us."

Metaphor affords one potentially common ground from which to approach the question of place (Arendt 1958; Buttimer 1981). In mainstream traditions of geographical thought, place has been regarded in a variety of ways: for example, place as *niche* within an organically-constituted universe, place as *contained space* within a world conceived as a mosaic of forms; place as *node* within networks of functionally-organized space in a world conceived as mechanical system; place as *arena of spontaneous events* in a world without fixed coordinates. These "root metaphors" allow for a succinct—if not exhaustive—inventory of major differences in geographical perspectives on place (Buttimer 1981, 1993).[1] Most creative writers play with varieties of these root metaphors, and Ernest Hemingway is no exception. For a geographer, at least, this approach offered one guiding torchlight on the works of a highly treasured body of literature.

To approach the work of Hemingway in terms of "place," however, already posed problems. In many ways, it seems that Hemingway and his characters felt claustrophobic with place as fixed form. The real business of life for them flowed through places—journey and flow, rhythm and cycle, seemed so much more interesting than rest or containment. An emphasis on "root metaphors" might also suggest a focus on form, on cognitive style, ideological orientation, and literary trope—all of which lay quite beyond the scope of geographical expertise. More familiar to the geographer are questions of material—biophysical—attributes of place, and, for human geographers at least, the human significance of particular constellations of these. Thus it seemed wise to open up the question of *matter* in Hemingway's imagination.

Throughout his short stories, environments (*milieux*) are described in terms of material elements and their emotional appeal. Sensory images of scorched earth, burned-down forests, clear, sunny skies and shorelines bespeckle his narratives. Within these metaphors of place, therefore, it would seem useful to cast a particular glance on *milieux* and ways in which the four elements—fire, air, earth and water—are treated in his stories. Particularly striking is the salience of water, of rivers and eventually of the ocean, in Hemingway's work. Waterscapes evoke at once streams of consciousness, mirrors of life and death, spaces of rescue, rejuvenation and healing, and also of Heraclitean flux. One never bathes twice in the same river because, as Bachelard noted, in his inmost recesses, the human being shares the destiny of flowing water (Bachelard 1983/1942: 6). While land-based stories tend to project an attitude of conquest and eccentricity (Plessner 1975), those involving water project images of a more reflective turn, with characters apparently pondering the challenges of a world transformed by modern science, technology and war, a placeless world.

PLACE AS NICHE IN A WORLD AS ORGANIC WHOLE

In childhood years Ernest Hemingway treasured summer holiday hunting and fishing escapades with his father in the Michigan woods and lakes—zones of freedom from the highly disciplined home place at Oak Park, Chicago. "Nature" appealed as an organic realm, and in "natural" settings human relationships could be organic as well, for example, relationships among brothers and between father and son. Leaving that summer home in 1920, he may indeed have felt cast out of Paradise. *The Garden of Eden* is a story filled with organic images of reality, romantic conceptions of village life and early marriage for David Bourne and Catherine, all unfolding at Grau du Roi (*The Garden of Eden:* 3):

> They could see the towers of Aigues Mortes across the low plain of the Camargue and they rode there on their bicycles at some time of nearly every day along the white road that bordered the canal. In the evenings and the mornings when there was a rising tide sea bass would come into it and they would see the mullet jumping wildly to escape from the bass and watch the swelling bulge of the water as the bass attacked.

Idyllic it may have been as honeymoon place, but as *niche* it displays also the intricate and sometimes harsh realities of biological process and interspecies competition. The honeymoon, too, is short-lived. David catches a large bass and becomes a sort of village hero. His wife ("the girl") runs down to join him and the crowd. He seems scarcely interested in her embrace and asks: "Did you see him?" (*The Garden of Eden:* 9–10). It was a very good catch, the story recalls, and the town was busy and happy. David gives the fish away to the village, thus respecting the natural order which he joins for the moment. Yet already the reader can sense the snake in the Garden of Eden—the gap between husband and wife. Catherine longs to become modern, change her hairstyle, drive the car, and has already been threatening to do something dangerous. With the powerful line "Don't call me girl" (*The Garden of Eden:* 17) she breaks his waking dream of happy villagers, fishing, and idealized sex. Even though he still loves her in some ways, his mind says "goodbye and good luck and goodbye."

Another potential organic relationship to place is experienced by Nick Adams in *The Big Two-Hearted River*. Standing in the stream, having lost the big trout, Nick takes the broken leader in his hand (*The Essential Hemingway:* 344):

> He thought of the trout somewhere on the bottom, holding himself steady over the gravel, far down below the light, under the logs, with the hook in his jaw. Nick knew the trout's teeth would cut through the snell of the hook. The hook would embed itself in his jaw. He'd bet the trout was angry. Anything that size would be angry. That was a trout. He had been solidly hooked. Solid as a rock. He felt like a rock, too, before he started off. By God, he was a big one. By God, he was the biggest one I ever heard of.

Healing comes also in that place (*The Essential Hemingway*: 344):

> Nick climbed out on to the meadow and stood, water running down his trousers and out of his shoes, his shoes squelchy. He went over and sat on the logs. He did not want to rush his sensations any.

Place as niche, that is, as local orchestration of *material elements*—fire, air, earth, water—in Hemingway's stories connotes the experience of reengagement with other living forms, a sensual experience of nature, however fleeting (*The Essential Hemingway*: 344):

> He sat on the logs, smoking, drying in the sun, the sun warm on his back, the river shallow ahead entering the woods, curving into the woods, shallows, light glittering, big water-smooth rocks, cedars along the bank and white birches, the logs warm in the sun, smooth to sit on, without bark, great to the touch; slowly the feeling of disappointment left him. It went away slowly, the feeling of disappointment that came sharply after the thrill that made his shoulders ache. It was all right now.

Concepts of place as niche are therefore discernible in Hemingway's short stories—places where characters can experience an orchestration of sensory and emotional immersion in particular places. But frequently there is evidence of tensions between *extasis* and *instasis*, between disciplined engagement and immersion, the controlling of or the belonging to place. Here one finds a replica of the tension which Husserl once called "the paradox of human subjectivity"—we, humans, are both objects in the world and subjects for the world, both actors in the play and theaters of experiences themselves (Husserl 1970: 178 ff.).

PLACE AS CONTAINED SPACE-TIME IN A WORLD AS MOSAIC OF FORMS

The Hemingway Oak Park home was a tidy one, with a place for everything and everything in its place. Such meticulous and at times loving attention to detail characterizes some of Hemingway's descriptions of place. There are vivid descriptions of the young couple's bedroom at Grau du Roi, of the bar, the beach and the cycle path in *The Garden of Eden*. While there was palpable delight in all of this, there was also a sense of repulsion at any prospect of remaining contained in any of these places for long.

To be constantly on the move, however, was not necessarily an attractive alternative. "Listen Robert," said Jake after repeated invitations to abscond to South America (in *The Sun Also Rises*), "going to another country doesn't make any difference. I've tried all that. You can't get away from yourself by moving from one place to another. There's nothing to that" (*The Essential Hemingway*: 16). There was one place which provided a sense of home—at least for a brief *séjour*—for Nick Adams. This was the one he made himself: the camp overlooking the "Big Two-Hearted River." Settling in for the night atop a hillock, he fixed cheesecloth across the mouth of the tent to keep out mosquitoes and crawled inside under the mosquito bar with various things from the pack to put at the head of the bed under the slant of the canvas, and here he felt at home (*The Essential Hemingway:* 337):

> Nick was happy as he crawled inside the tent. He had not been unhappy all day. This was different though. Now things were done. . . . He had made his camp. He was settled. Nothing could touch him. It was a good place to camp. He was there, in the good place. He was in his home where he had made it. Now he was hungry.

Here, alone, Nick feels "at home" in this particular place which he has made himself. This camping achievement displayed competence at both housekeeping and scouting.

Hemingway had a strong sense of duty and responsibility that, as journalist or freelance reporter, he should be writing. And in what kinds of places might writing be best done? *The Garden of Eden* illustrates well the imperative for David to have a special place where only writing was done. There are no doubt important connections between creativity and place (Buttimer 1983). These also vary from one individual to another, depending on childhood experiences, training, life goals and circumstances. But concepts of place as contained space—with given coordinates of space and time—seem to have been claustrophobic for Hemingway and his characters, unless they permitted some noteworthy achievement by way of mastery of the elements or of writing.

Here Hemingway attempts to combine mastery of space with an openness to sensory and emotional experience. One is reminded of the interpretation of Odysseus in Adorno and Horkheimer's *Dialectic of Enlightenment*—Odysseus as prototype of the (Western) quest for subjectivity and the struggle for liberation from the natural world, prototype of the autonomous subject who eventually succeeds in "disenchanting" the worlds. Each place visited is mapped out and named, and its peculiar demons and monsters are subdued. Odysseus as the prototypical cartographer—rationally surveying and organizing space, leaving no detail unnamed (Adorno and Horkheimer 1987/1944: 46 ff.)

By mid-century, physically definable places as contained spaces—whether in the Michigan peninsula or at Grau du Roi—were indeed becoming *passé*. The latter half of the twentieth-century world was moving to other rhythms—far more mechanical in temporal and spatial cadence—a world that was at once the fulfillment and the terror of Promethean (Faustian) dreams which also held strong appeal for Hemingway.

PLACE AS NODE IN A WORLD AS MECHANICAL SYSTEM

Standing in diametrical opposition to conceptions of place as niche in an organically harmonious world for Hemingway was the world of society. The latter represented societally engineered systems of cash flows and debits, of impending war and atrocities. He was also fascinated by the mechanical arts of bullfighting, fishing and game hunting (Hotchner 1966/1955). There seems to have been a profound ambivalence in his overall conceptions of the world: that of nature as organic whole on the one hand, and that of the (humanly created) world as mechanical system on the other. Some of Hemingway's characters are prototypically Promethean, ready to steal fire from the gods for the sake of humanity and, like his own father, hunt down wild animals and fish. Others, like Nick Adams, love the trout and bemoan the devastation of war and everything mechanically derived. Many of Hemingway's stories simply comment—often ironically—on the consequences of mechanistic sports such as the bullfight, a spectacle which at once fascinated and horrified him.

World War I may have really put an end to places as niches, as zones of security and havens of moral consensus. Landscape images are deployed to generate powerful impressions of anger and grief about the ways in which war had destroyed the places he remembers (*A Farewell to Arms*: 143):

> The mulberry trees were bare and the fields were brown. There were wet dead leaves on the road from the rows of bare trees and men were working on the road, tamping stone in the ruts from piles of crushed stone along the side of the road between the trees. . . . It did not feel like a homecoming.

A Farewell to Arms offers a uniquely striking condemnation of war and also an occasion to highlight the importance of movement to escape from war-ridden place. The hero is rescued by the river (*A Farewell to Arms:* 197–98). Exciting, indeed, and sensually engaging is this experience of the fast-flowing water. Then comes the transition from flowing water to the train, from organic to mechanical place (*A Farewell to Arms*: 202–03):

> Lying on the floor of the flat-car with the guns beside me under the canvas I was wet, cold and very hungry. Finally I rolled over and lay flat on my stomach with my head on my arms. My knee was stiff, but it had been very satisfactory. Valentini had done a fine job. I had done half the retreat on foot and swum part of the Tagliamento with his knee. It was his knee all right. The other knee was mine. Doctors did things to you and then it was not your body any more. The head was mine, and the inside of

> the belly. It was very hungry in there. I could feel it turn over on itself. The head was mine, but not to use, not to think with; only to remember and not too much remember.

In a mechanically organized world, places can only be momentary locations of transition. In many respects space, time and place were simply challenges to be mastered—externalities with which one should willy-nilly flow—rather than frameworks for dwelling or rest. Once again one recalls the story of Odysseus, or even more tangibly, that of Goethe's *Faust*, or Shelley's *Prometheus*. The challenge of mastery versus openness to experience unfolds even more dramatically.

PLACE AS ARENA IN A WORLD AS STAGE OF SPONTANEOUS EVENTS

A fourth conception of place, and the one which perhaps best corresponds to the worldviews of Hemingway and most of his heroes, is that of *arena*. Places are theaters of street ballet, of events, encounters, and above all movement. His very best "place" descriptions are those which evoke the dynamic settings through which he walked en route to somewhere, or those observed from café or terrace. In *The Sun Also Rises* Jake describes such a scene (*The Essential Hemingway*: 17–18):

> It was a warm spring night and I sat at a table on the terrace of the Napolitain after Robert had gone, watching it get dark and the electric signs come on, and the red and green stop-and-go traffic-signal, and the crowd going by, and the horse-cabs clippety-clopping along at the edge of the solid taxi traffic, and the *poules* going by, singly and in pairs, looking for the evening meal.

Here one finds perhaps the ideal encounter of geography and literature, descriptions of lived worlds, "place ballets" which vary across cultures and also by time of day. As all who have spent time in Paris know, the city's "geographies" change throughout circadian and seasonal cycles. Hemingway was obviously attuned to these realities. One morning, for example, Jake walked down the Boulevard to the Rue Soufflot for coffee and brioche and here is what he witnessed (*The Essential Hemingway*: 32–33):

> It was a fine morning. The horse-chestnut trees in the Luxembourg gardens were in bloom. There was the pleasant early-morning feeling of a hot day. I read the papers with the coffee and then smoked a cigarette. The flower-women were coming up from the market and arranging their daily stock. Students went by going up to the law school, or down to the Sorbonne. The Boulevard was busy with trams and people going to work.

These examples of "place ballet," to use a phrase from Jane Jacobs, yield splendid impressions of dynamic urban places (Jacobs 1961). Again in this case the account is quite explicitly construed in terms of the observer's own agenda. Paris in the early 1920s provides the ideal context for such impressions of place. The hero of *A Moveable Feast* carries around a notebook, a pencil and a pencil sharpener, ready for whenever inspiration might come. And it does once in Paris, in the corner of a café at the Place St. Michel. The experience is so dramatic that he even remembers the detail of the route from home to café and the "warm and clean and friendly" atmosphere where he immediately began to write (*A Moveable Feast*: 4–5):

> I was writing about up in Michigan and since it was a wild, cold, blowing day it was that sort of day in the story. I had already seen the end of fall come through boyhood, youth and young manhood, and in one place you could write about it better than in another. That was called transplanting yourself, I thought, and it could be as necessary with people as with other sorts of growing things. But in the story the boys were drinking and this made me thirsty and I ordered a rum St. James. This tasted wonderful on the cold day and I kept on writing, feeling very well and feeling the good Martinique rum warm me all through my body and my spirit.

So much "at home" does the writer feel that the "story was writing itself" and he is "having a hard time keeping up with it." The sight of a beautiful girl distracts for a while, and there is some chagrin when she has gone, but the writer is so much in place with his story that he scarcely grieves over her departure: "I've seen you, beauty, and you belong to me now, whoever you are waiting for and if I never see you again, I thought. You belong to me and all Paris belongs to me and I belong to this notebook and this pencil" (*A Moveable Feast*: 4–5). Then the writer returns to writing, enters far into the story and is lost in it. This café corner at the Place St. Michel allows the writer to transcend all material elements and even taken-for-granted passions such as another rum or gawking at a beautiful female profile. The writing project had assumed a life of its own and therefore this was a "good place."

One dominant impression of places in Hemingway's stories is that of *arena* (sand). Sand occurs geographically at the meeting point of land and water—a zone of instability, change, and unpredictable dune formation. *Arena* as metaphor could thus evoke, for example, in the work of Borges' *The Book of Sands,* a sense of the threshold between land and water, the shifting stage on which spontaneous and unpredictable events can occur. On terra firma characters often assume active roles—of writing or play, fighting or travel—whereas in water they tend to become more centered, more reflective. On land, places are simply stages on which events unfold: their physical or social features are relevant only to the actor's agenda. Hemingway's characters traverse places rather than dwell in them. In this sense one could ascribe to them the prototypically modernist attitude that Heidegger called *Gestell*—a view that regards the world as *Bestand*—a potential reservoir of resources for use by humans[2] (Heidegger 1954). In lake, river and stream, however, there is evidence of immersion and *Gelassenheit* (a letting be), a freedom to immerse oneself in the natural elements and flow (Heidegger 1957).

Arena (sand) might therefore be the most appropriate metaphor for sense of place in the writings of Hemingway. Sand allows scope for exposing the tensions between moral convictions and inner feelings. There is a link here which, to use Bachelard's expression, is "liquidity": it is only in fluid (liquid) places that conflicts are dissolved or transcended. "Liquidity," in Bachelard's view, "is the desire of language. Language needs to flow. It flows naturally" (Bachelard 1983/1942: 187). This liquidity is one gripping feature of Hemingway's language and prose style. And one wonders if he was not already sensitive to this when he intimated that famous "iceberg" theory of fiction—that 90 percent of the story remains hidden, felt by reader and writer alike, and only 10 percent is shown (Nagel 1984).

"STEADY IN THE CURRENT WITH WAVERING FINS"

"Water is the gaze of the earth," Paul Claudel wrote in *L'Oiseau noir*, "its instrument for looking at time . . ." (Claudel 1929: 229). Throughout various nuances of metaphor and matter, narrative and place, water symbolism seems to occupy a primary status in the stories of Ernest Hemingway. Water provides the locus for the struggles between dream and reality, ideals and indulgence, nature and society. Torn between two loves in *The Garden of Eden*, David Bourne plunges, barenaked, into the sea. For the hero in *A Farewell to Arms* water also affords the means of rescue, re-creation and even rebirth (*A Farewell to Arms:* 203–04):

> Anger was washed away in the river along with any obligation. Although that ceased when the carabiniere put his hand on my collar . . . I was not against them. I was through. I wished them all the luck. There were the good ones, and the brave ones, and the calm ones and the sensible ones, and they deserved it. But it was not my show any more.

Feared and loved, struggled against and yet submitted to, the river journeys through light and dark, life and death—this "double-hearted" river assumes a mantra role for Hemingway. Gazing over the bridge at Seney (*The Essential Hemingway:* 332):

> Nick looked down into the clear, brown water, coloured from the pebbly bottom, and watched the trout keeping themselves steady in the current with wavering fins. As he watched them they changed their positions by quick angles, only to hold steady in the fast water again. Nick watched them for a long time.

That trout—keeping itself steady in the current with wavering fins—was that not the ideal way of being-in-place? Nick Adams continued up the hill, his rucksack heavy, but he was happy—revived by the image of the trout in the flowing stream. This image brought serenity as he walked up the hill (*The Essential Hemingway:* 333):

> The road climbed steadily. It was hard work walking up-hill. His muscles ached and the day was hot, but Nick felt happy. He felt he had left everything behind, the need for thinking, the need to write, other needs. It was all back of him.

A meticulously detailed evening liturgy of tent assembly and food preparation follows. Then, after a nice cool smoke, comes sleep in the safety of the tent. Next morning (Part II) Nick Adams wakes up, emerges from the tent and "there was the meadow, the river and the swamp" (see Baker 1975: 150–59). This entire second day is devoted to fishing. There are numerous references to his thoughts—and to thoughts about not thinking—which combine to offer a sense that consciousness is a moving stream. This exploration into his own consciousness is an adventure which is both frightening and appealing.

The Big Two-Hearted River is a splendid orchestration of melodies, all to the tune of the river, the stream of consciousness, of time, of life. Fishing, as revealed later in *The Old Man and the Sea*, symbolizes other, perhaps more demanding, challenges. *The Big Two-Hearted River* symbolizes the hydrological cycle, the reciprocity of life and death. One heart is the active stream (brightness and goodness), the other is the swamp (darkness and evil). The story ends with a glimpse of the peace for which Nick was longing (*The Essential Hemingway:* 348):

> Nick stood up on the log, holding his rod, the landing net hanging heavy, then stepped into the water and splashed ashore. He climbed the bank and cut up into the woods, toward the high ground. He was going back to camp. He looked back. The river showed through the trees. There were plenty of days coming when he could fish the swamp.

Of all Hemingway's characters, Nick Adams has been regarded as the one who comes closest to being autobiographical. In his own life and thought Hemingway epitomized many of the dilemmas of twentieth-century America and its poignantly ambivalent attitudes toward nature, space and place: mobility and restlessness versus rootedness and identity; freedom versus belonging; adventure versus security; active mastery of environment versus awe-inspiring celebration of wilderness. Successive periods of war and peace, depression and boom, transatlantic travel and exposure to other civilizations also no doubt challenged taken-for-granted American identities. There was always, of course, the profoundly appealing work ethic, but here again one can note the tension between the thrill of completion and the sense that there was still more to do. Just as Nick Adams happily realized that there were "plenty of days coming when he could fish the swamp," so too his journey to "The Last Good Country" (1987) remained an unfinished one.

PLACE, POWER AND "PROGRESS"

Les-Saintes-Maries-de-la-Mer 1998 was quite a different place than the one I visited in 1966; Le Grau du Roi, Nîmes, Aigues Mortes and other places no doubt quite different from those visited by Hemingway. Why then, I wondered, the enormous appeal of these places for the Hemingway Society and Foundation? A romantic quest for something lost? What lessons could be gleaned from this encounter between geography and literature besides a shared sense of loss of those places where Hemingway had apparently drawn inspiration?

Returning by road along the *Languedocienne* and *Autoroute du Soleil,* I tried in vain to feel positive about the fact that this feat of highway engineering was affording a much more commanding view of landscape than was ever possible from the highways and byways of thirty years ago. The unrelenting procession of juggernaut trucks which filled the multilane highway screeched out the reality of commodity flows and the commodification of all matter through space—origins and destinations anonymous—riding roughshod over that mosaic of places about which classical regional monographs and novels were written. How, I wondered, might Hemingway have responded to this experience?

In many ways this whole pageantry along the *Autoroute* might have held as much appeal as the bullfight and the chasing of wild animals. Perhaps the artificiality of the situation would have been ignored—the sensory overload of those giant trucks transporting so many tons of stuff evoking other sensations—because in fact movement and flow mattered so much for him. Driving on this freeway demanded the prototypically "eccentric" attitude (Plessner 1975) of the "outsider" and spectator, overcoming "frictions of distance" rather than the empathetic gaze of someone eager to understand people's sense of place and home. In this land of Descartes, it seemed indeed ironic that the harvest of modern science could be so inimical to other human values such as dwelling or feeling at home in one's place. I suspect that this would have been the beacon light on Hemingway's horizon. In so many of his short stories, characters are described as genuinely confronting tensions between feelings of "home" and "homelessness," between inner feelings and moral convictions, rest and movement.

Here again one could ponder the ambivalence in Heidegger's reflections on technology and human progress. Could these freeway antics not be regarded as "local practices," and its rest stops as "gathering places" in the everyday lives of truck drivers and weary commuters? Hubert Dreyfus (1996: 13) surmised that Heidegger, in fact, might have found this *Autoroute du Soleil* to be exhilirating and hopeful. Heidegger indeed wrote:

> The highway bridge is tied into the network of long-distance traffic, paced as calculated for maximum yield. Always and ever differently the bridge escorts the lingering and hastening ways of men to and fro . . . [Each] bridge gathers to itself in its own way earth and sky, divinities and mortals (Heidegger 1971: 152–53).

Thirty years ago European geographers used terms such as sense of place, *pays*, *Heimat* to represent the lived experiences of people within particular localities. World War II exposed evidence of ways in which such catchwords had been exploited in the interests of political imperialism. History and literature have certainly made much noise about this. But it was surely the applied geographies of postwar times that ultimately threatened the salience of place, in geographical thought as well as in lived experience? The requestioning of disciplinary thought and practice has indeed been aided by a renewal of interest in dialogue with colleagues in literature.

From diverse corners come published works which illustrate the potentially fruitful common ground between geography and literature. This common ground contains not only landscape aesthetics (Schama 1995; Thompson 1995) and recollections of romantic literature, but also a firm commitment to eliciting concern about the integrity of places in all their human and biophysical reality (Ferrier 1998; Naess 1989; Ehrenfeld 1993; Berque 1996). At the approach of the third millennium philosophers now reflect on ways in which discourses on space have virtually drowned out knowledges of place (Casey 1997; Seamon and Zagonc 1998). On the wings of postmodernist prose, too, come resounding pleas for a rediscovery of meaning—symbolic, emotional and political—in the lived spaces of everyday experience. The dialogue between geography and literature remains an open horizon, just like the world to which they both attend. The world as horizon is a field which is always already there but which cannot be objectified, any more than it can be doubted (Husserl 1970: 142–43). Ernest Hemingway indeed beckons further encounters and mutually enriching exchanges on metaphor and matter, on place and narrative desire.

NOTES

I gratefully acknowledge help and critical commentaries graciously offered by Luke Wallin, University of Massachusetts, Ron Callan, Department of English and Tim Collins, Department of Philosophy, at University College Dublin in the preparation of this paper.

1. An overreliance on metaphor, too, could reiterate the empiricist claim that the world is meaningless in itself—something on which human intentions bestow significance. But the lived world, as Merleau-Ponty has pointed out, is not a metaphor; it is the seat or "homeland" of our thoughts (Merleau-Ponty 1962: 23–24).

2. The term *Gestell* involves three distinct meanings: *vorstellen* (to represent), *herstellen* (to fabricate), and *bestellen* (to order, enforce, set upon). See Tijmes (1998).

REFERENCES

Adorno, T. W., and M. Horkheimer. 1987/1944. *Dialectic of Enlightenment*, trans. J. Cumming. New York: Continuum Publishing Company.

Arendt, H. 1958. *The Human Condition*. Chicago: University of Chicago Press.

Bachelard, G. 1983/1942. *Water and Dreams: An Essay on the Imagination of Matter*, trans. Edith R. Farrell. Dallas: The Pegasus Foundation.

Baker, S. 1975. "Hemingway's Two-Hearted River." In *The Short Stories of Ernest Hemingway: Critical Essays*, ed. J. J. Benson. Durham, NC: Duke University Press.

Berque, A. 1996. *Etre humains sur la terre*. Paris: Gallimard.

Buttimer, A. 1971. *Society and Milieu in the French Geographic Tradition*. Chicago: Rand McNally. AAG Monograph No. 5.

———. 1981. "Musing on Helicon: Root Metaphors in Geography." *Geografiska Annaler* 64B (2): 89–96.

———. 1983. *Creativity and Context*. Lund, Sweden: Lund Studies in Geography Series B, No. 50.

———. 1993. *Geography and the Human Spirit*. Baltimore: Johns Hopkins University Press.

———. 1998. "Geography's Stories: Changing States of the Art." *Tidschrift voor Economische en Sociale Geographie* 89 (1): 90–99.

Buttimer, A., and D. Seamon, eds. 1980. *The Human Experience of Place and Space*. London: Croom-Helm.

Casey, E. S. 1997. *The Fate of Place. A Philosophical History*. Berkeley: University of California Press.

Claudel, P. 1929. *L'Oiseau noir dans le soleil levant*, 15th ed. Paris: Gallimard.

Cosgrove, D., and S. Daniels. 1998. *The Iconography of Landscape*. Cambridge: Cambridge University Press.

Dreyfus, H. 1996. "Being and Power." *International Journal of Philosophical Studies* 4 (1): 1–20.

Ehrenfeld, D. 1993. *Beginning Again. People and Nature in the New Millennium*. Oxford: Oxford University Press.

Ferrier, J.-P. 1998. *Le contrat géographique ou l'habitation durable des territoires*. Lausanne: Editions Payot.

Heidegger, M. 1954. *Vorträge und Aufsätze*. Pfüllingen, Federal Republic of Germany: Günter Neske.

———. 1957. *Gelassenheit*. Pfüllingen, Federal Republic of Germany: Günter Neske.

———. 1971. *Poetry, Language, Thought*, trans. and ed. Albert Hofstadter. New York: Harper Colophon Books.

Hemingway, E. 1964/1947. *The Essential Hemingway*. London: Jonathan Cape.

———. 1964. *A Moveable Feast*. New York: Charles Scribner's Sons.

———. 1966/1929. *A Farewell to Arms*. London: Jonathan Cape.

———. 1986. *The Garden of Eden*. London: Hamish Hamilton.

———. 1987. "The Last Good Country." In *The Complete Short Stories of Ernest Hemingway: the Finca Vigia Edition*, 504–44. New York: Charles Scribner's Sons.

Hotchner, A. E. 1966/1955. *Papa Hemingway: A Personal Memoir.* London: Weidenfeld and Nicolson.

Husserl, E. 1970. *The Crisis of European Sciences and Transcendental Phenomenology*, trans. D. Carr. Evanston, IL: Northwestern University Press.
Jacobs, J. 1961. *The Death and Life of Great American Cities*. New York: Vintage.
Kearns, M., and C. Philo, eds. 1993. *Selling Places: The City as Cultural Capital*. Oxford: Pergamon Press.
Keith, M., and S. Pile, eds. 1993. *Place and the Politics of Identity*. London: Routledge.
Ley, D., and M. Samuels, eds. 1978. *Humanistic Geography: Prospects and Problems*. London: Croom-Helm.
Linde-Laursen, A., and J.-O. Nilsson, eds. 1995. *Nordic Landscapes: Cultural Studies of Place*. Stockholm and Copenhagen: Nordic Council of Ministers.
Merleau-Ponty, M. 1962. *Phenomenology of Perception,* trans. Colin Smith. New York: Humanities Press.
Monod, G. 1894. *Les maîtres de l'histoire: Rénan, Taine, Michelet*. Paris: Calmann-Levy.
Naess, A. 1989. *Ecology, Community and Lifestyle.* Cambridge: Cambridge University Press.
Nagel, J., ed. 1984. *Ernest Hemingway: The Writer in Context*. Madison: University of Wisconsin Press.
Paasi, Anssi. 1996. *Territories, Boundaries and Consciousness: The Changing Geographies of the Finnish-Russian Border*. New York: John Wiley & Sons.
Plessner, H. 1975. *Die Stufen des Organischen und der Mensch*. Berlin: Walter de Gruyter.
Pocock, D. C. D., ed. 1981. *Humanistic Geography and Literature*. London: Croom-Helm.
Relph, E. C. 1976. *Place and Placelessness*. London: Pion.
Salter, C. L., and W. J. Lloyd. 1977. *Landscape in Literature*. Washington, DC: Association of American Geographers Resource Papers for College Geography No. 76–3.
Schama, S. 1995. *Landscape and Memory*. London: Fontana Library.
Seamon, D. 1979. *A Geography of the Lifeworld*. London: Croom-Helm.
Seamon, D., and A. Zagonc, eds. 1998. *Goethe's Way of Science: A Phenomenology of Nature*. Albany, NY: State University of New York Press.
Tall, D. 1993. *From Where We Stand: Recovering a Sense of Place*. New York: Knopf.
Thompson, P. B. 1995. *The Spirit of the Soil: Agriculture and Environmental Ethics*. London and New York: Routledge.
Tijmes, P. 1998. "Home and Homelessness: Heidegger and Levinas on Dwelling." *Worldviews: Environment, Culture, Religion* 2 (3): 201–13.
Tuan, Yi-fu. 1977. *Space and Place: The Perspective of Experience.* Minneapolis: University of Minnesota Press.
Vidal de la Blache, Paul. 1913. "Des caractères distinctifs de la géographie." *Annales de Géographie* 22 (124): 289–99.
Wagner, Philip L., and Marvin W. Mikesell, eds. 1962. *Readings in Cultural Geography*. Chicago: University of Chicago Press.
Yaeger, P., ed. 1996. *The Geography of Identity*. Ann Arbor: University of Michigan Press.

10

Cultural and Medical Geography: Evolution, Convergence, and Innovation

Charles M. Good

Cultural geography is at a strategic crossroads in its development. On the one hand, the field has been strengthened through its exposure to recent intellectual currents that have affected the social sciences and humanities in general. Paradoxically, however, reflecting the postmodern situation, cultural geography's collective identity has become rather less coherent than before. Overall, this is a most welcome, positive condition that invites reassessment and renewal.

To remain vital and innovative, visible and credible in the twenty-first century, cultural geographers must reexamine and strengthen their authentic links to the wider discipline. The conceptual, practical, and historical linkages between medical geography and cultural geography, in particular, have long ranked among the field's strongest. These intradisciplinary links have existed in the past and endure today (Hunter 1973; Good 1987; Roundy 1987; Kearns 1993; Haggett 1994; Gesler 1991, 1992, 1996; Lewis and Rapaport 1995). More than at any other time, the methods and perspectives of cultural and medical geography represent a rich medium for future scholarship that examines the conditions, and viability, of human existence on earth. The following themes suggest some of the commonalities and ideas for collaboration: (1) implications of persistence and change in national medical cultures; (2) globalization's role in the changing relations "on the ground" between biomedicine and "organized" indigenous medical systems; (3) the process of "re-placing," that is, how new immigrants to North America use culture in the transition to their new environments and in coping with novel health problems; (4) processes and health implications of ecologies and landscape changes in "silent" and "health-critical" environments; (5) geographical aspects of emerging infectious diseases; (6) "therapeutic" landscapes: the multiple meanings of place in health; (7) extending research beyond mortality and morbidity to encompass geographic analyses of cultural and social factors in the "health transition"; and (8) place and culture studies that explore the determinants of HIV/AIDS transmission and control in "developing" societies.

This chapter begins by examining factors that have influenced cultural geography's development over the past forty years. It considers the relationships between the "old" (traditional) and "new" (humanistic, structural, postmodernist) approaches, and identifies lines converging with medical geography. Finally, case studies from East Africa are used to illustrate opportunities within the emerging hybrid field of "cultural-medical" geography.

CONTINUITY AND CHANGE IN CULTURAL GEOGRAPHY

Arguably no single project in the field has superseded the intellectual influence of Wagner and Mikesell's *Readings in Cultural Geography* (1962). Their introductory essay and core themes, dealing with culture, cultural landscape, culture history, cultural diffusion, and cultural ecology, remain a coherent, if partial, framework of ideas, questions, and connections to other fields. A popular textbook, the anthology served to acquaint many readers with the legacy of Carl Sauer and the "Berkeley" school.

Shifting paradigms, including the ongoing engagement with social theory, have not pushed aside the "vintage-1962" themes. Rather, they continue to inform, and sometimes anchor, cultural geography's debates and contributions. Later reassessments of the field generally reaffirm the continuing value of Wagner and Mikesell's framework (Mikesell 1978; Foote et al. 1994).

Evolution and Internal Critique

Several aspects of cultural geography's trajectory deserve mention. First, Wagner and Mikesell's (1962: 1) definition of cultural geography as "*the application of the idea of culture to geographic problems*" was influential but incomplete. Wagner (1994: 7) more recently criticized their idea as unsystematic, shallow, and "faulty." The two scholars also asserted that "the cultural geographer is not concerned with explaining the inner workings of culture or with describing fully patterns of human behavior" (1962: 5). Immediately controversial (Brookfield 1964), this attempt to circumscribe the scope of inquiry probably drew more debate than any other position Wagner and Mikesell attempted to stake out.

Today, it is generally accepted that there is no intellectual rationale for limiting geographic inquiry. Linkages among cultural values, sexuality, and health status, for example, reflect newer concerns that previously were virtually taboo (Gould 1993). Such topics also exemplify the complementarity of cultural and medical geography (Good 1998).

Many cultural geographers can recall a growing restiveness in the field by the late 1970s. Duncan (1994: 401) asserts that by the early 1980s the discontent had become "civil war." There was a sense that the "new" cultural geography was "obligated" to overhaul the traditional approach. Accomplishing this goal meant embracing greater "theoretical rigor" and ideological commitment. There seemed little justification for continuing apolitical approaches to such themes as landscape change, cultural ecology, and historical geography. Other fundamental questions also begged answers. What does cultural geography propose to explain? Which among its basic premises, ideas, and methods constitute enduring strengths? What recent developments in theory and method in other social sciences and the humanities connect with scholarship by cultural geographers?

Several sources of change helped to energize cultural geography's "reimagining" process, including: (1) "critical" studies (for example, Harvey 1973, 1989); (2) social geography; (3) postmodernist thought; and (4) humanistic studies. "New" cultural and medical geographers in British, Irish, and Commonwealth universities were among the vanguard in this movement. Often grounded in a structuralist approach, this new work highlighted various ideological perspectives and promoted more "cross-scholarship." Topics of interest included "gendered" and health landscapes, inequality, oppression, and resistance. Cultural and historical geographers such as Meinig (1982) and Blaut (1993) offered reinterpretations of such overarching topics as imperialism, diffusion, and global inequality.

Simultaneously, Foucault's philosophy (1984) challenged the field's conventional "eyes" and influenced how some "new" cultural geographers understood their purpose. Others expressed skepticism about investing in ideas that were inflexible, contradictory, and colored by an exclusive ideology that gave birth to such notions as "postdisciplinarity."

Toward a "New" Cultural Geography

Cultural geography's internal discourse has been enlivened by access to ideas ranging from the postmodernist paradigm, including textual criticism and deconstruction, to political economy, gendered perspectives, and poststructuralism (Matless 1996). The humanistic tradition has also acquired new outlooks. Ideas about symbolic landscapes as social documents and texts have produced valuable connections to medical geography and other subfields (Gesler 1992, 1996).

Several other points bear emphasis. First, few cultural geographers have been able to ignore the debates and tensions affecting the field internally. Many scholars view "social theory" with both favor and skepticism. Their eclectic stance runs counter to the agendas of ideological deconstructionists who insist that cultural geography is splintering into shards of "plurality and difference" (Duncan 1994). Postmodernity studies also claim to draw greater attention to "the specific, the local, the place-bound"(Livingstone 1993). How strange that the "old" cultural geographers were once admonished for preferring to work in local places instead of popular spaces![1]

Second, by the 1980s cultural geographers no longer gave preeminence to rural and, particularly, non-Western environments and landscapes as research venues. Most would agree that this is a welcome, positive development.

On the debit side, the entire discipline has been threatened by the apparent decline of fieldwork and of regional expertise—both former staples in the life-long education of cultural geographers.

MUTUALITY OF CULTURAL AND MEDICAL GEOGRAPHY

Interest in the interdependence of cultural and medical geography picked up during the late 1970s, partly due to the expansion of medical geography in North America, the British Commonwealth, and Europe. An increase in funded research, joint international meetings, and participation in such professional journals as *Social Science & Medicine* encouraged scholarship and teaching. Medical geography began to draw more upon social theory and "critical" social geography, medical anthropology, and human ecology. Behavior, population, environment and spatial relations emerged as conceptual pillars of its ecological approach (Meade, Florin and Gesler 1988). Work on the "development of underdevelopment" theme criticized medical geographers for their failure to analyze health issues in political economy terms (Stock 1986). The inclusion of structuralist perspectives strengthened geographers' concepts of health care and disease.

Medical geographers endeavor to understand the complex interplay of cultural-ecological, political, and place-space processes and patterns in human health. Issues can range from the study of diseases promoted by human alterations of the environment, to spatial patterns of cancer, to the spiritual dimensions of health behavior. The hybrid term *cultural-medical geography* simply recognizes the reality of interdependence. A large tent, it covers more conceptual and topical ground than previous definitions.

Four groups of overlapping questions illustrate the irreducible connections in cultural-medical geography.

1. In what characteristic (and divergent) ways do societies perceive, define, value, and manage disease and health care? What is significant about how these factors vary in place and time, and over space?
2. Where, why, and how do specific cultural-behavioral, environmental, political-economic, and ethical factors affect disease exposure and experience (for example, poverty, forced migration, female circumcision, abortion, breast cancer, malaria, HIV)?
3. How does a society define its health resources, physically and symbolically,

and regulate access to them? What determines the use and efficacy of multiple systems of health care (Good 1987)?

4. What factors account for the relative success or failure of societies and cultures to cope over time with old and new health hazards in different places?

LANDSCAPE STUDIES

Cultural landscape studies have long been a distinguished staple of cultural geography (Sauer 1925). Medical geographers have recently been attracted to "humanistic" landscapes, with their invocation of such concepts as sense-of-place and symbolic analysis. Cultural-historical geographers have also contributed significantly to the cross-fertilization. Meinig's (1982) generic approach to the geography of imperialism, including social change on cultural frontiers, underscores the value of a historical perspective within cultural-medical-health studies. Whereas Meinig did not set out to highlight these particular interrelations, his "generic imperialism" scheme has broad application. The framework aids in discovering what happens to people's cultures, welfare, and cultural-symbolic landscapes when societies possessing vastly different worldviews and medico-religious systems meet.

Places and "health landscapes" are commonly "read," decoded, and interpreted differently among individuals and groups (Meinig 1979; Gesler 1991; Kearns 1993; Moon 1995). Insiders and outsiders may "read" and decode the same landscape text but reach dissimilar interpretations. In a "medical" context cultural landscapes reinforce awareness that health and illness are relative, nonrandom conditions. Landscapes may be "rich" or "poor," healthy or unhealthy; but they are never mute.

The journal *Health & Place* exemplifies the collaboration now established among British, European, Australian-New Zealand, North American, and other cultural-medical geographers. The editorial policies of this young journal explicitly support efforts to demonstrate how the perspectives of cultural and social geography enhance knowledge of the "health landscapes" of particular places and societies. Kearns' (1993) question: "What is the place of health in the health of place?" is much more than a play on words.

Cultural and landscape factors in health date back at least to Hippocrates' treatise *On Airs, Waters, and Places* from the fifth century B.C.E. However, systematic attention to geographical relationships of culture and health dates mainly to the second half of the twentieth century. Previously, the place of "culture" in medical geography rarely received systematic attention. Relevant human behavior was often left to medical anthropologists, historians, and others to sort out (Pugh 1996). Exceptions include extensive studies by May (1971) on disease ecology and dietary patterns and by Prothero (1965) on human mobility. Hunter (1966, 1973) conducted pioneering work on disease ecology. Roundy (1987) developed a useful model to assess relationships among human spatial and temporal behavior, environmental risk zones, and disease transmission patterns.

Gesler's synthesis, *The Cultural Geography of Health Care* (1991), illuminates the complementarity of the "old" and "new" cultural and medical geography. Traditional themes including culture systems, culture regions, cultural ecology, cultural evolution, cultural diffusion, and folk and popular culture are linked to medical geography. These themes have continuing value in a changing, hybrid discipline. Some essays spotlight the influence of social theory on recent work in cultural-medical geography. Others address social space, place and landscape, and structuration. Vignettes demonstrate interrelationships of culture, politics, and medical systems with health and disease in localities ranging from India and Detroit to East London, England.

RITUAL, THERAPEUTIC, AND SYMBOLIC LANDSCAPES: CASE STUDIES OF CULTURE, SEXUALITY, AND HEALTH

Rites of passage, sexuality and reproductive health, and traditional medical systems have not attracted much attention from geographers. Rather than being appreciated as universal features of human experience, such themes have been avoided because of their erotic or exotic auras. At best they have been seen as having peripheral importance for understanding places and regions. Squeamishness or narrowness produced a false trichotomy of culture, sexuality, and geography (Lange 1997). Along with sexuality, there has also been a sense that medico-religious systems and related themes lie remote from the "core" of the geographic enterprise. Sexual behavior has been curtained off and treated as something voyeuristic and professionally risky—more the province of anthropology, sociology, or public health. Notable exceptions include Symanski (1981). Recent work includes efforts to understand the politics and spatiality of gay and lesbian experiences (Bell 1994; Brown 1995).

Sexuality and its manifold health implications now seem positioned to become bona fide topics for scholarly and applied research in geography. The journey has been punctuated by stops at the sexual revolution and the HIV/AIDS pandemic, changing perceptions about reproductive health, and the emergence of gay and lesbian politics. While HIV/AIDS transmission appears basically unaffected by natural environmental factors, the virus's ubiquitous connections with human institutions and behavior are no longer news. This powerful pandemic thus offers broad scope for cultural-medical geographers to contribute to understanding and possibly mitigating its effects (Gould 1993; Good 1998).

Apart from sexuality and reproduction, cultural-medical geography also embraces the nexus of religion, space, and social relations. A topical and conceptual framework needs to be developed that incorporates indigenous medical systems, rites of passage, and therapeutic landscapes (ritual, symbolic and contested). Significantly, these concerns have important links to such conventional geographic interests as livelihood systems and resource use, human mobility, and urbanization.

CULTURE, HEALTH, AND LANDSCAPE: AFRICAN EXAMPLES

Health and health care, ritual and therapeutic symbolism, and human sexuality and reproduction are deeply interwoven features of any cultural landscape. Geographical expressions of these themes may be both direct and subtle, mundane and exceptional. Africa offers abundant historical and contemporary examples of them all. Africa's landscapes have stories to tell about cultural continuity, resistance to change, crisis, adjustment, and adaptation. Examples include disease ecologies and colonial and neocolonial social relations; spatiality of sexuality, including networking patterns, and epidemics of sexually transmitted diseases (STDs/HIV); collaboration of traditional healers and biomedical workers to minimize the spread and optimize prevention of STDs/HIV; and rites of passage, gender relations, and "missionary engineering" of human reproduction and demographic change.

The following case studies point to the mutuality of cultural and medical geography. They suggest how rites of passage, gender relations, and reproductive politics have influenced the formation and character of African landscapes. This evidence is based on field and archival work in Kenya and Britain. Labeling these case studies "old" or "new" cultural-medical geography offers little advantage, since they reflect important elements of both continuity and change.

RITES OF PASSAGE AND THE TRANSITION TO AFRICAN WOMANHOOD: FEMALE CIRCUMCISION

Awareness of female circumcision (FC) has spread widely and gained global notoriety during the 1990s. Great controversy has accompanied this growth in consciousness. A diverse and strongly regionalized institution, FC is practiced primarily in Africa north of the equator and in a few other regions. Millions of African girls and young women from specific ethnic groups remain subject to this rite of genital cutting wherever they may live, including North America (Nelson 1998) and Europe (Public Health Reports 1997).

The term *female circumcision* was coined in the setting of colonial overrule in such places as Kenya's former White Highlands. For generations, girls and women in hundreds of African societies have undergone distinct forms and degrees of cutting and related rituals as part of initiation to adulthood and as preparation for marriage.

Female circumcision has recently been socially reconstructed as female genital mutilation (FGM). The latter term commonly occurs in much of the Western-generated literature and rhetoric on the topic. Consciousness raising, reeducation, and political pressure from women's support groups in and outside Africa are contributing to important changes in attitudes and behavior toward FC (Inter-African Committee 1997). Many believe that abolition of dangerous forms of FC and acceptance of alternative, noninvasive rites of passage will eventually yield enormous benefits for women's mental and physical well-being.

In extensive areas of Egypt, Sudan, the Horn and parts of Kenya, 50 percent to over 90 percent of women undergo some form of FC. In West Africa, over half of the women undergo FC in Nigeria, Ivory Coast, Burkina Faso, Mali, Guinea Bissau, Guinea, the Gambia and Senegal. A similar frequency occurs among many ethnic groups in the rainforest and savanna zones of West Africa. The practice is also widespread in Chad and the Central African Republic, northern Cameroon, and the northern border peoples of Congo. In Kenya, the practice is customary among the Bantu-speaking Kikuyu, Embu, and Meru peoples and among some Nilotic peoples.

Efforts of non-African feminist spokespersons to shield African girls and women from abuse and the complications of chronic poor health have been well-meaning. However, there has been a tendency to overlook the desirability of focusing resources and support on women in ethnic groups practicing the most severe forms of cutting and trauma (for example, infibulation). In terms of public and personal health, lumping all rites together (there are six degrees of cutting in FC) makes little sense when they are not equally invasive, traumatic or dangerous to health. There are also geographically distinct patterns of FC, varied according to the relative severity of the customary procedures. Locational patterns of male circumcision also have health implications. Thus noncircumcision is now a recognized cofactor in HIV infection among African males (Caldwell 1995).

Genital surgery performed while one is in a conscious "observer" state can be extremely emotionally and physically traumatic. Wide variations in the amount of cutting also result in different risk levels for women. Tissue and nerve damage and pain levels may be insignificant or severe. Variations in blood loss, susceptibility to infection, healing time, degree of psychological trauma, and subsequent quality of life have major health and social consequences. Depending on the type of surgery, women may suffer for years from chronic pelvic infections, infertility, and related social problems, including divorce.

The surgical procedure customarily ranges from a small nick on the clitoris among some ethnic groups that is not life-threatening to the radical "Pharonic" infibulation practiced on Somali women in the Horn of Africa. The Pharonic procedure clearly presents the greatest risk of chronic complications. These may include keloid scarring, birthing difficulties, and delivery trauma. Complications of FC-induced vesico-vaginal fistula, which is generally correctable by surgery, may also subject women to social disgrace. Since it removes all external genital tissue and sutures the vagina shut (except for very small openings left for urination and menstruation), the infibulation procedure obviously causes excruciating pain and the potential for long-term discomfort. Immediately upon her marriage, an infibulated woman must be "opened" by her husband, which is typically painful, unpleasant, and sometimes drawn out over several days. Traditionally a Somali woman is resutured between the births of her children.[2]

In comparison, FC among the Nyeri Kikuyu, located at Mt. Kenya, reportedly involves minor cutting and far less trauma. In recent decades, girls between 12 and 14 years of age in this locality, perceiving enhancement of their own identity, have been known to seek out circumcision against the wishes of their elders.[3]

An ancient rite of passage for millions of African women,[4] FC has significant implications for the stability of extended family life and ethnic and class identity. It can determine the quality of sexuality, health, and spirituality (FGM Research Homepage). Beyond Africa, FC has often generated unfavorable images of indigenous cultures and of relations between African women and men.

Today, outsiders have appropriated some of the symbolism of FC/FGM, and spawned a massive popular "discourse" on the topic. Thus, in one important sense the issue no longer belongs completely to Africans. Female circumcision exemplifies a "constructed," highly charged, and now globally politicized practice. Its different interpretations emphasize African self-determination and identity, women's development and equity, and health and human rights. In the long run, noninvasive, substitute rituals seem a likely best compromise solution. However, African women remain divided over whether FC should be completely banned, or modified in line with modern sensibilities. These differences reflect its deeply rooted cultural symbolism and social implications. On the other hand, many African women have been of one mind "in condemning outsiders' intervention" in FC/FGM. They want the issue resolved in Africa, by Africans, and according to African values (Stock 1995: 244). One thing is certain: conflict will grow stronger because change is under way. Despite heated opposition from fundamentalist Islamic leaders, Egypt's highest court recently overturned a lower court decision and "upheld a ban on the genital cutting of girls and women." Violators face imprisonment for three years. Egypt's law sets a strong precedent for other national and local efforts aimed at eliminating FC (Crossette 1997).

For centuries, Africa has been defined by and for outsiders, frequently with tragic consequences for the indigenous peoples (Hochschild 1998; Gourevitch 1998), creating *The Africa That Never Was* (Hammond and Jablow 1992). Fifty years after decolonization began, a many-sided internal and international activism will surely change the character of female rites of passage. Inside and outside Africa the issue is increasingly framed by principles of individual choice, social responsibility, and human rights. Outsiders should not misinterpret Africans' ultimate resolve to enhance the health conditions and human rights associated with female initiation.

Cultural and medical geographers' frames of reference reveal a remarkable complementarity of subjects, concepts, and methods. Each subfield enhances the other's potential and contributes to the possibility of synergy. However, most writing and discussion about FC do not reflect the kinds of questions cultural-medical geographers would ask or anticipate. There is a weak sense of the character and relevance of place, of spatiality, and of time (all important "policy" domains). The historical case studies below emphasize the importance of context and provide a sense of continuity in the current contest over FC.

Colonial Kenya

Female circumcision in colonial Kenya meant quite different things to Africans and colonialists, especially missionaries. European doctors occasionally warned about the procedure's real dangers. Some district administrators counseled that FC was a deeply ingrained custom that would not easily yield, if ever, to outside manipulation. In short, it was best to let sleeping dogs lie. Colonial officers tended to show more interest in tackling problems when they believed they had a reasonable chance of realizing a positive outcome. They were more likely to spend their limited resources on such concerns as improving tax collection or reducing adolescent abortions than on efforts to abolish female circumcision.

Among Kenya's colonial missionaries, few seemed aware of the powerful bombshell that could drop if they challenged female circumcision, particularly among the Kikuyu-Embu-Meru peoples in the Central Province. Evidently until the last moment, they had little inkling about FC's power to mobilize cultural nationalism and challenge the authority of the European Christians in the African churches. These African-initiated forces ultimately translated into concrete physical and symbolic features of the cultural landscape. Resistance to subjugation by Europeans, independence of thought and action, and community support manifested themselves in new elements of local religious, educational, and therapeutic landscapes. Finally, who would have predicted the strong link between female circumcision and the rise of nationalism and ultimately Kenya's independence? Such developments leave no doubt about the importance of historical context to current discussions of FC.

More Than Cutting: The Contested Landscape of Female Circumcision

Historically, female circumcision in Kenya offers a well-documented case study of how early African nationalism manifested itself in breakaway resistance to coercive cultural change. By the late 1920s European Protestant missionaries and some African converts whose numbers varied by congregation and location had decided that baptism and FC could no longer coexist. A cultural crisis rapidly emerged, sustained in part by the Kikuyu Central Association's (KCA) goal of grassroots political mobilization against white domination (Annual Report

1929).[5] Attacks on FC further inflamed existing nationalist passions that were building with the growing conflict over African lands alienated to European settlers in the "White Highlands." A belief spread among an "overwhelming majority" of Africans that Europeans would try to marry uncircumcised girls in order to grab even more Kikuyu land and thus complete the destruction of their social order (Kenyatta 1962: 130).

Among the Kikuyu and related societies, FC linked the generations through a consultative process involving the mothers, fathers, grandfathers and other elders of the uncircumcised girls. Regardless of how little or how much genital cutting (*irua*) occurred, the operation was embedded in deeply symbolic rituals.

Differences also occurred among tribes in the required length of the initiation ceremonies, and to some extent in the cultural and social centrality of the FC ritual. Regardless of intergroup and place-to-place differences, FC was a virtual common denominator in that it was (and generally remains) a precondition for admission to adult society and to marriage. Female circumcision was typically accompanied by much singing and dancing, and it held profound significance for the initiates and their families (Strayer 1978). Each adult group thus had a degree of authority over the young girls entering maturity, and each had to approve their moving into this next stage of life (Presley 1992). Among the educated class, families sometimes arranged for their daughters to be circumcised in a hospital.

Female circumcision's centrality in the ritual life, marriage, and social organization of Kenya's Kikuyu, Embu and Meru peoples did not escape the missionaries' attention. Regardless, they concluded that even clitoridectomy, the "minor" form of cutting, was abhorrent and irreconcilable with church dogma.

The Church of Scotland Mission (CSM) conducted its evangelistic teaching and medical activities from four main stations ranging from Kikuyu, outside Nairobi, to Chogoria, 150 miles north on Mt. Kenya's east face. The CSM ultimately banned from church membership all individuals and families who supported clitoridectomy.[6] Other mission societies, including the Church Missionary Society (CMS) and Africa Inland Mission (AIM), but not the Roman Catholics, followed suit. CSM members had to mark their thumbprint on a document and swear their loyalty to the Presbyterian position against female circumcision, or else be dropped from church rolls. The CSM gambled on the power of public pressure and private sentiment to break down resistance to its theological imperialism. Those who refused to support the policy became known as *aregi* (defiers), while those who offered their thumbprint were *kirore* (loyalists). This period of intense crisis in the churches and communities—the *Kirore* era—extended from 1929 to the mid-1930s.

During *Kirore,* Kikuyu leaders in and out of the Kikuyu Central Association manipulated the power of indigenous and imported rituals and symbols, at considerable cost to personal and family life. They altered the geography of worship and education by establishing separatist churches and schools, enabling these functions to continue relatively free of missionary and colonial interference. Independent churches and schools formed the movement's primary weapons, and many of the latter eventually received limited government recognition until their closure at the outbreak of the Emergency (Mau Mau) in 1952. In the first month after the ultimatum, about 90 percent of the southern Kikuyu on the official membership rolls of the CSM at Kikuyu (near Nairobi) and of the AIM at Kijabe left the church. Over half of the children enrolled in CSM schools temporarily dropped off the rolls (Hyam 1990).

The ideological conflict and locational displacements that *Kirore* brought left some families painfully divided. Independent churches and schools, radical symbols of change and choice, created new senses of place identity and a refuge for a society whose core values and mores were under assault from two sides. Separate churches and schools also became living symbols of resistance to white hegemony, nurturing political awareness.

By 1929, FC had captured the attention of the British Parliament and the Colonial Office. Officials in London took a dim view of the missionaries' political naiveté and worried about their potential for stirring up even greater African discontent. Also in 1929, Jomo Kenyatta, the KCA president and leading Kikuyu activist (and later independent Kenya's first president), visited England. He proved to be a frustrating enigma to colonial authorities hoping to gather intelligence from him.[7] From *Kirore* onward, Kenyatta's remarkable political skills would become increasingly and distressingly evident to officials in London and to Kenya's hedonistic settler society (Hyam 1990).

Undeniably, the fallout from the *Kirore* events contributed decisively to the processes that led to Kenya's political independence in 1963. Predictably, missionary attempts to stamp out female circumcision ended in failure. The institution never came close to dying out, and by the 1940s it had been rejuvenated and was once again widespread. Kenya's white missionaries and the colonial government could not do much about it. In any case, the government had little enthusiasm for this particular form of coercive social engineering by the missions. Overall, the rift created between the missions and the Kikuyu people beginning in 1929 turned out to be a crucial turning point in Kenya's history. It was, Hyam (1990: 196) asserts, "comparable with Henry VIII's breach with the Church of Rome on the matter of his divorce. Both were seen as 'sexual' confrontations with an alien authority."

MICROCOSM OF A BROADER CULTURAL CONFLICT AND CHANGE ON MT. KENYA: CHOGORIA AND THE SCOTTISH PRESBYTERIAN CHURCH

Kenya's female circumcision crisis occurred principally in the Kikuyu, Embu, and Meru districts. Chogoria, situated at higher than 5,000 feet on eastern Mt.

Kenya and the northernmost CSM station, offers a fascinating case study of this phenomenon. Isolated and far from the cradle of the conflict near Nairobi, Chogoria's Mwimbi population was nevertheless deeply divided by the church's insistence that members renounce FC as a "pagan procedure." When presented with the *Kirore* ultimatum, or "promise paper,"[8] on September 20, 1929, the "overwhelming majority" of Africans (Mwimbi and neighboring Chuka peoples) in the church refused to provide their thumbprint (Irvine 1929). Instead they became *aregi.*

The missionaries had made a stunning miscalculation. Only fourteen out of about fifty Christian men and women at Chogoria took the oath (*Kikuyu News,* June 1930; Irvine 1929, 1972).[9] Many who broke away, called "the anti-Mission faction" by the CSM, received a sympathetic reception from local representatives of the nationalist Kikuyu Central Association. Again, this political linkage energized a movement to found independent churches and schools and thus the beginnings of a new (still incompletely deciphered) social map of Mwimbe.

At Chogoria the upheaval touched every aspect of mission life. Most African workers left the station: "all the staff" and "every hospital dresser deserted." Most teachers also left, and houseboys walked away from their employers (*Kikuyu News*, June 1930). The CSM responded to this repudiation by closing "all out-schools." Twenty-five Mwimbi did return to the church in the following month. "Only fear of persecution, ostracism, and ridicule keep a large number back," proclaimed Dr. Clive Irvine, the well-meaning but domineering Mission head (*Kikuyu News*, June 1930).

Soon after *Kirore*, itinerant foreign "prophets" arrived from Kikuyu districts in the south, 100 to 150 miles distant. Considered anti-European, they refused to adopt European clothing, and reportedly revived "an old scare that the missionaries and their followers were head-hunters." Prophets invoked the Holy Spirit through feverish shaking, and CSM claimed that they baptized followers "promiscuously." Within two years, local Mwimbi prophets had emerged as part of this changing scene (*Kikuyu News*, June 1931).

Although some *aregi* continued to drift back to the Chogoria Church, mission policies had, at least temporarily, damaged its credibility and divided the community. The CSM soon alleged that political opposition to CSM schools had withered. Yet within five years the "anti-Mission faction" had managed to establish seven independent schools. By 1939, reports claimed that the recession in church membership had been reversed throughout the CSM sphere. Bolstered by new converts and reentrants, Chogoria's membership rolls increased to 761 communicants (*Kikuyu News*, June 1939). Female circumcision continued in and out of sight.

In Mwimbe (and many other localities), FC was one part of a larger group of factors working to enlarge the cultural and demographic consequences of foreign religious and civil intervention. Alterations in traditional forms of social behavior created both subtle and pronounced changes in the social and cultural landscape, even at the household level. The CSM sought to introduce measures that would reshape the indigenous society by eliminating its least tolerable forms of sinful living.[10]

One major change the Scottish missionaries keenly wanted to promote from Chogoria was the cohabitation of married persons. The indigenous pattern for married Mwimbi was decidedly not Scottish, since it was customary for spouses to sleep under separate roofs. Younger married males, usually over 30 years old, and warriors customarily lived in communal bachelors' quarters. The all-male *gaaru* was a warriors' barracks centrally located in each settlement. Designed as a permanent guard to protect against raids, the social organization of the *gaaru* also contributed to the well-developed set of strategies the Mwimbi had evolved to regulate fertility. Men visited their wives for sexual relations but did not reside with them.

Unwilling to coexist with the prevailing mores, the CSM leadership decided to pressure married couples associated with the church to reject the old pattern of residence. Polygyny, of course, was also unacceptable. Several important consequences resulted from their meddling and social tinkering. Inadvertently aided by colonial administrative policies, the CSM cohabitation rule instigated changes in Mwimbe settlement and demographic patterns that reverberate to the present day. Consequences included a marked rise in fertility and, much later, a high level (for Africa) of modern contraception among local women.

Circumcision of girls occurred several years after the onset of puberty. Male circumcision did not take place until the mid-twenties. Intercourse and pregnancy before circumcision were forbidden and subject to powerful taboos and sanctions. Mwimbi spaced their babies at intervals of three to four years. Also, the custom of late marriage of warriors and their separate existence, living and sleeping in the *gaaru,* helped to depress fertility. For warriors, sexual intercourse served an instrumental purpose. In some cases a warrior who desired to sleep with his wife ("to seek a child") required the permission of his barracks' chief.

These customs began to erode under the British colonial administration and CSM policies. Taxation forced men to seek paid employment and sped the decline of barracks life. This process contributed to a relaxation of social controls, greater sexual license, and a steep increase in the pregnancy rate of uncircumcised girls. According to the Meru district commissioner, by 1933 abortion had become the locality's "most serious social problem." The administration responded by attempting to lower the circumcision age for both girls and boys to encourage earlier marriage.

As the price of remaining in good standing, the CSM expected its members to reject certain customary taboos and other behaviors that had profound cultural and demographic significance in traditional society. In return, the Scots offered the Africans Presbyterian doctrines, a doctor and hospital, teachers, and "opportunities" to observe Europeans and receive tutoring in "civilized" behavior. The two examples already mentioned—the CSM's insistence that families commit to stopping FC, and policies that succeeded in altering sexual behavior and reproductive outcomes—probably cannot be bettered as examples of coercive breaks with indigenous customs. As expected, the latter process also induced changes in the spatial and social structuring of spousal relations in Mwimbe.

Mission and secular influences also contributed to breaking down the customary prohibition on a couple resuming sexual intercourse with each other before weaning their child. This postpartum taboo obviously promoted birth-spacing and allowed for a long period of breast-feeding (generally not less than two years). Greeley (1988: 205) observed that the traditional cohabitation pattern in Mwimbe aimed to "achieve a desired small family size, and child spacing." The colonial government and the CSM both favored a degree of population expansion for economic reasons. However, the effects of European interference far exceeded expectations.

In the 1930s the effects of colonialist and CSM actions on population were reinforced by those of the Italian Catholic Mission, located several miles from Chogoria. The ultimate result was "a widesweeping pronatal effect" on local society. Ecological repercussions on the landscape closely followed the demographic changes. Examples include the clearing of formerly conserved forest lands, reductions in wildlife populations, and increases in "pest" forms of wildlife such as wild pigs (causing extensive crop damage). Other interventions in local environments included swamp drainage and the diversion for irrigation and subsequent drying up of streams on Mt. Kenya.

SEVENTY YEARS AFTER THE *KIRORE*

Today, discourse on female circumcision has been globalized and transformed into a human-rights movement aimed at eliminating such practices (Njambi 1997). According to one estimate, FC/FGM affects 114 million women throughout the world (FGM Research Homepage). The current debate reflects feminist activism and understandable fear and anger directed at patriarchy and male domination over women's bodies and their sexuality. Yet as the following discussion suggests, many women who reject FC are not passively sitting on their hands. Women and men wield different kinds of power in the struggle over FC. Based on the current situation and foreseeable landscape, men are more likely to be educated by and take their cues from women on FC than the other way around.

If, and where, FC is eradicated during the next 25 years, it will have relied on the leadership of African women and men. Only Africans will be able to lead the change and replace FC with other rituals.

In several African countries women have been steadily gaining political muscle sufficient to initiate organized, indigenous action against FC practices. There is now a vigorous FC eradication movement involving thousands of volunteers (government officials, traditional leaders, and young women in rural villages) and a network of affiliates in 26 African states and three European countries. It functions under an umbrella organization known as the Inter-African Committee on Traditional Practices Affecting the Health of Women and Children (IAC). Dozens of nongovernmental organizations and multilateral and bilateral donors provide financial assistance in support of the IAC's activities. It seeks to eradicate what it sees as harmful traditional practices (for example, both child marriage and FC) and instead to promote alternative customs.

CONCLUSION

Scholarly efforts to revitalize cultural and medical geography started to bear fruit during the 1970s and 1980s. The introduction of humanistic, structuralist, and postmodernist ideas into this process opened the door to "new" scholarship. Some cultural and medical geographers sought to "upgrade" the traditional and ostensibly more limited perspectives of their respective fields. Other scholars favored a radical break away from "orthodoxy." A minority has found it difficult to acknowledge their indebtedness to the history of ideas in cultural geography. Many have reconciled the politics of their philosophical differences by accepting the value of ideological diversity.

Cultural and medical geography have added conceptual richness and methodological sophistication during the past four decades. Together they form a hybrid with a future. Medical geography appears to have taken the lead in this relationship by systematically identifying areas of commonality, including common research themes. Such opportunities must now be intelligently exploited.

The case studies of FC and fertility change in Kenya testify to the value of historical and holistic methods when one attempts to explain health issues in landscapes. Female circumcision's centrality in many African societies enables it to exercise distinct ideological influence in the development of place and space relations. It is part of the nexus of society, ecology, landscape, and health.

Politically, the missionary assault on FC in Mwimbi provoked several forms of African resistance that affected the tangible cultural landscape, including parallel independent schools and churches. African Presbyterians forced out of the churches also lost access to missionary medicine.[11] They began to rely more openly on traditional healers. The *kirore* crisis kindled cultural nationalism at the local level and ultimately made a significant contribution to the rise of Kenyan nationalism and political independence. Missionary intervention in sexual behavior and residential patterns contributed to a demographic transformation on Mt. Kenya. The results included increases in fertility and juvenile abortion, as well as environmental interventions and land use changes, which no one even tried to imagine in advance.

This chapter outlines recent developments that have spurred convergence and innovation in cultural and medical geography. The African case materials help to illustrate this point by focusing attention on how the subtle implications of culture, behavior, and place influence the physical and psychic health of individuals and entire societies (Haggett 1994; Lewis and Rapaport 1995; Good 1998).

NOTES

1. This criticism began in the early 1960s, in keeping with a movement to redefine geography as a spatial and quantitative science.
2. Female Somali informant: A. Abubakar.
3. Personal communication, Mrs. W. Karanja (pseudonym).
4. By a common estimate, over 114 million women and girls have recently undergone some form of FC/FGM, up from 80 million in 1982.
5. KCA president Johnson Kenyatta made a highly publicized visit to England to lobby for Kikuyu rights, including FC.
6. In some locations excision included the labia minora and part of the labia majora. Europeans "socially constructed" the custom into "minor" (excision of clitoris) and "major" forms. Unlike the southern Kikuyu sphere, the Mwimbe and neighboring Chuka people of eastern Mt. Kenya leave a hole large enough to receive half a golf ball (Hyam 1990: 190).
7. By 1929 Kenyatta was already seen as "the tool of men not friendly to Britain" (Hyam 1990: 198–99, note 35).
8. The CSM (Chogoria) version of *Kirore* was: "I promise to have done with everything connected with the circumcision of women, because it is not in agreement with the things of God, and to have done with the Kikuyu Central Association, because it aims at destroying the Church of God" (Irvine 1929: 14).
9. Rev. Kabii and Rev. Bundi identified eighteen people who signed the oath (Kabii and Bundi 1991).
10. By 1939, the CSM Chogoria had been hugely successful at undermining the local culture. Referring to the "vacuum" this created, Dr. Irvine, the mission head, candidly admitted that "African life is largely social," that "the village and clan are strong unities," and people "are friendly and sociable." Regarding the church's ban on many clan-based social functions and behavior, he stated that "we are doing little to replace this social side. [The] . . . social life of the Church is a poor disconnected thing. . . . Now the parents cannot control the young, especially their evening activities" (*Kikuyu News*, March 1939: 222–23).
11. Notably at the CMS Chogoria Hospital and its outreach clinics.

REFERENCES

Annual Report. 1929. Kikuyu Province. PC/CP/4/1/1. Kenya National Archives (KNA).

Bell, D. J. 1994. "In Bed with the State: Political Geography and Sexual Politics." *Geoforum* 25 (4): 445–52.

Blaut, James M. 1993. *The Colonizer's Model of the World: Geocentric Diffusionism and Eurocentric History*. New York: Guilford Press.

Brookfield, H. 1964. "Questions on the Human Frontiers of Geography." *Economic Geography* 40: 283–303.

Brown, Michael. 1995. "Ironies of Distance: An Ongoing Critique of the Geographies of AIDS." *Environment and Planning D: Society and Space* 13: 159–83.

Caldwell, J. C. 1995. "Lack of Male Circumcision and AIDS in Sub-Saharan Africa: Resolving the Conflict." *Health Transition Review* 5 (1): 113–17.

Crossette, Barbara. 1997. "Court Backs Egypt's Ban on Mutilation." *New York Times*, December 29.

Duncan, J. 1994. "After the Civil War: Reconstructing Cultural Geography as Heterotopia." In *Re-Reading Cultural Geography*, ed. K. Foote et al., 401–08. Austin: University of Texas Press.

FGM Research Homepage. *FGM in Africa: Statistics I*. <http://www.hollyfield.org/ ″xasture/>.

Foote, Kenneth E., Peter J. Hugill, Kent Mathewson, and Jonathan M. Smith, eds. 1994. *Re-reading Cultural Geography*. Austin: University of Texas Press.

Foucault, Michel. 1984. *The Foucault Reader*, ed. Paul Robinson. New York: Pantheon Books.

Gesler, Wilbert M. 1991. *The Cultural Geography of Health Care*. Pittsburgh: University of Pittsburgh Press.

———. 1992. "Therapeutic Landscapes: Medical Issues in Light of the New Cultural Geography." *Social Science & Medicine* 34 (7): 735–46.

———. 1996. "Lourdes: Healing in a Place of Pilgrimage." *Health & Place* 2 (2): 95–105.

Good, Charles M. 1987. *Ethnomedical Systems in Africa: Patterns of Traditional Medicine in Rural and Urban Kenya*. New York: Guilford Press.

———. 1998. *Mobilizing Traditional Healers to Prevent STDs and Combat the Spread of HIV/AIDS in Tanzania: A Baseline Study*. CMG/USAID P.O. 621-0177-P-00-7097. April.

Gould, Peter. 1993. *The Slow Plague: A Geography of the AIDS Pandemic*. Cambridge: Blackwell.

Gourevitch, Philip. 1998. *We Wish to Inform You That Tomorrow We Will Be Killed with Our Families: Stories from Rwanda*. New York: Farrar, Straus, and Giroux.

Greeley, E. H. 1988. "Planning for Population Change in Kenya: An Anthropological Perspective." In *The Anthropology of Development and Change in East Africa*, ed. D. Brokensha and P. Little, 201–16. Boulder: Westview Press.

Haggett, Peter. 1994. "Geographical Aspects of the Emergence of Infectious Diseases." *Geografiska Annaler* 76B: 91–104.

Hammond, Dorothy, and Alta Jablow. 1992. *The Africa That Never Was*. Prospect Heights, IL: Waveland Press.

Harvey, David. 1973. *Social Justice and the City*. London: Arnold.

———. 1989. *The Condition of Postmodernity: An Enquiry Into the Conditions of Cultural Change*. Oxford: Blackwell.

Hochschild, Adam. 1998. *King Leopold's Ghost*. Boston: Houghton- Mifflin.

Hosken, Fran. 1993. *The Hosken Report: Genital Sexual Mutilation of Females*, 4th ed. Lexington, MA: Win News.

Hunter, John M. 1966. "River Blindness in Nangodi, Northern Ghana: A Cyclical Hypothesis of Advance and Retreat." *Geographical Review* 56: 398–416.

———. 1973. “Geophagy in Africa and the United States: A Culture-Nutrition Hypothesis.” *Geographical Review* 63: 170–95.
Hyam, R. 1990. *Empire and Sexuality: The British Experience*. Manchester: Manchester University Press.
Inter-African Committee (IAC) on Traditional Practices Affecting the Health of Women and Children. 1997. *Newsletter* No. 21.
Irvine, C. 1929. “Chogoria Letter, 28th September.” *Kikuyu News*, No. 110 (December): 14–15.
———. 1972. “Jubilee at Chogoria 1922–1972.” n.p.: Presbyterian Church of East Africa. [Kenya National Archives File: The Church at Chogoria]
Kabii, Rev., and Rev. Bundi. 1991. “Centennial History and Report of PCEA Chogoria Presbytery.” 25/3/91. Mimeo.
Kearns, Robin. 1993. “Place and Health: Towards a Reformed Medical Geography.” *Professional Geographer* 45 (2): 139–47.
Kenyatta, Jomo. 1962. *Facing Mt. Kenya*. New York: Vintage Books.
Kikuyu News. No. 112 (June, 1930): 34; No. 116 (June, 1931): 31–32; No. 147 (March, 1939): 222–23; No. 148 (June, 1939): 256.
Lange, Caroline A. 1997. “Review of *Mapping Desire: Geographies of Sexualities*, ed. D. Bell and G. Valentine. London: Routledge, 1995.” *Professional Geographer* 49 (1): 143–44.
Lewis, Nancy, and M. Rapaport. 1995. “In a Sea of Change: Health Transitions in the Pacific.” *Health & Place* 1 (4): 211–26.
Livingstone, D. 1993. *The Geographical Tradition*. Oxford: Blackwell.
Matless, D. 1996. “New Material? Work in Cultural and Social Geography, 1995.” *Progress in Human Geography* 20 (3): 379–91.
May, Jacques. 1971. *The Ecology of Malnutrition in Seven Countries of Southern Africa and Portuguese Guinea*. New York: Hafner.
Meade, Melinda, John Florin, and Wilbert Gesler, eds. 1988. *Medical Geography*. New York: Guilford Press.
Meinig, Donald W. 1979. “The Beholding Eye: Ten Versions of the Same Scene.” In *The Interpretation of Ordinary Landscapes*, ed. D. Meinig, 33–48. New York: Oxford.
———. 1982. “Geographical Analysis of Imperial Expansion.” In *Period and Place: Research Methods in Historical Geography*, ed. A. Baker and M. Billinge, 71–78. Cambridge: Cambridge University Press.
Mikesell, Marvin W. 1978. “Tradition and Innovation in Cultural Geography.” *Annals of the Association of American Geographers* 68: 1–16.
Moon, Graham. 1995. “(Re)placing Research on Health and Health Care.” *Health & Place* 1 (1): 1–4.
Nelson, Jill. 1998. “Review of Fauziya Kassindja, *Do They Hear You When You Cry?*” *New York Times*, April 5: 13.
Njambi, Wairimu Ngaruiya. 1997. “Colonizing Bodies? Representations of Genital Mutilation, Bodies, and Sexuality.” Ph.D. dissertation proposal, Science and Technology Studies, Virginia Tech (Dec. 22).
Presley, Cora. 1992. *Kikuyu Women, the Mau Mau Rebellion, and Social Change in Kenya*. Boulder: Westview Press.
Prothero, R. M. 1965. *Migrants and Malaria*. Pittsburgh: University of Pittsburgh Press.

Public Health Reports. 1997. "Female Genital Mutilation/Female Circumcision: Who is at Risk in the U.S.?" *Public Health Reports* 112 (5): 368–77.

Pugh, Judy. 1996. "Many Medicines, Many Voices." *Research in Anthropology* 25 (3): 167–74.

Roundy, Robert. 1987. "Human Behavior and Disease Hazards in Ethiopia: Spatial Perspectives on Rural Development." In *Health and Disease in Tropical Africa: Geographical and Medical Viewpoints*, ed. Rais Akhtar, 261–78. Chur/London: Harwood.

Sauer, Carl. 1925. "The Morphology of Landscape." *University of California Publications in Geography* 2 (2): 19–53.

Stock, Robert. 1986. "Disease and Development or the Underdevelopment of Health." *Social Science & Medicine* 23: 689–700.

———. 1995. *Africa South of the Sahara: A Geographical Interpretation*. New York: Guilford.

Strayer, Robert W. 1978. *The Making of Mission Communities in East Africa*. London: Heinemann.

Symanski, Richard. 1981. *The Immoral Landscape: Female Prostitution in Western Societies*. Toronto: Butterworths.

Wagner, Philip L. 1994. "Foreword: Culture and Geography: Thirty Years of Advance." In *Re-Reading Cultural Geography*, ed. K. Foote et al., 3–8. Austin: University of Texas Press.

Wagner, Philip L., and Marvin W. Mikesell, eds. 1962. *Readings in Cultural Geography*. Chicago: University of Chicago Press.

11

Language and Identity in Russia's National Homelands: Urban–Rural Contrasts

Chauncy D. Harris

If one compares an older with a newer cultural geography, at least as exemplified in the pioneering work *Readings in Cultural Geography* (Wagner and Mikesell 1962) and in the more recent collection *Re-reading Cultural* Geography (Foote et al. 1994), one might gain the impression that the newer cultural geography is devoting less attention to ethnicity, languages, and religions. The earlier volume contained significant articles on cultural boundaries and ethnographic maps (Weiss 1962), on geography of language (Delgado de Carvalho 1962), and on geography of religions (Fickeler 1962), whereas the later volume contains no articles devoted to these fundamental characteristics of culture and society. But, as James Parsons (1994: 284) observes in the latter volume, "The persistence of cultural grouping based on ethnicity, language, religion, territory, and a common history and way of life is emerging as a powerful new political reality."

Marvin Mikesell has displayed a persistent interest in ethnicity in his teaching, conversations, and publications. It has been my great privilege during the last forty years to be a colleague of Mikesell's and to have had an office near his. In hundreds of informal conversations our most common topic has been some aspect of ethnicity, language, and religion as complex regional and cultural phenomena closely tied to other characteristics in history, politics, and society. This interest found fruitful expression in Mikesell's seminal articles "The Myth of the Nation State" (1983) and "Culture and Nationality" (1985) as well as in a joint paper "A Framework for Comparative Study of Minority Group Aspirations" (Mikesell and Murphy 1991). His student Alexander B. Murphy also wrote an exemplary study on the cultural geography of language in Belgium (Murphy 1988).

Furthermore, ethnicity, language, and religion have moved to center stage in cultural and political geography as factors in the breakdown or reunion of countries. In Europe alone the 1990s witnessed the breakup of several countries (Harris 1993c). The former Yugoslavia splintered into five countries. Czechoslovakia split into two countries. The former Soviet Union shattered into 15 separate countries (Harris 1993a, 1993b, 1994). Within the Russian Federation itself 32 non-Russian national republics or similar units based on ethnicity and associated language and religion were recognized (Harris 1993d). Germany was reunited (Harris 1991).

American cultural geographers, however, have devoted little attention to the former Soviet Union or to the current Russian Federation. Perhaps this reflects in part the absence of a field of cultural geography in the former USSR, where geographical studies have focused mainly on economic and physical geography. Some American geographers, notably Robert Kaiser (1994; Chinn and Kaiser 1996), Ralph Clem (1980), and Ronald Wixman (1980, 1984), have contributed significantly to ethnic studies in this area, but, with the exception of Wixman, they do not identify themselves as cultural geographers. Special mention should be made of Kaiser, who has developed the concept of the interactive contingent nature of the emergence of nationalism in the former Soviet Union and the contemporary Russian Federation.

Among numerous ethnic studies of this area, mention may be made particularly of works by geographer V. Pokshishevskiy (1969); ethnographers such as Viktor Kozlov (1988) and M. Guboglo (1991, 1992–1993, 1993, 1994); political scientists such as Jerry Hough (1996), David Laitin (1995, 1996, 1998), Gail Warshofsky Lapidus (Lapidus and Zaslavsky 1992), Brian Silver (1974a, 1974b, 1976, 1986), Mark R. Beissinger (1995), Ian Bremmer and Raymond Taras (1997); historians such as Ronald Grigor Suny (1995); sociologists and demographers such as Barbara Anderson (Anderson and Silver 1990), Michael Paul Sacks (1995) and Susan Goodrich Lehmann (1997, 1998); and others (Drobizheva 1994; Drobizheva et al. 1996).

THE FORMER SOVIET UNION AND THE RUSSIAN FEDERATION

In its ethnic complexity the former Soviet Union was matched only by India. In the last census taken in the USSR in 1989, Russians represented only 50.8 percent of the population, or 145 million out of a total population of 286 million. Politically the country was a federal union made up of 15 union republics based on 15 major nationalities or ethnic groups. These union republics possessed a territorial base, a legal political administrative status, and a titular national group, which had an evolved sense of nationality. The titular group represented a clear majority of the population in each union republic, except in Kazakhstan, in which Russians then outnumbered Kazakhs but neither group had a clear majority. On December 31, 1991, the Soviet Union disintegrated into 15 independent countries.

The new political arrangement reduced the ethnic complexity greatly, but as Marvin Mikesell first called to my attention, the famous George Orwell (1968, vol. 4:282)[1] noted that "Whenever A is oppressing B, it is clear to people of goodwill that B ought to be independent, but then it always turns out that there is another group C, which is anxious to be independent of B." In the case of the successor states to the former Soviet Union, more than 25 million Russians overnight were transformed from being members of a dominant nationality in the Soviet Union into members of a minority in 14 separate independent non-Russian countries (Harris 1993a; Shlapentokh, Sendich, and Payin 1994; Kolstoe 1995; Zevelev 1996; Bremmer and Taras 1997; Laitin 1998).

The other major complexity is that within the Russian Federation itself 27 million non-Russians constitute a substantial minority. The more prominent of these minorities, which are concentrated in particular territories, have been given a degree of cultural and political autonomy in 32 ethnic homelands but not outside of them. Two types of minorities within Russia do not have such a territorial-administrative base nor autonomy. Some with large numbers are dispersed members of the groups that were the titular nationalities of the former union republics, now independent countries; the most numerous are Ukrainians (4.4 million), Belorussians (1.2 million), Kazakhs (0.6 million), and Armenians (0.5 million). Others also have no such homeland in Russia such as the Germans, gypsies, and Koreans. The Jews have a nominal symbolic homeland, but it is remotely located in the Far East. Far removed from the main concentrations of Jewish population, this region is home to only about 1 percent of the Jewish population in Russia, and Jews constitute a mere 4 percent of its population.

Most studies of nationalities in the former Soviet Union have been devoted to those non-Russian nationalities that were formerly the titular nationalities in union republics and are now the dominant titular nationality in 14 different independent countries. Far fewer studies have been devoted to the minorities within the Russian Federation itself, which this chapter examines. The relations between Russians and non-Russians is one of the most urgent problems facing the Russian Federation today (Drobizheva 1994; Radvanyi 1996; Stadelbauer 1996). Kaiser (1995) suggests that ethnicity is so powerful a force that it must be considered in an analysis of prospects for the disintegration of the Russian Federation. Sacks (1995) demonstrated that ethnicity is associated with different levels of participation in the workforce and in educational attainment. Kaz'mina and Puchkov (1994) have surveyed the ethnodemography of the Russian Federation and Guboglo (1991) and associates have analyzed ethnic processes.

URBAN–RURAL DIFFERENCES

Virtually no attention at all has been paid to the sharp urban–rural differences in ethnic composition (Harris 1945). This chapter attempts to fill this gap by an analysis of urban–rural differences in the ethnic composition of the population within the Russian Federation, especially within its 32 ethnic homelands, and of the significance of these differences.

Specifically, the degree of urbanization of a national or ethnic minority is an indicator of a way of life, of the relative participation of that group in socioeconomic development, of the relative access of the group to higher-status and higher-income occupations and a higher standard of living, but also of social and cultural development, education, better health care, integration into a larger community, acculturation through knowledge of the Russian language, and even potential assimilation (Clem 1980: 23; Stadelbauer 1996: 198). Pokshishevskiy (1969) long ago noted relationships between both function and size of cities and their ethnic composition. Ruble (1989) called for more attention to the role of the city in heightened ethnic awareness, noting that city size is an important variable determining ethnic and social status. Kaiser (1994: 195–225) analyzed urbanization of various ethnic groups in the former Soviet Union, particularly trends over time in successive censuses (1959–1989), and Silver (1986) summarized the census sources of ethnic data. Kozlov (1988) addressed ethnic aspects of urbanization, noting particularly factors in urban–rural differences in ethnic composition. Guboglo (1993) studied the ethnographic and language structure of the capitals of the former union republics of the USSR. Still relatively little attention has been paid to the capitals of republics within the Russian Federation.

NATIONAL HOMELANDS OF THE RUSSIAN FEDERATION

The Russian Federation, one of the 15 successor states of the former Soviet Union, has within its own boundaries 32 major ethnic homelands that now have a legal-territorial basis. These are organized into 21 republics, 10 autonomous okrugs, and one autonomous oblast. Together they cover 53 percent of the total land area of the country in the vast but sparsely settled expanses in northern Siberia, but including also densely settled areas in the Middle Volga region, in the North Caucasus, and in parts of southern Siberia. Their locations and boundaries are depicted in Figure 11.1 and their ethnic composition in Table 11.1. The table includes only 31 areal units, since the separate Chechen and Ingush republics are grouped together in the table (area 14 in Table 11.1), as they were at the time of the 1989 census. They were separated only in 1992, and because of unsettled political and military conditions the boundary between them has not yet been demarcated. Data are thus not available for them separately.

The present study is focused on census data for 1989. It is primarily a statistical analysis: (1) of each ethnic homeland in the Russian Federation and the percentage which each major nationality forms of the rural and urban population (Table 11.1); and (2) of the urban–rural and homeland–outland contrasts in degree of shift to use of Russian as mother tongue for each non-Russian minority

Figure 11.1 The Ethnic Homelands in the Russian Federation.

Table 11.1 Ethnic Groups within Homelands of the Russian Federation, Percentage of Rural and Urban Population, 1989.

Political Unit / *Ethnic Group*	*No.*	*Percent Rural*	*Percent Urban*
Homelands in North European Russia			
1. Karelian R	790,150	100	100
Karelians	78,928	21	8
Russians	581,571	58	77
Belorussians	55,530	11	6
2. Mordvin R	963,504	100	100
Mordva	313,420	46	22
Russians	586,147	47	72
3. Chuvash R	1,338,023	100	100
Chuvash	906,922	86	55
Russians	357,120	9	40
4. Mari-El R	749,332	100	100
Mari	324,349	70	26
Russians	355,973	23	63
5. Tatarstan R	3,361,742	100	100
Tatars	1,765,404	66	42
Russians	1,575,361	23	51
6. Udmurt R	1,605,663	100	100
Udmurts	496,522	57	20
Russians	945,216	38	68
Tatars	110,490	3	8
7. Bashkortostan R	2,943,113	100	100
Bashkirs	863,808	35	15
Tatars	1,120,602	33	26
Russians	1,548,291	18	51
8. Komi-Permyak AOk	158,526	100	100
Komi-Permyaks	95,415	64	51
Russians	57,272	32	46
9. Komi R	1,250,847	100	100
Komi	291,542	51	14
Russians	721,780	37	64
Homelands in the North Caucasus			
10. Adygey R	432,046	100	100
Adygey	95,439	31	14
Russians	293,640	60	75
11. Karachay-Cherkess R	414,970	100	100
Karachay	129,449	42	19
Cherkess	40,241	13	6
Russians	175,931	26	60
Abaza	27,475	8	5

(Table 11.1—*Continued*)

Political Unit Ethnic Group	No.	Percent Rural	Percent Urban
12. Kabardino–Balkar R	753,531	100	100
Kabards	363,494	71	34
Balkars	70,793	10	9
Russians	240,750	14	43
13. North Osset R	632,428	100	100
Ossets	334,876	61	49
Russians	189,159	19	35
Ingush	32,783	10	3
14. Chechen R and Ingush R	1,270,429	100	100
Chechens	734,501	74	35
Ingush	163,762	14	11
Russians	293,771	8	45
15. Dagestan R	1,802,188	100	100
Avars	496,077	34	20
Dargins	280,431	19	11
Kumyks	231,805	12	14
Lezgins	204,370	12	10
Laks	91,682	3	7
Tabasarans	78,196	5	3
Russians	165,940	2	18
16. Kalmyk–Khal'mg Tangch R	322,579	100	100
Kalmyks	146,316	42	49
Russians	121,531	34	42
Homelands in Siberia			
17. Altay R	190,831	100	100
Altay	59,130	38	12
Russians	115,188	52	82
Kazakhs	10,692	7	2
18. Khakass R	566,861	100	100
Khakass	62,859	26	5
Russians	450,430	64	86
19. Tuva R	308,557	100	100
Tuvinians	198,448	85	41
Russians	98,831	14	53
20. Ust'-Orda Buryat AOk	135,870	100	100
Buryats	42,298	37	31
Russians	76,827	55	63
21. Buryat R	1,038,252	100	100
Buryats	249,525	33	17
Russians	726,165	62	76
22. Aga Buryat AOk	25,134	100	100
Buryats	42,362	66	33
Russians	31,473	31	52
23. Jewish AO	214,085	100	100
Jews	8,887	1	6
Russians	178,087	81	84
Ukrainians	15,921	11	5

(Table 11.1—*Continued*)

Political Unit *Ethnic Group*	*No.*	*Percent Rural*	*Percent Urban*
24. R. Sakha (Yakutiya)	1,094,065	100	100
Yakuts	365,236	75	13
Russians	550,263	15	68
Ukrainians	77,114	2	9
Homelands of Peoples of the North			
25. Nenets AOk	53,912	100	100
Nentsy	6,423	29	2
Russians	35,489	49	76
Komi	5,124	16	6
26. Yamal-Nenets AOk	494,844	100	100
Nentsy	20,917	17	1
Khanty	7,247	6	0
Russians	292,808	45	63
Ukrainians	85,022	13	18
Tatars	26,431	5	6
27. Khanty-Mansi AOk	1,282,396	100	100
Khanty	11,892	7	0
Mansi	6,562	3	0
Russians	850,297	66	66
Ukrainians	148,317	8	12
Tatars	97,689	4	8
28. Taymyr (Dolgano–Nenets) AOk	55,803	100	100
Dolgans	4,939	24	1
Nentsy	2,446	12	1
Russians	37,438	44	79
Ukrainians	4,816	7	10
29. Evenki AOk	24,769	100	100
Evenki	3,480	16	10
Russians	16,718	65	73
Ukrainians	1,303	5	6
30. Chukotsk AOk	163,934	100	100
Chukchi	11,914	24	1
Russians	108,297	50	72
Ukrainians	27,600	13	18
31. Koryak AOk	39,940	100	100
Koryaks	6,572	21	9
Chukchi	1,460	6	0
Russians	24,773	54	74
Ukrainians	2,896	7	8

R = Republic; AO = Autonomous oblast; Aok = Autonomous okrug.

Source: Compiled by the author from data in Goskomstat RSFSR, *Natsional'nyy Sostav Naseleniya RSFSR po Dannym Vsesoyuznoy Perepisi Naseleniya 1989 g.* (Moscow: Respublikanskiy Informatsionno-Izdatel'skiy Tsentr, 1990), table 11, 102–53.

(table 11.2). Table 11.1 is arranged by political-administrative territorial units and Table 11.2 is arranged by ethnic groups. The data have been calculated by the author from information in the 1989 census, variously published (Goskomstat RSFSR 1990; Goskomstat SSSR 1991; Goskomstat SNG 1992–1993; Tishkov 1994).

In order to aid regional analyses, the ethnic homelands and national minorities in Tables 11.1 and 11.2 have been arranged by four areal groups: North European Russia and the Middle Volga, the North Caucasus, Siberia, and Peoples of the North.

Failure to take account of urban–rural differences is dramatically illustrated by Russian military operations in the Chechen Republic. Here Groznyy, the capital, had a majority Russian population (53 percent in the 1989 census), whereas the rural population was overwhelmingly non-Russian. The proportion of the Chechen rural population is doubtless well over 90 percent with the percentage of Russians less than 10 percent (exact figures are not available because of boundary changes since the census). Control of the city of Groznyy, with its majority Russian population, does not ensure control of the rural areas, where Chechen fighters are protected by the native Chechen population in the villages. Russian attacks on Groznyy may well have killed more Russian than Chechen civilians. Russian troops were not able to establish control over the Chechen rural areas during the bitter military operations in the civil war of 1991–1997. The political and economic background of the problems of the republic has been discussed by Bond and Sagers (1991).

Most of the ethnic homelands have complex ethnic structures in which the titular nationalities, though present in significant numbers, do not generally represent a clear majority of the total population. Indeed the titular nationalities constitute more than 50 percent of the population in only seven out of the 32 homelands: Chuvash and Komi-Permyaks in the Middle Volga area; Ossets, Chechens, and peoples of Dagestan in the North Caucasus; Buryats in Aga Buryat Autonomous Okrug (AOk); and Tuvinians, both in the southern part of Siberia. Dagestan is a special complex area in which Russians constitute only 2 percent of the population, but the other 98 percent includes an amazing variety of peoples. What these seven areas have in common is the absence of resources to attract large numbers of Russians. They are economic backwaters on the periphery of the country and lack developed mineral resources or major industry.

The most significant generalization to be drawn from the data in Table 11.1 is that minorities are much more rural than the Russians. In the 32 ethnic homelands the national groups form a higher percentage of the rural than of the urban population; two exceptions reflect very special circumstances as noted below (Table 11.1). This typical and widespread pattern can be illustrated by figures for selected republics. In northeast Siberia, in Sakha Republic (Yakutiya), for example, Yakuts form 75 percent of the rural population but only 13 percent of the urban population. In the Tuva Republic in Eastern Siberia on its southern border with Mongolia, Tuvinians make up 85 percent of the rural population but only 41 per-

Table 11.2 Percentage of Each Ethnic Group in the Russian Federation Using Russian as Mother Tongue, 1989

	Inside Homeland		*Outside Homeland*	
Ethnic Group	*Rural*	*Urban*	*Rural*	*Urban*
Peoples of European Center and North				
1. Karelians	35	56	40	66
2. Mordva	4	23	24	50
3. Chuvash	2	31	18	40
4. Mari	5	23	15	36
5. Tatars	1	5	8	25
6. Udmurts	12	39	26	50
7. Bashkirs	1	10	9	27
8. Komi-Permyaks	11	36	45	52
9. Komi	11	42	42	55
Peoples of the North Caucasus				
10. Adygey	1	3	8	18
11a. Karachay	0	2	6	16
11b. Cherkess	0	5	9	26
12a. Kabards	0	2	13	26
12b. Balkars	0	3	19	30
13. Ossetians	1	2	19	33
14a. Chechens	0	1	2	10
14b. Ingush	0	1	3	9
15a. Avars	0	2	5	17
15b. Darghins	0	2	2	16
15c. Lezghins	0	2	11	22
15d. Kumyks	0	1	3	15
15e. Laks	0	3	9	21
15f. Tabasarans	0	2	7	15
16. Kalmyks	2	6	22	36
Peoples of Siberia				
17. Altay	9	26	33	45
18. Khakass	8	26	47	50
19. Tuvinians	0	2	13	13
20–22. Buryats	—	—	20	38
20. Ust-Orda Buryat AOk	9	14	—	—
21. Buryat R	3	20	—	—
22. Aga Buryat AOk	1	4	—	—
23. Jews	87	88	75	91
24. Yakuts	2	15	36	29
Peoples of the North				
25, 26, 28a. Nentsy	—	—	39	54
25. Nenets AOk	34	69	—	—
26. Yamal-Nenets AOk	3	16	—	—
28a. Taymyr AOk	16	52	—	—

Ethnic Group	Inside Homeland		Outside Homeland	
	Rural	Urban	Rural	Urban
27a. Khanty	38	59	23	52
27b. Mansi	54	76	46	69
28b. Dolgans	7	34	23	43
29. Evenki	22	41	24	47
30. Chukchi	25	52	22	51
31. Koryaks	43	57	36	64

Sources: Calculated by the author from data in the following publications of the 1989 census: Goskomstat RSFSR, *Natsional'nyy Sostav Naseleniya RSFSR po Dannym Vsesoyuznoy Perepisi Naseleniya 1989 g.* (Moscow: Respublikanskiy Informatsionno-Izdatel'skiy Tsentr, 1990), table 11, 102–53, and table 12, 160–737. Goskomstat SSSR, *Natsional'nyy Sostav Naseleniya SSSR po Dannym Vsesoyuznoy Perepisi Naseleniya 1989 g.* (Moscow: Finansy i Statistika, 1991), 34–49. Goskomstat SNG, *Itogi Vsesoyuznoy Perepisi Naseleniya 1989 goda.* (Minneapolis: East View Publications, 1992–1993), vol. 7, part 1, table 3, 66–73, 78–83, 90–95, and table 4, 104–683.

Note: The figures outside the homeland have been calculated by subtracting the numbers of each ethnic group inside the homeland from the total number of each ethnic group in the entire Russian Federation and then calculating the percentages. Only the resulting percentages are presented in this table.

cent of the urban population. In the middle Volga area in European Russia four republics also illustrate these differences. In Tatarstan, Tatars make up 66 percent of the rural population but only 42 percent of the urban. Bashkirs form 35 percent of the rural population but only 15 percent of the urban population in Bashkortostan. In the nearby Mari-El Republic the Maris make up 70 percent of the rural population but only 26 percent of the urban. In the Udmurt Republic, Udmurts constitute 57 percent of the rural population but only 20 percent of the urban. In the Komi Republic in extreme northeast European Russia, Komis form 51 percent of the rural but only 14 percent of the urban population.

The two exceptions to this persistent and widespread pattern of minority concentration in rural areas are the Jewish Autonomous Oblast in Eastern Siberia and the Kalmyk-Khal'mg Tangch Republic near the mouth of the Volga River in the northwest Caspian Sea.

In the Jewish Autonomous Oblast the Jews were not significant rural residents before the creation of this symbolic homeland. Not only are the Jews predominantly urban throughout Russia, but those in this homeland were actually immigrants in the Soviet period from other areas; thus Jews form only 6 percent of the urban population and only 1 percent of the rural population.

In contrast in dry steppe lands of the Kalmyk-Khal'mg Tangch Republic, the Kalmyks, a Mongolian people, mainly Buddhists, are traditionally herdsmen. How did they become more urbanized? Their 1989 greater relative concentration in urban areas reflects tragic events during the Soviet period. Great losses of life took place in the rural areas during the Civil War and during collectivization, mainly as a result of their deportation to Siberia during World War II in 1943. Those who remained were allowed to return to Kalmykia only after 1956 and then mainly to cities, as other peoples had been settled on their farms and pasturelands. Thus the percentage of Kalmyks in Kalmykia dropped from 76 percent in 1926 to 35 percent in 1959. During this same period the percentage of Russians rose from 11 to 56 percent, though thereafter the proportion of Kalmyks rose somewhat and the proportion of Russians declined.

Thus, in 12 of the homelands the titular nationality constitutes a majority of the rural population, but in only two is the group for which the region is named actually a majority of the urban population.

TYPES OF URBAN AREAS

In areas with a substantial non-Russian population, both rural and urban, which have been incorporated into the Russian state over the course of history, one might expect on the basis of general theory to find a regular gradation with decreasing percentages of indigenous population from rural areas, to smaller urban centers, to the largest city, typically the capital. The Russian unifying imperial power, originally external, bringing new political organization and economic development, would have its strongest impact on the capital, less on smaller urban centers, and least in rural areas. Thus the proportion of Russians would be expected to be highest in the capital of each ethnic homeland, at the top of the central-place hierarchy, and to be less well represented in the smaller urban centers, with a lower order in the central-place hierarchy.

The expected patterns are well marked in areas with a high density of non-Russian population and with a well-established settlement pattern antedating the Russian conquest, as in the North Caucasus. For example, the percentages of the titular nationality in the population of the Adygey Republic decreased regularly from 31 percent in rural areas to 20 percent in urban population outside the capital to 11 percent in the capital, Maykop. Similar percentages for Karachay in the nearby Karachay-Cherkess Republic and its capital (Cherkessk) were 42–34–8, for Ossets in North Ossetia, 61–56–46 (Vladikavkaz), and for Chechen in the former Chechen-Ingush Republic 74–49–31 (Groznyy).

Similar regular steps characterize also the titular nationalities in republics of old dense non-Russian settlements of the Middle Volga, Mari-El 70–29–23 (Yoshkar-Ola), Mordovia 46–25–20 (Saransk), Bashkortostan 35–17–11 (Ufa), and Udmurt Republic 57–23–17 (Izhevsk), and in Northwest European Russia: Karelia 21–9-5 (Petrozavodsk).

The relative proportion of Russians and non-Russians in cities is related not only to size but also to function. In general the proportion of Russians increased and the proportion of titular nationalities decreased with city size. But industrial cities and smaller unifunctional towns (settlements of urban type) had relatively high proportions of Russians and low proportions of titular nationalities. These smaller Russian cities and towns set in non-Russian rural areas are typically mining and industrial cities developed in the Soviet period.

This entirely different picture emerges in areas of Russian expansion into sparsely settled areas without a dense pre-existing settlement pattern, as in the vast forests of the north. The Komi Republic in northeast European Russia is a good example. Komi form 51 percent of the rural population, only 8 percent of the urban population outside the capital, and 34 percent in the capital. The vigorous Soviet thrust for minerals has produced a whole series of small noncentral places, specialized mining towns in which Russians form the bulk of the population, overtopping the percentage in the political capital, Syktyvkar, which lies in the southern part of the republic peripheral to the main centers of production of coal, oil, and natural gas.

Even more striking is the pattern produced by Soviet development of the oil and gas of the northern part of West Siberia. Here the indigenous population is very small and widely scattered, forming only a tiny part even of the rural population. All of these peoples are virtually absent from the oil and natural gas producing towns. Russians, together with Tatars and Ukrainians, form the bulk of the urban population of the mining towns.

In a number of other ethnic units of Siberia and its Northlands, minorities also form a lower percentage of the population in the specialized mining, electricity-generating, and manufacturing towns than in the more diversified capitals. Thus the percentage of Khakass in the Khakass Republic is 26 percent in rural areas, only 3 percent in smaller towns, and 9 percent in the capital (Abakan). Similar percentages characterize the Buryats in Ust'-Orda Buryat Autonomous Okrug (AOk) and Aga Buryat AOk. Particularly significant are the figures for the Yakut Republic, where the Yakuts form 75 percent of the rural population, only 9 percent for smaller mining towns, but 25 percent in the capital (Yakutsk).

DEGREE OF URBANIZATION OF MINORITIES INSIDE AND OUTSIDE THEIR NATIONAL HOMELANDS

On the basis of theory one would expect that members of an ethnic group who have left their homeland would be more urban than those who remain in the homeland. Ethnic groups in their homeland retained a substantial rural base even though during the Soviet period these groups experienced substantial urbanization. Those who have migrated to areas outside the homeland in search of economic opportunities would be expected to be predominantly urban because economic development in the Soviet period took place mainly in the cities.

Of the 37 ethnic groups in the ethnic homelands, 32 have a higher percentage of the ethnic urban population outside the ethnic homeland than inside it. The five exceptions are all small groups with only slightly higher figures outside of the ethnic homeland.

For the numerically prominent peoples in the Middle Volga region, the Bashkir are 42 percent urban in Bashkortostan and 64 percent urban outside the homeland, and the Mordva 38 percent urban inside the Mordov Republic and 58 percent outside. For the little-developed Komi-Permyaks the corresponding figures are 25 and 67 percent.

The peoples of the densely settled North Caucasus have similar contrasting percentage figures for the urban population inside and outside the homeland. The Adygey are 33 percent urban inside and 70 percent urban outside. Karachay, the Cherkess, the Kabards, and the Balkars have similar figures. For Siberian peoples the figures are 11 and 62 percent for the Altay, 30 and 80 percent for the Tuvinians, and 26 and 75 percent for the Yakuts.

URBAN–RURAL CONTRASTS IN DEGREE OF SHIFT TO USE OF THE RUSSIAN LANGUAGE BY NON-RUSSIAN MINORITIES

Ethnic minority parents face many language problems, as, for example, in choice of schools, whether in Russian or the national language, one emphasizing economic and social mobility, the other preservation of the national heritage. Language usage may be very complex. An individual may use the national language at home and on social occasions but use Russian in shopping and at work. In rural areas there may be little occasion to use Russian, as one's indigenous neighbors may all be members of the same nationality. In cities with mixed ethnic structures, in advanced schools, and in the workplace, Russian may be the predominant language. In mixed marriages Russian usually becomes the language of the home. In the political processes in national homelands one may use Russian or the national language depending on the circumstances.

Urban–rural contrasts exist among many economic, political, social, and educational characteristics of non-Russian minorities in Russia, as well as among the Russians themselves. Soviet census data are particularly detailed on language usage, which may serve as a useful surrogate for many characteristics associated with the participation of national minorities in the processes of economic development. Ability to speak Russian is crucial to participation in the higher levels of education, to high-status occupations, to higher wages and salaries, and the ability to work and function throughout the country. One Soviet ethnographer asserted that "full integration into urban life required not only a knowledge of the Russian language, but mastery of it . . . as the working language for the majority of qualified occupations, by which the normal process of urban social and professional mobility was determined." (Kozlov 1988: 56). Knowledge of Russian is a key to both social and geographic mobility. As Anderson and Silver (1990) have emphasized, the various processes of economic development, acculturation, and assimilation are complex and interrelated. Silver (1974b; 1976: 412) has measured the effect of religion on assimilation; in general ethnic groups that are Orthodox assimilate at much higher rates than those that profess Islam.

Data on knowledge of Russian as a second language are highly subjective, but data on use of Russian as the language spoken at home (the mother tongue) are likely to be far more reliable. Members of ethnic minorities in the Russian Federation generally develop some knowledge of Russian through education or participation in the workforce. An advanced stage of knowledge of Russian is a very significant step in adopting Russian rather than the language of the ethnic group as the language spoken at home. Taking all non-Russian minorities within the Russian Federation together, the proportion who have shifted from the language of the minority to Russian as the mother tongue is 14 percent among the rural population and 37 percent among the urban population (Harris 1993d: 572). Thus among the ethnic minorities two and a half times as high a proportion of the urban members as of the rural members have shifted to speaking Russian at home. A shift to the use of Russian as the language of the home is a significant indicator of Russian acculturation and a step toward possible assimilation.

On the basis of theory one would expect the level of usage of Russian as the language of the home among non-Russians to be positively associated with both urbanization and location outside the ethnic homeland. In a pioneering study based on the 1959 census, Silver (1974a) measured these relationships. Three relationships in the 1989 census as measured in the current study may be noted: (1) inside the ethnic homeland in all 41 ethnic groups recorded in Table 11.2, Russian language usage is higher in urban than in rural areas, and usually is much higher; (2) outside their ethnic homelands, all these groups also have higher rates of Russian acculturation in urban than in rural areas, except for the Yakuts; and (3) the 37 non-Russian ethnic groups for which data are presented in Table 11.2 have higher rates of home use of Russian outside the ethnic homeland than within it, except for four peoples of the North and the Jews, all with very small numbers in their homelands overwhelmed by a flood of Russians. The rates of home use of Russian are usually much higher outside the homeland than within it, generally two to five times as high among peoples of the Volga and European North and about ten times as high among the peoples of the North Caucasus.

In general there is a stepwise increase in percentage of home usage of Russian among the non-Russian ethnic groups from rural homeland to urban homeland to rural areas outside the homeland to urban areas outside the homeland with the extreme contrast between rural within homeland to urban outside.

Thus among Tatars, the most numerous group and one with a long history of contact with Russians, the percentages using Russian as the home language rises from 1 percent among the rural population within Tatarstan, to 5 percent among the urban population within this area, to 8 percent for rural Tatars outside Tatarstan, to 25 percent for urban Tatars outside Tatarstan. Language usage in cities of

Tatarstan has been studied by Hough (1996: 99–104) on the basis of an interview survey.

In contrast, particularly high figures for the shift to the Russian language are recorded for the Karelians, who have been within the Russian Empire for a long time. High figures also characterize other peoples of the Volga and the European North. The peoples of Siberia and of the North generally have rather high figures.

All the North Caucasus peoples, conquered much later, have very low figures except for urban dwellers outside the homelands. Among the Chechens, currently struggling for independence, the percentages are zero for rural Chechenya, 1 percent for urban Chechenya, 2 percent for rural Chechens outside the homeland and 10 percent for urban Chechens outside the homeland. Among the Tuvinians, who joined the Soviet Union only in 1944, the corresponding figures are also low: 0, 2, 13, and 13 percent. The effects of urbanization and residence outside the homeland are particularly strong when they are combined. Thus the main spatial contrast within the Russian Federation is between the peoples of the North Caucasus who stoutly resist Russian acculturation, especially within their homelands, and the minority ethnic groups in most other parts of the Russian Federation with much higher percentages of adoption of the Russian language.

Urban–rural and homeland–outland contrasts are remarkably persistent over time as revealed by a comparison between data in the 1959 and 1989 censuses.

The complexities in the Tatarstan and Bashkortostan republics on the Middle Volga (numbers 5 and 7 in Table 11.1 and Figure 11.1), by far the two most populous national republics within the Russian Federation, deserve special attention. They have been investigated in detail in studies by Guboglo (1992–1993, 1994) and Hough (1996). The usage of Russian, Tatar, or Bashkir, for example, shows remarkable diversity in education, work, social discourse, newspapers read, radio listened to, and television watched. Attitudes toward the status of the titular language also differ by nationality.

Urban–rural differences are well illustrated in the six republics on the Middle Volga. Paired percentages of each national group within its homeland speaking Russian as mother tongue for rural areas and for urban areas are 4 and 23 percent for the Mordva, 2 and 31 percent for the Chuvash, 5 and 23 percent for the Mari, 12 and 39 percent for the Udmurts, 1 and 5 percent for the Tatars, and 1 and 10 percent for the Bashkirs (Table 11.2).

Figures for the Middle Volga republics also reveal the profound effect of religion. In the first four cases, all Russian Orthodox in religion, the percentages of the urban dwellers who have shifted to Russian as the mother tongue are 23, 31, 23, and 39 percent, whereas for the last two, the Tatars and Bashkirs, who profess Islam, the corresponding figures are much lower, 5 and 10 percent. Lehmann (1997, 1998) has recently discussed such contrasts.

Use of the Russian language among minorities shows an inverse relationship to age. The younger a person, the more likely he or she is to use Russian in the home.

CONCLUSION

Ethnic composition and language retention among the non-Russian ethnic groups of the Russian Federation exhibit clear spatial patterns. These ethnic groups are to a considerable extent localized in specific places, recognized by political-administrative units, generally called republics, where the groups have special rights in language usage, education, and many social, political, and cultural activities that constitute a distinctive way of life.

Within ethnic homelands, sharp urban–rural differences are characteristic, with the titular ethnic group regularly forming a higher percentage of the rural than of the urban population. Among urban settlements in these areas, two types are recognized. In areas already densely settled at the time of Russian imperial conquest, a gradation exists with the titular ethnic group, regularly increasing in proportions of the population from capital city to other urban settlements to rural areas. In areas that were and are sparsely settled, however, the titular ethnic groups represent a lower percentage of the population in non-capital cities and towns than in the capital, since in such areas these smaller urban settlements generally are industrial or mining towns dominated by Russian laborers who have come into the area as part of an industrialization process.

Shifts among non-Russian ethnic groups to the use of the Russian language in the home, a form of acculturation and a possible step toward assimilation, show a regular pattern, being highest in capital cities, lower in other urban centers, and lowest in rural areas. Such shifts also vary spatially over the country, being significantly higher among members of the ethnic group living outside the homeland area than those within their own republic. A religious factor is also revealed in that language shifts are much higher among groups that have become Orthodox Christian than among those who profess Islam.

The process of language shift has been unfolding over time as revealed by the regular increase in the proportions of members of non-Russian ethnic groups who have shifted to the use of Russian with each increasingly younger cohort group. As effectively described by Kaiser (1994), ethnic self-identity and assertiveness result from an interactive process in the contact and conflicts among different ethnic groups.

The possibilities in the new cultural geography in the analysis of language shifts are well illustrated by the imaginative research by Laitin (1995, 1996, 1998), Hough (1996), and others of attitudes, expected economic returns, perceived social status, and local and job requirements, combined with concrete case studies of individual families. These studies did not, however, directly investigate urban–rural differences, contrasts between minorities resident inside or outside the designated ethnic homelands, or differences between central-place capital cities and smaller special-function mining and industrial towns within the Russian Federation. This chapter has attempted to suggest some of the significant features of these variables.

The Russian Federation, with its many peoples, offers a rich field for research on national identity, language usage, and religion in association with a wide variety of cultural, social, political, legal, and economic attributes, and through a rich assortment of methodologies. The relations of its ethnic groups are sure to change in the future, as they have in the past, with the contending pressures for greater autonomy, independence, and cultural preservation on one hand and for acculturation, assimilation, and mobility on the other.

NOTE

This statement originally was published in Orwell's weekly column, *Tribune,* February 2, 1947.

REFERENCES

Anderson, Barbara A., and Brian D. Silver. 1990. "Some Factors in the Linguistic and Ethnic Russification of Soviet Nationalities: Is Everyone Becoming Russian?" In *The Nationalities Factor in Soviet Politics and Society,* ed. Lubomyr Hajda and Mark Beissinger, 95–130. Boulder: Westview Press.

Beissinger, Mark R. 1995. "The Persisting Ambiguity of Empire." *Post-Soviet Affairs* 11(2): 149–84.

Bond, Andrew R., and Matthew J. Sagers. 1991. "Checheno-Ingushetia: Background to Current Unrest." *Soviet Geography* 32 (10): 701–06.

Bremmer, Ian, and Raymond Taras, eds. 1997. *New States, New Politics: Building the Post-Soviet Nations.* New York: Cambridge University Press.

Chinn, Jeff, and Robert Kaiser. 1996. *Russians as the New Minority: Ethnicity and Nationalism in the Soviet Successor States.* Boulder: Westview Press.

Clem, Ralph S. 1980. "The Ethnic Dimension in the Soviet Union." In *Contemporary Soviet Society: Sociological Perspectives,* ed. Jerry Pankhurst and Michael Sacks, 11–62. New York: Praeger.

Delgado de Carvalho, C. M. 1962. "The Geography of Languages." In *Readings in Cultural Geography*, ed. Philip L. Wagner and Marvin W. Mikesell, 75–93. Chicago: University of Chicago Press.

Drobizheva, Leokadia M. 1994. "Etnicheskiy faktor v zhizni rossiyskogo obshchestva k seredine 90-kh godov" (The Ethnic Factor in the Life of Russian Society in the Middle of the 1990s). In *Konfliktnaya Etnichnost' i Etnicheskiye Konflikty*, ed. Leokadia M. Drobizheva, 4–14, 145–153. Moscow: Institut Etnologii i Antropologii im. N. N. Miklukho-Maklaya Rossiyskoy Akademii Nauk.

Drobizheva, Leokadia M., Rose Gottemoeller, Catherine McArdle Kelleher, and Lee

Walker, eds. 1996. *Ethnic Conflict in the Post-Soviet World: Case Studies and Analysis.* Armonk, NY: M. E. Sharpe.

Fickeler, Paul. 1962. "Fundamental Questions in the Geography of Religions." In *Readings in Cultural Geography*, ed. Philip L. Wagner and Marvin W. Mikesell, 94–117. Chicago: University of Chicago Press.

Foote, Kenneth E., Peter J. Hugill, Kent Mathewson, and Jonathan M. Smith, eds. 1994. *Re-reading Cultural Geography.* Austin: University of Texas Press.

Goskomstat RSFSR (Russian Soviet Federal Socialist Republic). 1990. *Natsional'nyy Sostav Naseleniya RSFSR po Dannyim Vsesoyuznoy Perepisi Naseleniya 1989 g.* (National Composition of the Population of the RSFSR Based on Data of the All-Union Census of Population 1989). Moscow: Respublikanskiy Informatsionno-Izdatel'skiy Tsentr.

Goskomstat SSSR (Union of Soviet Socialist Republics). 1991. *Natsional'nyy Sostav Naseleniya SSSR po Dannyim Vsesoyuznoy Perepisi Naseleniya 1989.* (National Composition of the Population of the USSR Based on Data of the All-Union Census of Population 1989). Moscow: Finansy i Statistika.

Goskomstat SNG (CIS/Commonwealth of Independent States). 1992–1993. *Itogi Vsesoyuznoy Perepisi Naseleniya 1989 goda.* (Results of the All-Union Census of Population 1989). Minneapolis: East View Publications. 12 vols. in 24 parts, 144 microfiches. (vols. 1–6 issued by Goskomstat SSSR in 1992; vols 7–12 by Goskomstat SNG in 1993).

Guboglo, M. N., ed. 1991. *Natsional'nyye Protsessy v SSSR* (National Processes in the USSR). Moscow: Institut Etnologii i Antropologii, Tsentr po Izucheniyu Mezhnatsional'nykh Otnosheniy).

———, ed. 1992–1993. *Etnopoliticheskaya Mozaika Bashkortostana: Ocherki, Dokumenty, Khronika* (The Ethnopolitical Mosaic of Bashkortostan: Essays, Documents, Chronicle). Moscow: Institut Etnologii i Antropologii, Tsentr po Izucheniyu Mezhnatsional'nykh Otnosheniy, 3 vols.

———. 1993. "Etnodemograficheskaya i Yazykovaya Situatsiya v Stolitsakh Soyuznykh Respublik SSSR v kontse 80-kh—nachale 90-kh godov." (The Ethnodemographic and Language Situation in Capitals of the Union Republics of the USSR at the End of the 1980s—the Beginning of the 1990s). *Otechestvennaya Istoriya* l: 53–65.

———, ed. 1994. *Bashkortostan i Tatarstan, Paralleli Etnologicheskogo Razvitiya.* (Bashkortostan and Tatarstan: Parallel Ethnopolitical Development). Moscow: Institut Etnologii i Antropologii, 4 vols.

Harris, Chauncy D. 1945. "Ethnic Groups in Cities of the Soviet Union," *Geographical Review* 35 (3): 466–473.

———. 1991. "The Unification of Germany in 1990." *Geographical Review* 81 (1): 170–82.

———. 1993a. "The New Russian Minorities: A Statistical Overview." *Post-Soviet Geography* 34 (1): 1–27.

———. 1993b. "Ethnic Tensions in Areas of the Russian Diaspora." *Post-Soviet Geography* 34 (4): 233–39.

———. 1993c. "New European Countries and Their Minorities." *Geographical Review* 83 (3): 301–20.

———. 1993d. "A Geographic Analysis of Non-Russian Minorities in Russia and Its Ethnic Homelands." *Post-Soviet Geography* 34 (9): 543–97.

———. 1994. "Ethnic Tensions in the Successor Republics in 1993 and Early 1994." *Post-Soviet Geography* 35 (4): 185–203.

Hough, Jerry F. 1996. "Sociology, the State and Language Politics." *Post-Soviet Affairs* 12 (2): 95–117.

Kaiser, Robert J. 1994. *The Geography of Nationalism in Russia and the USSR*. Princeton: Princeton University Press.

———. 1995. "Prospects for the Disintegration of the Russian Federation." *Post-Soviet Geography* 36 (7): 426–35.

Kaz'mina, O. Ye., and P. I. Puchkov. 1994. "Etnodemograficheskiye Protsessy v Rossiyskoy Federatsii." (Ethnodemographic Processes in the Russian Federation). In *Osnovy Etnodemografii* (Fundamentals of Ethnodemography), 97- 117. Moscow: Nauka.

Kolstoe, Paul. 1995. *Russians in the Former Soviet Republics*. Bloomington: Indiana University Press.

Kozlov, Viktor. 1988. "Ethnic Aspects of Urbanization." In Viktor Kozlov, *The Peoples of the Soviet Union*, trans. Pauline M. Tiffen, 49–70. Bloomington: Indiana University Press. (Translation of his *Natsional'nosti SSSR*. Moscow, 1982).

Laitin, David D. 1995. "Identity in Formation: The Russian-Speaking Nationality in the Post-Soviet Diaspora." *Archives Européenes de Sociologie* 36 (2): 281–316.

———. 1996. "Language and Nationalism in the Post-Soviet Republics." *Post-Soviet Affairs* 12 (1): 4–14.

———. 1998. *Identity in Formation: The Russian-Speaking Populations in the Near Abroad*. Ithaca: Cornell University Press.

Lapidus, Gail Warshofsky, and Victor Zaslavsky, eds. 1992. *From Union to Commonwealth: Nationalism and Separation in the Soviet Republics*. New York: Cambridge University Press.

Lehmann, Susan Goodrich. 1997. "Islam and Ethnicity in the Republics of Russia." *Post-Soviet Affairs* 13 (1): 78–103.

———. 1998. "Inter-Ethnic Conflict in the Republics in Russia in Light of Religious Revival." *Post-Soviet Geography and Economics* 39 (8): 461–93.

Mikesell, Marvin W. 1983. "The Myth of the Nation State." *Journal of Geography* 82: 257–60.

Mikesell, Marvin W., and Alexander B. Murphy. 1991. "A Framework for Comparative Study of Minority Group Aspirations." *Annals of the Association of American Geographers* 81 (4): 581–604.

Murphy, Alexander B. 1988. *The Regional Dynamics of Language Differentiation in Belgium. A Study in Cultural-Political Geography.* Chicago: University of Chicago, Department of Geography, Research Paper No. 227.

Orwell, George. 1968. *The Collected Essays, Journalism and Letters of George Orwell*, ed. Sonia Orwell and Ian Angus. London: Secker & Warburg.

Parsons, James J. 1994. "Underrepresented Nations." In *Re-reading Cultural Geography*, ed. Kenneth E. Foote et al., 281–284. Austin: University of Texas Press.

Pokshishevskiy, V. V. 1969. "Etnicheskiye Protsessy v Gorodakh SSSR i Nekotoryye Problemy ikh Izucheniya." (Ethnic Processes in the Cities of the USSR and Some Problems of Their Study). *Sovetskaya Etnografiya* September-June: 3–30.

Radvanyi, Jean. 1996. "Russes et non-Russes: un fragile équilibre à préserver." In *La Nouvelle Russie. L'après 1991: un nouveau "temps" des troubles. Géographie écono-*

mique, régions et nations, géopolitique, ed. Jean Radvanyi, 73–75. Paris: Masson/Armand Colin.

Ruble, Blair. 1989. "Ethnicity and Soviet Cities." *Soviet Studies* 41 (3): 405, 407.

Sacks, Michael Paul. 1995. "Ethnic and Gender Divisions in the Work Force of Russia." *Post-Soviet Geography* 36 (1): 11–12.

Shlapentokh, Vladimir, Munir Sendich, and Emil Payin, eds. 1994. *The New Russian Diaspora: Russian Minorities in the Former Soviet Republics.* Armonk, NY: M. E. Sharpe.

Silver, Brian D. 1974a. "The Impact of Urbanization and Geographical Dispersion on the Linguistic Russification of Soviet Nationalities." *Demography* 11 (1): 89–103.

———. 1974b. "Social Mobilization and the Russification of Soviet Nationalities." *American Political Science Review* 68 (1): 45–66.

———. 1976. "Bilingualism and Maintenance of the Mother Tongue in Soviet Central Asia." *Slavic Review* 35 (3): 406–24.

———. 1986. "The Ethnic and Language Dimensions in Russian and Soviet Censuses." In *Research Guide to the Russian and Soviet Censuses,* ed. Ralph S. Clem, 70–97. Ithaca: Cornell University Press.

Stadelbauer, Jörg. 1996. *Die Nachfolgestaaten der Sowjetunion*. Darmstadt: Wissenschaftliche Buchgesellschaft.

Suny, Ronald Grigor. 1995. "Ambiguous Categories: States, Empires and Nations." *Post-Soviet Affairs* 11 (2): 185–96.

Tishkov, V. A., ed. 1994. *Narody Rossii Entsiklopediya* (Peoples of Russia. Encyclopedia) Moscow: Nauchnoye izdatel'stvo Bol'shaya Rossiyskaya Entsiklopediya.

Wagner, Philip L., and Marvin W. Mikesell, eds. 1962. *Readings in Cultural Geography.* Chicago: University of Chicago Press.

Weiss, Richard. 1962. "Cultural Boundaries and the Ethnographic Map." In *Readings in Cultural Geography*, ed. Philip L. Wagner and Marvin W. Mikesell, 62–74. Chicago: University of Chicago Press.

Wixman, Ronald. 1980. *Language Aspects of Ethnic Patterns and Processes in the North Caucasus.* Chicago: University of Chicago, Department of Geography, Research Paper No. 191.

———. 1984. *The Peoples of the USSR: An Ethnographic Handbook.* Armonk, NY: M. E. Sharpe.

Zevelev, Igor. 1996. "Russia and the Russian Diasporas." *Post-Soviet Affairs* 12 (3): 265–84; comments by Gail W. Lapidus: 285–87.

12

Sharing Sacred Space in the Holy Land

Chad F. Emmett

When Moses came unto "the mountain of God" the Lord commanded him: "Draw not nigh hither . . . for the place whereon thou standest is holy ground" (Exodus 3:1, 5). In this instance, deity designated Horeb as holy ground. For other sacred sites, it is mere mortals who offer the designation of sacredness. In both instances the designation of being holy has at its root the notion that access to that place is limited, a place beyond the profane. Tuan (1978: 84) stresses this notion of separateness in his associating the word "sacred" with "apartness and definition" as suggested by the Latin word *sacer*, which means an area that stands apart or has limited access because of its association with deity. Similar in meaning is the Greek *templos*, which stems from a root meaning "to cut out," and the Hebrew root *k-d-sh*, which means holy and has connotations of being set apart. Isaac (1984: 28) makes a similar point in noting that on "almost all levels of culture there are segregated, dedicated, fenced, hallowed spaces." He also suggests that these places "are frequently thought to possess, or be possessed by the power of the holy."

Humans also seek power in these places. The mortal desire to identify, control and even limit access to sites that are sacred is at its simplest a desire to honor and revere those places where identifying events of one's religious beliefs have taken place. However, at a more complex level it can also be part of a human desire to control territory. Sack (1986: 93) makes this clear when applying his theory of territoriality to that of the Catholic Church. He writes: "Church buildings, properties, holy places, parishes, and dioceses are elements in the visible Church. These are not simply things located in space. They are places set apart by boundaries and within which authority is exerted and access is controlled. In other words they are territories." And as territories, Sack maintains, churches then become both a component and a physical symbol of power.

Associating power with religious structures is not limited to just Christianity. In his analysis of the Muslim Dome of the Rock, Grabar (1973: 66) suggests that the Quranic inscriptions found within the dome "forcefully assert the power and strength of the new faith and the state based on it." He then concludes that the dome was built as "a symbol of the conquering power or faith within the conquered land" which showed "the Jewish and especially the Christian worlds that the new faith was their successor in the possession of the one revealed religion and that its empire had taken over their holiest city" (Grabar 1973: 67).

The association of sacred space with power is not limited to any one period, region or religion. Whether it be the Ottoman Empire's turning the Byzantine basilica of Haigha Sophia into a royal mosque, Spain's building of Catholic churches atop Aztec and Incan temples, or Israel's demolishing an Arab residential quarter to enlarge the plaza in front of the holy Western Wall, the desire to control sacred space is based on more than just piety. Politics and power are often part of the formula.

Most sacred sites fall within the domain of a single religious tradition and therefore lack competition for control. However, as religions have expanded territorially and divided theologically, some sacred sites and holy lands have become more than just the sole claim of a single religion. Of these jointly claimed sites, some are peacefully shared while others are hotly contested. At one extreme are the more than 2,500 Muslims and Hindus killed during riots which followed the 1992 destruction by Hindu nationalists of a mosque in Ayodha India in order to rebuild a long-gone temple at the site of the birth of the Hindu god Ram (Appleby 1994).

A much less confrontational approach can be found in Independence, Missouri, among the various "Mormon" factions who look to Joseph Smith as their founder. These groups view Independence as the New Jerusalem and the site of a latter-day temple. In 1831 Joseph Smith dedicated a 63-acre parcel of land for the building of a temple. Smith and his followers were forced from the land before the temple could be built, but the desire to build it has persisted. Over time, the original temple site has gradually fragmented into parcels controlled by the various sects: the Reorganized Church of Jesus Christ of Latter-Day Saints with its auditorium and new nautilus-shaped temple, the Church of Jesus Christ of Latter-Day Saints with its visitor center, and the Temple Lot Church of Christ with its chapel. Campbell (1993: 520) notes: "As the New Jerusalem develops, it looks increasingly and eerily like the old one."

Old Jerusalem and the many other sacred sites of the Holy Land can indeed be contentious as religions compete for control. These conflicts over control of sacred space stem from the fact that three great religious traditions all view the land of Israel/Palestine as holy. For Jews, this is the promised land, a land of patriarchs, prophets and kings. It is also the land of temples, past and future. For Christians, it is the land not only of the Old Testament prophets, but the land where Jesus walked and taught and to where he will one day return. For Muslims it is where prophets extending from Abraham to Jesus and on to Muhammad taught the truths of God and where the final judgment will one day take place. Some sites associated with these religious events are shared peacefully, while others seem destined for continual conflict.

In this chapter I will illustrate how these sites are shared or, in the case of a few, not shared. I will begin with those that are most peaceful and conclude with those that are most contentious. Why some sites have found accommodation while others still only know conflict can in part be understood through the Mikesell and Murphy (1991) formula for minority group aspirations. Although this formula fits best within a multination-state model, it can also work for smaller-scale issues of territorial control. In general, sacred sites are peacefully shared because the various claimants recognize that there are others who also have the right of access and the right to participate in religious worship at the site. Contended sites are those where such accommodations have not been approved or enforced and as a result each group seeks for separate control.

Once the various patterns for sharing sacred space have been established, I will then present those patterns as possible models that could be used by Israel and the Palestinians in sharing a land both deem holy. Finding a workable model is not an easy task, since both nations use differing criteria to justify their claims. As with other nationalist competitions for territorial sovereignty, there are always competing claims and conflicting views.

Trying to comprehend these cultural attachments to land is not an easy task. Mikesell (1994: 438) challenges cultural geographers to "accept that we are trying to explain meanings that may not be comprehensible to the people entangled in the webs and that cannot be comprehended fully by those who are not." The sacred places of the Holy Land evoke strong emotional attachments among all of its inhabitants. As an outsider I cannot claim to comprehend those feelings fully. But, by being an outsider, I may have the advantage of seeing and understanding both sides and thereby being able to offer an acceptable middle ground.

SHARED SACRED SPACES

The peaceful sharing of sacred space most often occurs when there is more than a single site for a sacred event. In Nazareth, there are two claimed sites for where the angel Gabriel made his annunciatory visit. In Greek Orthodox tradition, Gabriel first spoke to Mary at the village well as she filled her pitcher. She then returned home, where the angel stood before her and told her she would bear a son whom she should name Jesus. In Roman Catholic tradition, Gabriel's only visit was to Mary at her home. The Greek Orthodox are happy to commemorate the annunciation at their Church of the Annunciation, which is centered over the well, while Catholics honor the event at their Basilica of the Annunciation, which is centered over the house of Mary (Emmett 1995: 81).

Nazareth also has two possible sites for the home of Joseph where Jesus would have been raised. The traditional location of the home of the holy family is the Franciscan-controlled Church of St. Joseph. However, just down the street, the Sisters of Nazareth, while not wanting to contend with their Catholic brothers, still quietly claim that excavations below their convent are indeed the childhood home of Jesus based on compliance with the earliest known descriptions of the site (Livio 1982).

In Jerusalem, Protestants, with their Garden Tomb and nearby Golgatha, have not had to compete with the claims of other Christians who revere the Church of the Holy Sepulchre as the site of the crucifixion and resurrection.

A sign posted at the entrance to the town of Beit Sahour, near Bethlehem, points the way to Catholic, Orthodox and Protestant sites for the fields where heavenly hosts visited the shepherds. While the various Christian sects may contend over claims to authenticity, they at least are not contending over control of sacred space. They each have their own. There are plenty of fields to go around.

Some sacred sites are jointly claimed and peacefully shared. Those that have known the greatest peace comprise areas large enough to accommodate dual claims. Mount Sinai is such a place (see Figure 12.1). A Christian pilgrim to the mountain in 1912 wrote: "Here on the mountain, sacred to Christian and Mohammedan alike, in silent friendliness, chapel and mosque lie side by side, as if ignorant or fearful of the antagonisms of their servants in this world" (Hobbs 1995: 169). The expansive summit of Mt. Tabor, one of two traditional sites of the Transfiguration, is neatly divided by a rock wall. The Franciscans control the southern half and the Greek Orthodox the northern half (Hoade 1984: 710).

The much more localized Sanctuary of the Ascension atop the Mount of Olives involves various Christian communities as well as Muslims who have long controlled the site (see Figure 12.2). A small octagonal edicule (once enclosed within a large Crusader church) enshrines a footprint in stone that is attributed to where Jesus stood before ascending into heaven. Since 1187, the shrine has functioned as a Muslim place of prayer with a *mihrab* or niche on the southern wall indicating the direction of prayer to Mecca. Surrounding the domed structure is a circular walled yard with several stone altars ready for Greek Orthodox, Armenian, Coptic and Syrian services during the Feast of the Ascension (Hoade 1984: 257).

Armenians and Greek Orthodox share joint control of the Tomb of the Virgin at Gethsemane. The Latin Catholics at one time had exclusive possession of the church, but lost their claim in a *firman* or decree from the Ottoman sultan in 1757. The Armenians and the Greeks each have their own altars, but they share the main altar at the tomb. Copts and Syrians are allowed to hold weekly services at the Armenian altars (Cust 1980: 34–36).

Figure 12.1 Church and Mosque atop Mt. Sinai.

Figure 12.2 Dome of the Ascension.

More complex is the sharing of the Church of the Holy Sepulchre.[1] Throughout the centuries, different religious communities have struggled for control of the church. At one point, ten different Christian sects shared rights in the church. However, rights often changed. For example, in the 1400s, possession of Calvary changed hands between the Georgians and Armenians five times in 30 years (Hintlian 1989: 42). Latins had greatest control during the Crusades, while during Ottoman rule the Greeks and Armenians, as subjects of the sultan, gained significant rights, most notably in the 1757 *firman*.

This and other *firmans* were meant to bring peace to the holy places, but more often than not, the *firmans* with their contradictory decrees only served to exacerbate the problem of control. Complicating the matter was the geopolitical reality that Russia promoted Orthodox gains in the holy places, while France backed the Catholics. In an attempt to appease religious subjects as well as political rivals and thus settle centuries of competition, Ottoman sultan Abdul Majid decided finally to resolve the issue of control of the holy places. In 1852 he established the existing state, or Status Quo, which meant that the current system of having to share sacred space would in a sense be codified and that there would be no more changes in control of holy places. Since the issuance of the Status Quo, the Church of the Holy Sepulchre has been shared by five religious communities in a complex mosaic of scattered sovereignty (see Figure 12.3).[2] The three main powers are the Greek Orthodox, who control the greatest share; the Armenians; and the Franciscans, who are the designated Catholic custos in the Holy Land. Egyptian Copts and Syrians have limited claims. Ethiopian control is restricted to a monastery on the roof.

Common areas shared jointly by the Greeks, Armenians and Latins include, among others, the entrance courtyard, the entry (the keys of which have long been held by two neutral Muslim families), the edicule which houses the tomb, and the rotunda which rises above the edicule. Calvary is divided into side-by-side Latin and Greek chapels. The Armenians, who lost their rights to Calvary in the 1400s, now control the Armenian Gallery as well as the Chapel of St. Helena, which is reached by descending stairs from the Greek-controlled ambulatory. Beyond this chapel and down another flight of stairs is the Latin-controlled Grotto of the Invention of the Cross. The Latins also control two other chapels. Greek control includes the central Katholikon, three chapels on the west side of the courtyard, two altars along the ambulatory, the Chapel of Adam underneath Calvary, and most of the small storerooms surrounding the central rotunda. The Copts possess a small chapel attached to the backside of the tomb.

Control of the Chapel of St. Nicodemus to the west of the edicule is disputed between the Syrian Orthodox (who worship there each Sunday) and the Armenians. This dispute stems from Ottoman times when the Armenians, as the designated representative to the sultan of other eastern sects, allowed the Syrians to worship in the chapel. Also in dispute is the north transept or Seven Arches of the Virgin, which is claimed by both the Latins and Greeks. The Status Quo did

Figure 12.3 Church of the Holy Sepulchre.

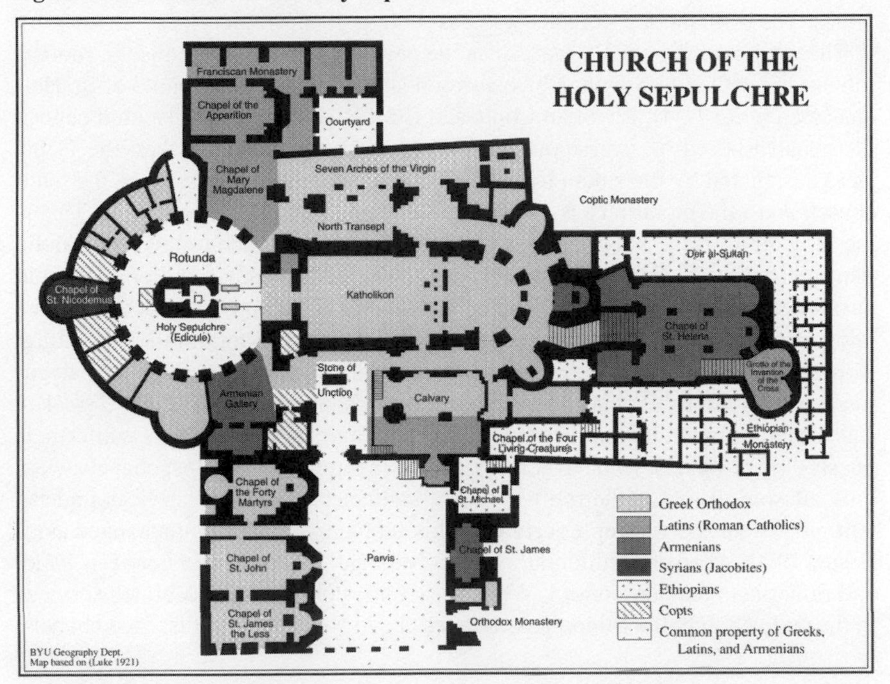

not identify rights to these areas, and so none of the claimant sects has been able to repair and furnish them (Cust 1980).

Within the church no borders or lines of demarcation exist. Only the lamps, candles and icons give a clue as to who controls what sections. Religious leaders and pilgrims can freely move throughout the church. On holy days the order of procession and prayer is regulated by very specific schedules. Tensions do exist, but relations continue to increase in amicability. Everyone knows their place and that nothing will change. As a result, religious communities can be more concerned with the daily cycles of worship than worrying about losing territory or rights. Communities still keep watch for violations of the Status Quo, but only occasionally do problems arise.

Evidence of continually increasing cooperation is that the Greek, Armenian and Latin churches finally agreed to jointly restore the long-disputed rotunda. The agreement to adorn the dome with a sun motif acceptable to the varying artistic traditions of the three churches was reached in 1994 with a historic signing by the Armenian and Greek patriarchs and the Latin custos of the Holy Land. Armenian patriarch Manogian called the agreement "a turning point for all Christendom." He then stated: "The very fact that it was signed at all provides telling evidence of the new spirit of ecumenical rapprochement that is sweeping not only the Western Christian world, but [the Eastern] region as well" (Anonymous 1995). The brilliant new rotunda was unveiled in 1997.

The section of the church that defies the peace of the Status Quo is the rooftop monastery of Deir al-Sultan, which surrounds the dome of the Chapel of St. Helena (see Figure 12.4). Egyptian Copts and Ethiopians have long disputed control of this area based on overlapping claims from Ottoman times when the Copts were designated by the sultan to represent the Ethiopians. Access from the main church up to the monastery is via two small chapels and some stairways. This is the only direct route between the Church of the Holy Sepulchre and the Coptic Patriarchate, which lies to the north of Deir al-Sultan. In 1838 a cholera epidemic nearly decimated the Ethiopian community of Deir al-Sultan. The Copts used this opportunity to regain control by petitioning a sympathetic Ibrahim Pasha, ruling from Egypt. To solidify their claim the Copts even burned "contaminated" documents which the Ethiopians had used to justify their claim (Pedersen 1988: 40). The Copts thus gained control of the chapels and the monastery, while their guests the Ethiopians, who never relinquished their claim to the lost chapels, were only allowed to live in the rooftop rooms and celebrate Easter services under a tent erected in the rooftop courtyard. This situation remained unchanged until Easter 1970, when the Ethiopians took advantage of strong ties between Israel and Ethiopia to reassert control. While the Coptic community celebrated services in the main church, the Ethiopians changed the locks to the two disputed chapels.

Figure 12.4 Deir al-Sultan with Ethiopian Monks and the Dome of the Chapel of St. Helena.

Israeli police who were stationed around the monastery did nothing to prevent the change. The Israeli courts first ruled in favor of the Copts, but then backed down and decided to take the matter under consideration. No further action has been taken, and so the Ethiopians are still in control (Elsner 1978). With nationalist pride at stake, the Egyptian government continues to discourage its Christians from making pilgrimages to Israel until the Ethiopians relinquish control to the Copts.

On Jerusalem's Mt. Zion, the sharing of sacred space is complicated by the unusual fact that Muslims, Jews and Christians all claim it as holy, particularly the structure housing David's tomb and the Room of the Last Supper. The upper room of the structure (which only dates back to the Crusader period) has traditionally been celebrated by Christians as the site of the Last Supper. In 1333 the King of Naples purchased the room from the Mamluk sultan of Egypt and then entrusted control to the Franciscans. In 1552 the Franciscans were ousted by the Muslims and the shrine was converted into the Mosque of the Prophet Da'ud based on the tradition that the lower level housed the Tomb of David. During Jordanian control of the Old City, the Tomb of David rose in prominence among Jerusalem's Jews due to the fact that Jordan denied them access to the more holy Western Wall.[3] Since the structure was never designated a part of the Status Quo, Israel's Ministry of Religious Affairs has assumed the ruling role as a neutral agent (Rabinovich 1986; Shapiro 1986b). Meanwhile Franciscans claim control of the building, which technically remains a Muslim religious endowment or *waqf* that is used as a place of prayer by Jewish and Christian pilgrims.

More problematic, even though it is only claimed by Christians, is the Church of the Nativity in Bethlehem (see Figure 12.5). Built during the Byzantine period, the structure is the oldest standing Christian church in the Holy Land. Tradition holds that it survived the Persian conquest of 614 because the invaders were impressed with the depiction of Persian Magi in a mosaic on the front of the church (Freeman-Greenville 1993: 18). Long revered as the site of the birth of Jesus, the Church of the Nativity has a complex history of competition for control among the various Christian sects. One of the most noted encounters occurred in the mid-nineteenth century when Greeks and Latins jostled for expanded control of the jointly shared, but hotly contested, church. During 1847, Greek monks beat Latins with staves in an attempt to drive them from holy sites in Bethlehem. Meanwhile Latins took possession of a tapestry which they claimed marked a portion of the church that they deemed theirs. Later a Greek bishop was wounded in a clash with members of the Latin clergy. This was followed by the mysterious disappearance of the silver star in the Grotto of the Nativity, which marked the site of the birth. Latins accused the Greeks of stealing the star out of fear that Latin inscriptions on the star would be used as justification by the Catholics to expand their territorial rights in the church (Broadus 1979).

What may now seem to be a rather trivial contest was at the time a very serious conflict, for it involved not only the local clergy but also world powers with stra-

Figure 12.5 Basilica of the Nativity with Latin (left), Armenian (center), and Orthodox (right) Steeples Indicating the Three Main Claimants.

tegic interests in the region. Catholic France had long claimed the role of protector of the Roman Catholics in the Ottoman Empire, while its rival, Orthodox Russia, backed the more numerous Greek and Arab Orthodox communities under Ottoman control. During the Easter seasons of 1847 and 1848, France sent an armed ship to the port of Jaffa as a way of backing Latin claims to sacred sites. Meanwhile, Russian diplomats in Istanbul and emissaries in Jerusalem sought to further Orthodox claims. After several years of political maneuverings by both France and Russia, the Ottomans finally allowed the Latins to replace the silver star on December 22, 1852. As this was viewed as a setback to Greek claims in the church, Russian representatives hinted at retribution. Russian troops were mobilized in the south of Russia, and ships in the Black Sea were put on alert. Russia demanded that it be recognized as the protector of all Orthodox subjects of the sultan and that Latin claims in the Holy Places be revoked. Fearing Russian expansion into the Ottoman Empire, Britain and France began to mobilize troops and navies. Negotiations between the European powers over Ottoman–Russian relations were unfruitful, and by October 1853 war had broken out between Russia and the Ottoman Empire (Broadus 1979). Rivalry over holy places, and in particular the silver star of the Church of the Nativity, did not cause the Crimean War, but it certainly added to the already tense geopolitics of the region.

While not as cataclysmic, there are still disputes over control of certain portions of the Church of the Nativity. These disputes center on cleaning and repairs, acts which serve to solidify claims. During the annual 1983 and 1984 Christmas cleanings of the church, up to 50 Armenian and Greek priests came to blows using brooms, chairs and ladders as weapons as they fought for the right to clean a disputed six-meter-by-fifteen-centimeter stretch of wall above the north entrance to the grotto (Shapiro 1984). In an attempt to prevent a similar brawl in 1985, the two sects agreed to leave the disputed wall as an uncleaned sort of no-man's-land. The agreement worked, and it has allowed priests annually to clean the rest of the church in peace (Shapiro 1986a). In 1989 Latins and Armenians protested Greek attempts to repair the leaky roof of the church because it was seen as an attempt by the Greeks to assert their control of the roof, something that the Latins and the Armenians consider to be jointly controlled by all three sects (Shapiro 1989). Eventually the Israeli government came in as a neutral fourth party and fixed the leaks (Shapiro 1990).

While not all of the above sacred sites know true peace, they at least know the absence of the intense conflicts that once plagued many of them. Minor disputes may still arise, but Ethiopia and Egypt are not about to wage war over Deir al-Sultan, and it is highly unlikely that another Crimean War would be fought over the star in the Church of the Nativity. Joint claimants of these holy sites have learned that they can coexist, however tenuously, by sharing. Unfortunately, there are other sites in the Holy Land that lack favorable formulas for the sharing of sacred space. With strong nationalist underpinnings, three sacred sites claimed by both Muslims and Jews have the potential of inciting future conflicts.

CONTENDED SACRED SPACES

In Hebron, Muslims and Jews both claim the tomb where the patriarchs Abraham, Isaac and Jacob are buried. It is known as the Tomb of the Patriarchs or Cave of Macphelah by Jews and as the Ibrahimi (Abraham) Mosque by Muslim Arabs, who see it as their fourth most holy site after Mecca, Medina and Jerusalem. Since both Jews and Arabs claim descent from Abraham, and since both Jews and Muslims revere Abraham, Isaac and Jacob as prophets, this site has become particularly contentious. In many ways control of the tomb has become symbolic of the attempt by Israelis to expand control into what they refer to as Judea and of the desire of Palestinians to establish a state of Palestine that would include what they refer to as the occupied West Bank.

The imposing structure was built over two thousand years ago during the Hasmonean period and is often attributed to King Herod. During the Byzantine period, a hall for Christian prayer was located around the cenotaphs of Isaac and Rebekah while rooms for Jewish worship were near the cenotaphs of Jacob and Leah. A pilgrim visiting the tombs in 570 noted that at that time there was sharing of this sacred place. He writes: "The Christians enter from one side and the Jews from the other, and they burn a great deal of incense and distribute gifts and candles to the attendants and to all those praying there. They come there in multitudes from near and far" (Klein and Roman 1989: 85). The church hall became a mosque following the Muslim conquest of 638 and then a church again during the Crusader period. With the demise of the Crusaders, the sanctuary once again became a mosque, and for over seven hundred years, from 1266 to 1967, non-Muslims were forbidden entry. Jews were only allowed to pray on the first seven steps of the southwestern stair landing. Following the Israeli conquest of the West Bank, and with the approval of the military government, Jews were permitted first just to enter the sanctuary and then on Yom Kippur in 1968 to pray in the sanctuary (Klein and Roman 1989: 85). Since then Jewish rights in the mosque have continually expanded (see Figure 12.6). In 1971 an ark for Torah scrolls was placed in the hall of Abraham, which then began to function as a synagogue. In 1975 the hall of Jacob and the interior courtyard were designated as additional areas where Jews could worship. In 1979 Jewish settlers forcibly entered the large hall of Isaac in an attempt to further expand areas in which Jews could pray. After months of deliberations, the Israeli government authorized Jewish services in part of the hall.

The ongoing Israeli encroachment within the Mosque of Abraham and within the city of Hebron has resulted in very tense relations between Jews and Arabs. In 1968 a hand grenade exploded in front of the mosque among a crowd of Jewish visitors. In 1976 rare and valuable Qurans in the mosque were stolen by a Jewish resident of the nearby settlement of Kiryat Arba. Angry Muslims retaliated by expelling Jewish worshippers and setting their scriptures and ceremonial objects on fire. Torah scrolls are now housed in a fireproof safe. During 1980 a yeshiva student and six settlers were murdered by Arabs, and in 1983 a Jewish settler was stabbed to death in broad daylight while walking through the Arab market. "These and other incidents were always followed by reprisals by Kiryat Arba residents," including the 1983 murder of three Arab students at the Hebron Islamic college (Romann 1985: 9).

By 1982 a sort of status quo emerged in which hours and areas of worship for Muslims and Jews were specifically defined and militarily enforced. For example, on Fridays during the day the entire mosque was reserved exclusively for Muslims to pray and to gather for their weekly sermon, but on Friday evenings precisely 15 minutes before the lighting of Sabbath candles, Jews were permitted to enter and for the rest of the Sabbath they could pray in designated areas. At other times during the week, Jew and Arab could be seen praying side by side in front of Abraham's tomb. All of this was done under the watchful eye of the Israeli military, complete with a military commander assigned to the mosque/synagogue (Halevi 1995: 12).

This tenuous but workable status quo solution for sharing lasted until 1994. On an early Friday morning of February 24, 1994, Baruch Goldstein, a Jewish emigrant from New York City to Kiryat Arba, entered the Ibrahimi Mosque and started firing his machine gun into an assembled crowd of 600 men and boys and

Figure 12.6 Tomb of the Patriarchs.

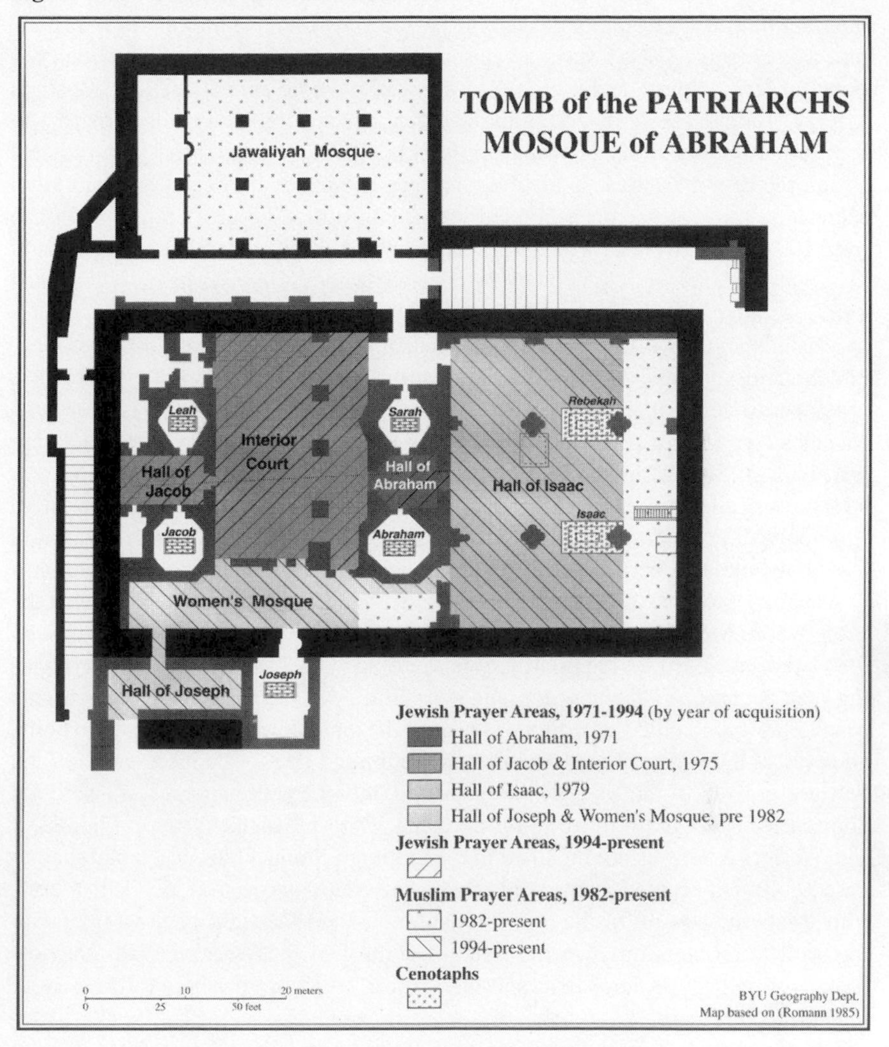

100 women as they bowed in prayer during the fasting month of Ramadan. Before being subdued and then beaten to death by the worshippers, Goldstein had killed 21 men and 8 boys (*The Massacre in Al-Haram Al-Ibrahimi Al-Sharif* 1994). Goldstein's tomb in Kiryat Arba has become a place of pilgrimage for those Jews who admire his willingness to assert Jewish rights to the Tomb of the Patriarchs and his efforts to try to thwart the peace process and the Israeli troop redeployment from Hebron.

Since the massacre, the Tomb of the Patriarchs has been rigidly divided between Jews and Muslims. Jews have turned the halls of Abraham and Jacob and the interior courtyard into a synagogue, while Muslims use the rest of the structure as a mosque. For ten days a year on Jewish holidays the whole sanctuary is open for Jews to pray, and for ten days a year the whole area is open for Muslims to pray (Halevi 1997: 20). Multiple security checks for all visitors now try to prevent any more violence within the shrine. The city of Hebron also requires strong security measures. In an effort to keep the peace between Abraham's descendants, the city has been divided into a Palestinian-controlled sector for the over 100,000 Arabs who live in Hebron and an Israeli-controlled sector around the Tomb of the Patriarchs and nearby enclaves where 500 Jews have settled since 1967.

Muslims and Jews also compete for control of their holy shrines in Jerusalem, which include the Western Wall and, towering above it, the Temple Mount as it is known to Jews or the Haram al-Sharif or Noble Sanctuary as it is known to Muslims (see Figure 12.7). The Western (or Wailing) Wall is part of the retaining wall built in 20 B.C. under the direction of King Herod to support the massive structures of the Temple Mount above. When Romans destroyed the temple in 70 A.D., only the retaining wall remained. As the last remaining part of Judaism's most holy site, the Western Wall then became Judaism's most holy site (Murphy-O'Connor 1986: 87). The wall is also holy to Muslims who honor the site as the place where Muhammad tied up his winged steed Buraq before ascending from the Haram al-Sharif on his night journey into heaven. The wall and its surrounding area became a Muslim *waqf*, and over time Muslim migrants from Africa's Maghreb region settled in *waqf* houses near the wall. Jews were allowed to come and pray at the wall, although numbers were limited due to the narrowness of the corridor in front of the wall, which measured only 4 by 28 meters. Jews were not allowed to repair the wall, to bring in chairs, or to blow the shofar on holidays. The Western Wall was not included in the Ottoman Status Quo. When nationalist-based conflicts erupted at the wall during the Mandate period, the British reaffirmed Islamic control of the area but acknowledged the right of Jews to pray at the wall. So contentious was the issue of control of the Western Wall that riots broke out in 1929 in Jerusalem and then spread to Hebron, where 69 Jews were massacred by Arabs.

Restricted Jewish rights at the wall all changed in 1967 when, within days of Israel's gaining control of the old city, the 650 Arab Muslim residents of the Maghrabi quarter were expelled and their 135 houses destroyed to make way for increased access to the wall. While Muslims still claim the land as *waqf*, the area is now administered by the state of Israel. The large plaza in front of the wall is used as a central gathering place for both religious and nationalist functions (*Memorandum on the Israeli Violation of the Religious Status Quo at the Wailing Wall* 1968).

The adjacent Temple Mount/Haram al-Sharif with its two mosques, the Dome of the Rock and al-Aqsa, could also have come under Israeli control in 1967, but

Figure 12.7 Western Wall (below) with the Haram al-Sharif and Dome of the Rock (above).

the government withdrew and allowed the Islamic *waqf* to continue to administer Islam's third most holy site. Without the temple and not wanting to tread upon the sacred ground where the Holy of Holies would have been located, observant Jews have shown less interest in the mount than in the wall. Some Jews still insist on their right to pray on the mount, while others plan to one day build a temple there. On Easter Sunday of 1982, a follower of the extremist Kach group opened fire at the Dome of the Rock in an attempt to "liberate" the Temple Mount from Muslim control. Two people were killed in the shooting and at least 184 people injured when riots broke out in and around Jerusalem (Wasserstein 1982). In October 1997, the Temple Mount Faithful laid a cornerstone for the third temple just outside the walls of the old city of Jerusalem ("Temple Mount Faithful Lay Cornerstone for Third Temple" 1998). In 1996 an archaeological tunnel running along the western perimeter of the mount from the Western Wall to the Via Dolorosa was completed. Muslims saw this as another attempt by Israel to excavate underneath the Haram al-Sharif and its adjoining buildings in anticipation of a day when the Temple Mount would once again be under Jewish control. Protests and violence ensued, including an attack on Israeli soldiers guarding the isolated Tomb of Joseph in Nablus. Other indications of Jewish efforts to gain control of the Temple Mount and build the temple include a model of the third temple, which is prominently displayed at a temple museum in the adjacent Jewish Quarter, and posters of the hoped-for new temple, which are hung in Jewish establishments and are then used by Muslims in their own posters to warn about the need to protect the Haram al-Sharif.

While these sites are the scene of intense interreligious conflict, they are also the site of intense intrareligious competition. The Dome of the Rock has both religious and political symbolism for Muslim Arabs in Jordan, Saudi Arabia and Palestine. The Hashimite Kingdom of Jordan has long been the official caretaker. This caretaking is done with great fervor, partly as consolation for the Hashimites having lost control of Mecca and Medina to the rival Sa'ud family in 1919. The Saudis, on the other hand, as designated protectors of the holy sites in Arabia, would also like to gain influence at another holy site. When the Dome needed repairs and new gold plating, the wealthy Saudis offered to pay for its restoration, but instead, King Hussein of Jordan sold his London mansion to raise the necessary funds to ensure continued Jordanian upkeep and therefore Jordanian influence (Friedland and Hecht 1996: 474). Added to the Saudi–Hashimite rift is the renascent Palestinian nation, which has coopted the Dome as a symbol of its hoped-for state. Palestinians want not only for Israel to relinquish control of the Arab portions of Jerusalem, including the Haram al-Sharif, but also for Jordan (which Israel recognizes as caretaker) to relinquish its caretakership of the Dome.

At the Western Wall there is a growing rift between the Reform and Orthodox Jewish communities over how Jews should be able to worship at the wall. The Orthodox call for strict segregation by gender, while Reform Jews readily mix by gender when praying. In 1997 there were several outbreaks of violence at the Wall as Orthodox Jews sought to break up the mixed prayer gatherings of Reform Jews at the perimeter of the plaza (Shapiro 1997).

MODELS FOR SHARING SACRED SPACE

Within and without the discipline of geography there is an increasing awareness and interest in "cultural politics" (Mikesell 1994). This is not to say, however, that the correlation between culture and politics has not been noted before. Broek, for example, recognized during World War II that the decolonization process in Southeast Asia could result in an "Asiatic Balkans" due to the cultural diversity of the region based in part on the "divergent interests" that evolved during the various forms of colonial rule. He hoped, however, that common cultural characteristics, like that of the Malay-speaking peoples, would influence a non-European path (at the time engaged in a great war among nations) in which some form of regional cooperation was established (Broek 1962).

A few years later, Gottman (1964: 27) noted that the "stubborn regionalisms" of the Old World "have evolved toward nationalism in modern times, partitioning the lands more and more." He then stated that partitioning political space "would be much easier to settle if nations were less attached to their pride, their past, their way of life, their culture—all that we round up in the term 'national spirit.' "

Knight (1982: 514), in addressing the influence of cultural identity on territorial attachments, stated: "In some general works by political geographers lip service is given to the emotional bonds of group to politico-territorial identities, generally defined loosely as the nation or nation-state, but remarkably little has been done beyond this." He then concludes his article with the suggestion that "by focusing on the issue of group territorial identities, we can come into contact with the age-old problem that remains very much alive, that of how best to give political recognition to these identities" (Knight 1982: 526).

Many articles have followed, perhaps motivated by the rise of a new world order in which boundaries continue to evolve and new states emerge. Mikesell and Murphy (1991: 601) note, for example, that "Much of the cultural discord in the world is probably an inevitable consequence of the uncomfortable containment of too many peoples in too few states." In a similar vein, Nietschmann (1994: 242) notes: "It is apparent, then, that a new political architecture is developing globally, formed from the cultural boundaries of nations. States come and go—nations remain." Thus nationalisms among the many non-state nations of the Fourth World continue to rise.

Noted among the many nationalist movements is the competing claim by the Israeli and Palestinian nations for control of the land of Israel–Palestine. That this land is deemed holy by both groups means that the desire to control sacred space extends beyond specific sites to include the entire Holy Land. To ascertain degrees of holiness is an exercise in futility. It is holy to all, and all have justified rights to the land. Israelis and Palestinians must figure out a way to share the land they love. Sharing the land of Israel–Palestine will not be easy, but perhaps there are lessons to be learned from some of the already discussed examples of sharing smaller parcels of sacred space.

The first model for sharing sacred space is the Nazareth model, whereby competition for control of territory is ameliorated by the fact that there are two sites of the Annunciation. This ideal plan would work if history could be rewritten so that either there were dual Holy Lands or the Holy Land was holy to only one group of people. This cannot be, for the multiple claims are irreversible and feelings of attachment are equally strong for both peoples. In a 1989 conversation with an Israeli/Palestinian/Arab woman from Nazareth, I asked if she would move to a state of Palestine if it were ever established in the West Bank. Without even needing to think about it, she quickly responded "never." Nazareth was her home, for it is here that her family has lived for generations. She has very strong ties to that particular place, not to some distant Arab country or even a closer state of Palestine. Nazareth is the place for her, even if it means remaining as an Arab citizen of Israel. One Israeli, after visiting with a Palestinian man from the West Bank, described the man's inability to vocalize his feelings for the land as an "inarticulate, silent love that couldn't be relayed in any rhetoric." The Israeli then explained similar feelings in which, during the past two thousand years, Jews had "become expert pornographers with a highly symbol-wrought intellectualized yearning for this land" (Shehadeh 1984: 86). Upon arrival in Tel Aviv, one recent Jewish immigrant to Israel from England replied: "I feel like I'm coming home" (Cohen 1998: 8). These intense feelings for place cannot be denied either Arab or Jew. They are real and they are strong. There is no getting around the fact that both peoples are on the single Holy Land to stay. There will never be an alternative Holy Land.

The second model is the Mt. Sinai model. It calls for an equitable sharing of the land through the establishment of a secular democratic state in which Jews as well as Arabs have the same freedoms and rights. Such a plan has ancient precedents. Before entering the promised land, Moses admonished the Israelites, saying: "The stranger that dwelleth with you shall be unto you as one born among you and thou shalt love him as thyself for ye were strangers in the land of Egypt" (Leviticus 19:34). Anyone who has stood atop Mt. Sinai in Egypt to watch the sunrise has probably done so with a host of strangers, both pilgrims and partyers, from such diverse countries and religious traditions as Egyptian Muslims, Israeli Jews, Korean Presbyterians, German Lutherans, and American Catholics. As a sacred site to all three monotheistic religious traditions, and without a single religious sovereign, Mt. Sinai allows all who come to revel in the grandeur and glory of the place. A new state, perhaps call it Canaan, could similarly allow for all residents in the land to participate in a true secular democracy. The one problem with this otherwise fair and just solution is that it would mean the demise of Israel as a Jewish state, and until Jews can feel secure and accepted wherever they may live, they will feel the need for a Jewish state as a guaranteed place of refuge.

A third model is the Hebron model. It calls for a single sovereign state, either Jewish or Arab, to be established throughout the entire land. This plan is what the minority Israeli and Palestinian extremists want. They want it all. Terrorism in the form of Palestinian hijackers and suicide bombers and ethnic cleansing in the form of forced expulsion of the Arabs in 1948 and the continued demolition of Palestinian homes by the Israeli government are methods used to maintain or to create such exclusionist states. Such exclusion will not last when there are such strong ties to the land. Hebron's Tomb of the Partiarchs/Mosque of Abraham was first Christian, then Muslim, then Christian, then exclusively Muslim for 700 years, and yet beginning in 1967 Jews have gradually moved in and now control a significant portion of the structure. No single religion has been able to maintain control of the Tomb of Abraham, and no single state has been able to maintain control of what was anciently known as Canaan. Roots run deep and memories endure. Jews dreamed for nearly two thousand years before finally being able to return to their ancestral land. Palestinians talk of steadfastly enduring on the land under Israeli rule knowing that in this region of the world, things will change. The Hebron model is not a solution; it is a prescription for endless conflict.

The final and most feasible solution, from my perspective, is the Church of the Holy Sepulchre model, for it is here that, after centuries of feuding and fighting, multiple claimants to a very holy place have learned to share. That sharing is in the form of scattered sovereignty in which each group has control over its designated places. The idea of partitioning place is not new in the Middle East. When there was strife between the herdsmen of Abraham and the herdsmen of his nephew Lot, Abraham said to Lot: "Let there be no strife, I pray thee, between me and thee, and between my herdsmen and thy herdsmen, for we be brethren. Is not the whole land before thee? Separate thyself, I pray thee, from me: if thou wilt take the left, then I will go to the right, or if thou depart to the right hand, then I will go to the left" (Genesis 13:8–9).

For centuries under the reign of Islam, minorities coexisted in cities through the establishment of religious, ethnic and family quarters. Muslims, Jews, Christians and Armenians lived together peacefully in their separate four quarters of Jerusalem. Likewise Orthodox, Latin and Muslim quarters in Nazareth, Armenian quarters in Beirut and Aleppo, Coptic quarters in Cairo, and Jewish quarters from Morocco to Yemen all helped to protect and preserve religious minorities living in the House of Islam.

The Status Quo division of territory in the Church of the Holy Sepulchre and other sacred sites as well as the establishment of quarters in the cities of the Middle East are not an immediate or lasting guarantee of peaceful coexistence, but separation sometimes seems to be the only solution. Such must be the case for Israelis and Palestinians. Both need a state. In 1947, the United Nations voted to partition the land. Jews, who made up only one-third of the population, welcomed the plan and were happy to settle for even a small fragmented state. Arabs, on the other hand, who were the majority of the population and controlled the vast majority of the land, rejected partition, for it gave them only 43 percent of the land of Palestine. Fifty years later things have reversed. Many Israelis now reject the idea of partition, while most Arabs would be happy to settle for a small fragmented state comprised of just the West Bank, Gaza and East Jerusalem, a fraction of what they consider to be Palestine. How to create and maintain those states is not an easy task since the two peoples are very intertwined in terms of history and settlement patterns. It will require compromise and forgetting on both sides. It can, however, and should be done. Through the establishment of two states, Israel and Palestine, with Jerusalem as a shared capital, the sacred space of the Holy Land might finally be shared in peace.

NOTES

1. For a more detailed description of how the church is shared, see "The Status Quo Solution for Jerusalem" (Emmett 1997).

2. The five sites governed by the 1852 Status Quo are the Church of the Holy Sepul-

chre, its rooftop monastery of Deir al-Sultan, the Sanctuary of the Ascension, the Tomb of the Virgin, and the Church of the Nativity (Cust 1980).

3. The veneration of tombs among Jews has changed since biblical times. Japhet (1998: 58) writes: "nowhere does the Bible ever suggest that a tomb is a sacred place, in any respect whatsoever."

REFERENCES

Anonymous. 1995. "Christian Leaders' Accord on Sepulcher Hailed as 'Turning Point.' " *Christians and Israel* 4 (2): 3.

Appleby, R. Scott. 1994. *Religious Fundamentalisms and Global Conflict*. Ithaca: Foreign Policy Association.

Broadus, John R. 1979. "Church Conflict in Palestine: The Opening of the Holy Places Question During the Period Preceding the Crimean War." *Canadian Journal of History* 14 (3): 395–416.

Broek, Jan O. 1962. "Diversity and Unity in Southeast Asia." In *Readings in Cultural Geography*, ed. Philip L. Wagner and Marvin W. Mikesell, 170–85. Chicago: University of Chicago Press.

Campbell, Craig S. 1993. "Images of the New Jerusalem: Latter Day Saint Faction Interpretations of Independence, Missouri, 1830–1992." Ph.D. dissertation, University of Kansas.

Cohen, Aryeh Dean. 1998. "Making Israel Home—Despite Everything." *Jerusalem Post International Edition*, August 15: 8–9.

Cust, L. G. A. 1980. *The Status Quo in the Holy Places.* Jerusalem: Ariel Publishing House.

Elsner, Alan. 1978. "Room at the Top." *Jerusalem Post*, November 24.

Emmett, Chad F. 1995. *Beyond the Basilica: Christians and Muslims in Nazareth.*Geography Research Paper No. 237. Chicago: University of Chicago Press.

———. 1997. "The Status Quo Solution for Jerusalem." *Journal of Palestine Studies* 26 (2): 16–28.

Freeman-Greenville, G. S. P. 1993. *The Basilica of the Nativity in Bethlehem.* Jerusalem: Carta.

Friedland, Roger, and Richard Hecht. 1996. *To Rule Jerusalem.* Cambridge: Cambridge University Press.

Gottman, Jean. 1964. "Geography and International Relations." In *Politics and Geographic Relations,* ed. W. A. Douglas Jackson, 20–51. Englewood Cliffs, NJ: Prentice-Hall.

Grabar, Oleg. 1973. *The Formation of Islamic Art.* New Haven: Yale University Press.

Halevi, Yossi Klein. 1995. "Side by Side in Hatred." *The Jerusalem Report*, June 15: 12–14.

———. 1997. "The Next Minefield." *The Jerusalem Report*, May 15: 20–22.

Hintlian, Kevork. 1989. *History of the Armenians in the Holy Land.* Jerusalem: Armenian Patriarchate Printing Press.

Hoade, Eugene. 1984. *Guide to the Holy Land.* Jerusalem: Franciscan Printing Press.

Hobbs, Joseph J. 1995. *Mount Sinai.* Austin: University of Texas Press.

Isaac, Erich. 1984. "God's Acre." *Landscape* 14 (2): 28–32.

Japhet, Sara. 1998. "Some Biblical Concepts of Sacred Space." In *Sacred Space: Shrine, City, Land,* ed. Benjamin Z. Kedar and R. J. Zwi Werblowsky, 55–72. New York: New York University Press.

Klein, Yigal, and Yadin Roman. 1989. "The Legacy of Abraham." *Eretz Magazine*, Summer: 34–87.

Knight, David B. 1982. "Identity and Territory: Geographical Perspectives on Nationalism and Regionalism." *Annals of the Association of American Geographers* 72 (4): 514–31.

Livio, Jean-Bernard. 1982. "Les Fouilles chez les Religieuses de Nazareth." *Le Monde de la Bible* 16: 28–36.

Luke, H. C. 1921. "The Christian Community in the Holy Sepulcher." In *Jerusalem 1918–20*, ed. C. R. Ashbee, 46–50. London: Council of the Pro-Jerusalem Society.

The Massacre in Al-Haram Al-Ibrahimi Al-Sharif: Context and Aftermath. 1994. Jerusalem: Palestine Human Rights Information Center.

Memorandum on the Israeli Violation of the Religious Status Quo at the Wailing Wall, Jerusalem. 1968. Beirut: Institute for Palestine Studies.

Mikesell, Marvin W. 1994. "Afterword: New Interests, Unsolved Problems, and Persisting Tasks." In *Re-reading Cultural Geography*, ed. Kenneth E. Foote et al., 437–44. Austin: University of Texas.

Mikesell, Marvin W., and Alexander B. Murphy. 1991. "A Framework for Comparative Study of Minority-Group Aspirations." *Annals of the Association of American Geographers* 81 (4): 581–604.

Murphy-O'Connor, Jerome. 1986. *The Holy Land.* Oxford: Oxford University Press.

Nietschmann, Bernard. 1994. "The Fourth World: Nations Versus States." In *Reordering the World: Geopolitical Perspectives on the 21st Century*, ed. George J. Demko and William B. Wood, 225–42. Boulder: Westview Press.

Pedersen, Kirsten. 1988. "Deir Es-Sultan: The Ethiopian Monastery in Jerusalem." *Quaderni di Studi Etiopici* 8–9: 34–47.

Rabinovich, Abraham. 1986. "New Era on Mount Zion." *Jerusalem Post*, June 6.

Romann, Michael. 1985. *Jewish Kiryat Arba Versus Arab Hebron.* Jerusalem: West Bank Data Project.

Sack, Robert David. 1986. *Human Territoriality: Its Theory and History*. Cambridge: Cambridge University Press.

Shapiro, Haim. 1984. "Monks Battle with Sticks in the Church of the Nativity." *Jerusalem Post*, December 28.

———. 1986a. "Peaceful Cleaning for Church of Nativity." *Jerusalem Post*, January 1.

———. 1986b. "Silence Rules the Cenacle." *Jerusalem Post*, May 30.

———. 1989. "Roof Repair Sparks Bethlehem Church Row." *Jerusalem Post*, January 13: 2.

———. 1990. "Churches Agree to Let Government Repair Church of Nativity Roof." *Jerusalem Post*, November 23: 2.

———. 1997. "Haredim Attack Non-Orthodox at Western Wall." *Jerusalem Post*, June 21: 5.

Shehadeh, Raja. 1984. *Samed: Journal of a West Bank Palestinian.* New York: Adama Books.

"Temple Mount Faithful Lay Cornerstone for Third Temple." 1998. *Washington Report*, Jan/Feb: 40.

Tuan, Yi-Fu. 1978. "Sacred Space: Explorations of an Idea." In *Dimensions of Human Geography*, ed. Karl W. Butzer, 84–99. Chicago: University of Chicago, Department of Geography, Research Paper No. 186.

Wasserstein, Bernard. 1982. "Trouble on the Temple Mount." *Midstream* 28(7): 5–9.

13

An Absence of Place: Expectation and Realization in the West Bank

Shaul E. Cohen

As geographers continue to devote energy to their long-standing interest in the amalgam of place, space, and landscape, their subjects are diversifying and becoming more complex. Though cultural geography still roots itself in etymologies (O'Sullivan 1992) and traces back to representations in the arts, new challenges and forms of inquiry make a regular appearance. Thus, we have new sites of "place" to investigate (as in cyberspace; Adams 1997; Taylor 1997; Starrs 1997), new types of people to query (with notions of hybridity and subalterns), and a simultaneity of perspectives to explore (the postmodern and beyond). Accompanying (driving) this proliferation of opportunity for cultural geographers is the changing world itself, with its ever more complex patterns of society and space. New and emerging conditions necessitate an updating of the generalizations that we make through a monitoring of the particular. If these ongoing inquiries do not always lead to new paradigms, they at least test the veracity of standard definitions and perspectives. Such a process maintains the intellectual continuity of our endeavor, though it may not contribute to efforts to chart a purely linear progression.

It seems logical that geographers have an abiding curiosity about place, as both object and subject. My disciplinary chauvinism argues that no one is better equipped for such an interest, nor more attuned to the plethora of stimuli in the environment that confront all but the most obtuse observer. At the same time, we are, in the first instance, students, learning from anyone and anything that can teach us. The field component, therefore, is an essential element of learning about a place; direct observation and experience is something that geography cannot lose sight of, even if it straddles a complex line between science and art. This is particularly so in some of the more volatile regions of the world, where place can change in short order—both temporally and spatially.

One such place is the so-called West Bank,[1] where history and geography are formed on a week-to-week, if not a daily, basis. History has a role, of course, in constructions of space and place, both at the individual and the communal level, and as both internal and external forces. Indeed, one component of place is history, in terms of the process that has led to the development of a place, and the way that history is used by those creating and experiencing a place, both consciously and unconsciously. Place exists as an overlapping phenomenon, both spatially and temporally, with links between and across the two. In this spirit, a term that students of place like to use, and one that has been applied with some frequency to Palestine, is palimpsest.

It is not always sufficient, however, to study the uppermost layer of place, even with attention paid to the past. In the case of the Palestinians of the West Bank, it is necessary to include the past *and* the future in order to appreciate their sense of place in the present, and one needs to be aware of how this temporal bracket has evolved over the last fifty years. This is because, in addition to memory of place, anticipation, too, can play a part in experience and perception. Palestine is a place that is subsumed by history and that is also "in progress," being shaped by and generating ideologies if not always ideals. Yet it has been suggested that for Palestinians place is in abeyance, and Edward Said has argued that "in a very literal way the Palestinian predicament since 1948 is that to be a Palestinian at all has been to live in a utopia, a *nonplace*, of some sort" (Said 1980: 124, emphasis in the original). Other Palestinians make a case that place is portable, as in the statement by refugee author Fawaz Turki (1988: 36) that "As we grew up, we lived Palestine every day. We talked Palestine every day. For we had not, in fact, left it in 1948. We had simply taken it with us."

In this chapter I argue that the very appreciation of place, the self-conscious attention given to the concept and the political manipulation of place in popular culture, has engendered a situation in which yearning for place is frustrated rather than fulfilled, even as it is being realized. One can trace this problem back to Woodrow Wilson's offer of self-determination, if not earlier, and the suggestion that self-determination is the sine qua non of nationhood. In such a construction, place can be conflated with notions of unbridled sovereignty, with the absence of the latter impinging on an appreciation of the former. Place may then be more fulfilling in the abstract than in the harsh realities of contested space, as suggested by Turki's experience.

Is place, in fact, fixed? It has certainly been argued that place is an infinitely varied experience, with common elements for groups, but shaded to the personal by each individual (Relph 1980). Can one be at home but in a nonplace, or away from home yet with an acute sense of place? The Palestinian experience, ranging from citizenship in Israel, noncitizenship residence in East Jerusalem, at home but occupied in Gaza and the West Bank, displaced internally or as a refugee, to exile in its various manifestations, provokes an examination of many of the elements of place that intrigue geographers and, increasingly, others as well.[2] It is obvious that this case is not typical; yet, at the same time, place in extremis is becoming more common. Exile, frustrated national ambitions, truncated territory, rapid landscape change motivated by exogenous forces, all of these are experienced, separately or in conjunction, in manifold conflicts around the world. For this reason, place may not be something that people can take for granted, at least not those caught up in the political and spatial shuffling that has characterized the twentieth century and will succeed it into the next. The Palestinian case is therefore an important example of place for people in conflict. More accurately, it demonstrates how one's expectations of place can come to detract from or inhibit the experience of place. It shows how a blend of myth and anticipation of place can foster a maximalist dynamic, wherein anything short of the ideal is tainted, and has the capacity to create dissonance and disappointment (Figure 13.1).

Both the past and the future are components of the present for Palestinians and their sense of place. Inasmuch as the machinations of the current "peace process" are stimulating my reflection on place in the West Bank, I will explore the representations of the past and the future (now eclipsed in time by current events) that have shaped place in Palestine. To do so, I am confined to standard tools of literature, interview, and landscape, though these all present problems that require

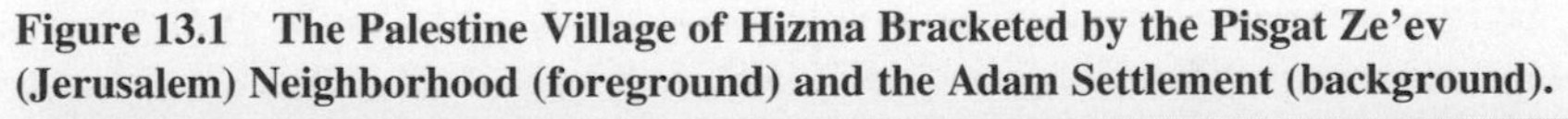

Figure 13.1 The Palestine Village of Hizma Bracketed by the Pisgat Ze'ev (Jerusalem) Neighborhood (foreground) and the Adam Settlement (background).

careful navigation. For the past I look to poetry and literature, while acknowledging that this is a limited and privileged optic—one that is often distant from Palestinians whose class or cultural orientation is the subject of writing that they have not themselves produced (Khalidi 1988). Similarly, I draw upon the promises of Palestinian and Arab elites, who sought to convey a future that they would deliver to the masses. As with the notions of the past, these promises became premises, ingrained in the culture of politics and place. After exploring Palestinians' conceptions of place in the past and in the anticipated future, I turn to their current situation on the ground, and probe the challenges to those conceptions that are generated by their collision with the political structures imposed by both Israel and the Palestinian Authority. My examination of the past draws upon written material that tried to convey a sense of place. For the present, I explore the complex map that fractures daily Palestinian life, posing a real challenge to those who hold to images of an idyllic past and an ideal future.

A PALESTINIAN PAST

> Do you know Jerusalem? You were probably too young when the Zionist monster gobbled up the most beautiful half of the most beautiful city in the world. It is said that Jerusalem is built on seven hills, but I have walked up and down all its hills, among its houses built of stone—white stone, pink stone, red stone—castle-like houses, rising high and low along the roads as they go up and down. You'd think they were jewels studding the mantle of the Lord. Jewels remind me of the flowers in its valleys, of spring, of the glitter of its blue skies after spring showers.[3]

So writes the Palestinian novelist Jabra Jabra in his novel *The Ship* (1985: 20). The passage is rich in the mechanics of place making, at least through literature. It is studded with superlative, its narrative reveals intimacy with and possession of the land, and it reflects the type of oral tradition indicated by Turki above. The telling of Palestine to the young, and old as well, is characteristic in this way of Palestinian poetry in the years following their dispossession (Ashrawi 1978; Shinar 1987).

The lost landscape of Palestine is depicted in ideal terms, often Edenic in its quality and texture. Jabra focuses on Jerusalem in the passage above, striking a powerful chord with Palestinians from all walks of life, both Muslim and Christian. While Jerusalem is seen as a common birthright by Palestinians, other writers develop rural images, or of particular cities or villages that are dear to them, or places associated with episodes in the national or historic Arab narrative. Palestinian authors draw out the texture of the land, and play to the connection between land holding and honor that is an icon of Palestinian society. Trees, flowers, and crops of various kinds are spoken of in appreciative and longing terms. Parmenter (1994: 79) notes that Palestinian writers "enlist nature in general, and the land in particular, as their last and strongest ally." Flora and fauna are often active figures in Palestinian poetry, substituting for the defeated and dislocated people themselves. The mechanism of idealization is not limited to the poets and authors, however. Indeed, the lens of nostalgia is common in Palestinian society, and has produced maps of what was—showing the location of Palestinian villages destroyed in 1948—while denying what is, the creation of Israeli communities in their place (Cohen 1998; Cohen and Kliot 1992). Historiography and geography have also been invoked in the reification of the Palestinian past (Khalidi 1992; Falah 1996), as scholars inventory what was lost and attempt to preserve or revivify it. The creation and mobilization of an idealized memory is both a cultural and political act, and Withers (1996) notes its power in terms of national identity and struggle in his work on Scotland.

Even as the past was being codified in literature and scholarship, Palestinians continued to experience daily life in and around the places that were lost to them. While many were displaced from their homes, some remained in or near their villages in Israel (those that were not destroyed in the war). Others lost their homes but became proximate refugees, at times within sight of Israel. Indeed, in the years immediately after the war many Palestinian exiles were close enough to reach home with a moderate walk, and initially there was no border fence to prevent them from doing so.[4] Thus, alongside the immense disruption and suffering that came with the creation of the state of Israel, traditional Palestinian life continued to exist in its basic forms in many areas of the West Bank. There, and in other areas too, the traditions and genres that were being memorialized/idealized continued to play out as daily life. They did not, however, earn the veneer of romanticism applied to that which was lost. Peasant life remained peasant life, poverty was still the norm, and Palestinians struggled to cope with their new conditions.

A critical transformation was that, in fundamental respects, Palestinians went from being insiders to being outsiders in many of their own places (Relph 1980). Though their national identity was not yet fully formed (Khalidi 1997), it was obvious that their aspirations as a people had been thwarted, and their homeland was controlled by others who would give it an alien character (Figure 13.2). By taking control of the land, Israel invested it with even greater meaning for the Palestinians (Parmenter 1994), even as Israel and Jordan (in the West Bank and East Jerusalem) began to stamp their imprint on the land and interdict the evolution of Palestinian place making. Thus, even in situ, Palestinians were responding to the loss of place. The loss was related to issues of control of the land. For if land ownership was closely connected to Palestinian honor, the loss of the land, at least in political terms, had a profound impact on self-image (Committee on International Relations 1978).

The imposition of Israeli and Jordanian rule made government more invasive, bringing outside power to the heart of Palestinian society, both urban and rural (Aburish 1988: 154). The inability to circulate freely also was a radical change

Figure 13.2 A Mosque in the Village of Beit Iksa, with the Jerusalem Ramot Neighborhood in the Background.

for Palestinians. Within Israel, Palestinians were subject to military control until 1965, and their movement within the state was conditioned on approval from military authorities. Palestinians in the West Bank were cut off not only from Israel but from the Gaza Strip as well. In Jerusalem, formerly Palestinian neighborhoods were tantalizingly close, but beyond their reach. Disruption of trade patterns and communication routes compounded the sense of dislocation and upheaval that filtered into society. The construction of a well-guarded armistice line slicing through the countryside (and Jerusalem) served as a tangible message in the landscape that Palestine was divided. In some places, Israel deliberately undertook projects adjacent to the border as a statement to the Palestinians of Israel's tangible and ongoing existence (Cohen 1993). Other efforts, whether deliberate or incidental, subtle or explicit, have operated in and on the landscape in ways that challenge Palestinian attachment to and sense of place (Falah 1996, Baxter 1991).

LIBERATING PALESTINE

Palestinians did not necessarily subscribe to notions of Israeli permanence, and depictions of lost Palestine were not solely laments. There was a suggestion, however muted at first, that the land and its beauty could be reclaimed, restored to its Palestinian inhabitants. The link between the possibility of recovery and the immaculate descriptions that posited a Palestine that never existed but was then created through a combination of memory, nostalgia, and construction helped to nationalize the land through the building of its attributes (Gruffudd 1995). By the time the Palestinian national movement began to organize and move toward the political stage in the mid-1960s, what had been lost to Palestinians had already become mythic. What was promised by the nascent national movement was nothing less than the "return" of that mythic past, and the symbols that Palestinians identified with were eagerly taken up by the political organizations. "Trim the leaves of youthful olive trees/But the roots of the olive/ Return and stretch in the depth/Of the soil. . ./Roots of olive, you have become the model/And man is competing, imitating your root,/Olive of the land" (Mahmoud Awad Abbas cited in Ashrawi 1978: 91).[5]

Baxter's (1991) ethnographies of Palestinian life during the *Intifada* include descriptions of organized "field trips" around the West Bank and inside Israel that were designed to give Palestinians some familiarity with the landscapes of the past, with the suggestion that possession was to shift in the future. Similarly, field surveys that I conducted with Palestinians during the late 1980s and early 1990s were a mix of "what was" and "what will be" when Israeli rule comes to an end (Cohen 1993, 1994). One concrete manifestation of the orientation toward the future was the construction of housing subsidized by the Palestine Liberation Organization (PLO). Individuals could secure loans and grants for building in areas that were vulnerable to expropriation by Israel, and such dwellings were often constructed well in advance of their actual use as domiciles. Built during the occupation, they were viewed as homes for a better day, when the anticipated future was delivered through the defeat of Israel.

In ideological terms, the Charter of the Palestine Liberation Organization (1968) served as the guiding principle for struggle against Israel until the program it outlined was eclipsed by greater pragmatism in the early 1990s. The document called for armed conflict that would lead to the destruction of Israel and the creation of a Palestinian state in all of Palestine.[6] It aspired to more than that in seeking to turn the clock of history backward for the Palestinian people. Article 6 of the Charter states that "The Jews who had normally resided in Palestine until the beginning of the Zionist invasion will be considered Palestinians" (Moore 1974: 706). This phrasing amended an earlier (1964 Article 7) version of the Charter which suggested that "Jews of Palestinian origin shall be considered Palestinian if they desire to undertake to live in loyalty and peace in Palestine" (Moore 1974: 700). The later version, which came after the Six Day War and the Israeli occupation of the West Bank, called in effect for the disenfranchisement of all but the oldest members of the Jewish community, inasmuch as the "Zionist invasion" began in the nineteenth century. It also precluded the possibility of rights by indigenous birth that were technically possible in the earlier version. This element of the Charter reflects not only Palestinian resentment over the immigration of Jews and the usurping of political rights, but also the dissonance created by the presence of Jews, at least politically and militarily dominant Jews, in Palestinian daily life.

The combination of political and cultural rejection is evident in the refusal to use the phrases "Israel" and "Israeli," which marked the Palestinian lexicon until the 1980s, preferring "Zionist entity" for the state and "Zionists" or more commonly "Jews" instead of Israelis.[7] Negation of the presence of Jews in Palestine and their impact was accompanied by positive reinforcement of connection to the land. Article 7 of the PLO Charter states that "All means of information and education must be adopted in order to acquaint the Palestinian with his country in the most profound manner, both spiritual and material, that is possible" (Moore 1974: 707). The acquaintance, of course, was intended for Palestine as a whole, and emphasis was placed on what was lost, and the promise of recovery. Indeed, much of the Charter relates to liberating and liberated Palestine, as the future is anticipated by the political movement (Figure 13.3).

THE PRESENT TENSE PLACE

With the advent of the peace process in the 1990s, Palestinians have shifted the course of their future to a negotiated outcome, conceding, perhaps irrevocably, that they are not sole masters of their fate, nor of the territory that they call their own. Thus, although initial strides made toward ending the occupation have been taken, the domination by Israel persists. Paradoxically, even as Israel begins to withdraw its military from portions of the West Bank and Gaza, the landscape of occupation becomes more complex. Whereas prior to the peace process the occupation was relatively ubiquitous in its imprint throughout the West Bank, today it is a maze of adjacent and overlapping jurisdictions of various hybrid characters, existing in remarkable proximity (Figure 13.4). The map of authority in the West Bank is fodder for political geography, but its impact cannot be gauged solely in political terms. Palestinian sense of place is shaped by the juxtapositions of outright occupation and shades of autonomy that mark the passage of their day as they move from home to school, the store, work, and so on. In order to demonstrate the pervasiveness of the peace-process map, some description is required.

One of the main mechanisms of the peace process generated by the Oslo Accords of 1993 is a phased transfer of authority over parts of the West Bank and Gaza Strip from Israeli to Palestinian hands, with variations of rule and responsibility for the territories as a whole subsumed to ultimate Israeli control. The maximum extent of Palestinian power is autonomy in daily life and self-policing within proscribed (and circumscribed) areas. Such areas, given the alphabetical identification "A," comprise the smallest of the territorial types in the West Bank, confined to the major cities and towns and a few stray villages. A lesser

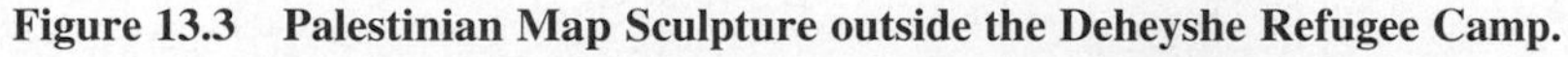

Figure 13.3 Palestinian Map Sculpture outside the Deheyshe Refugee Camp.

measure of latitude is accorded to Palestinians in Area B—those which have joint Israeli-Palestinian policing, but autonomy in daily administration and bureaucracy related to infrastructural issues. Area B status applies to the remaining villages of the West Bank. Area C, full Israeli control (but not sovereignty), is applied to all Jewish settlements, a small portion of the city of Hebron (given a special designation H1), and all parts of the West Bank not identified as either Area A or B.[8]

Both the military and the cultural landscapes reflect the zonation, and Israeli checkpoints at the perimeter of Area A, with Palestinian checkpoints just inside

Figure 13.4 Map of Zones of Authority in the West Bank.

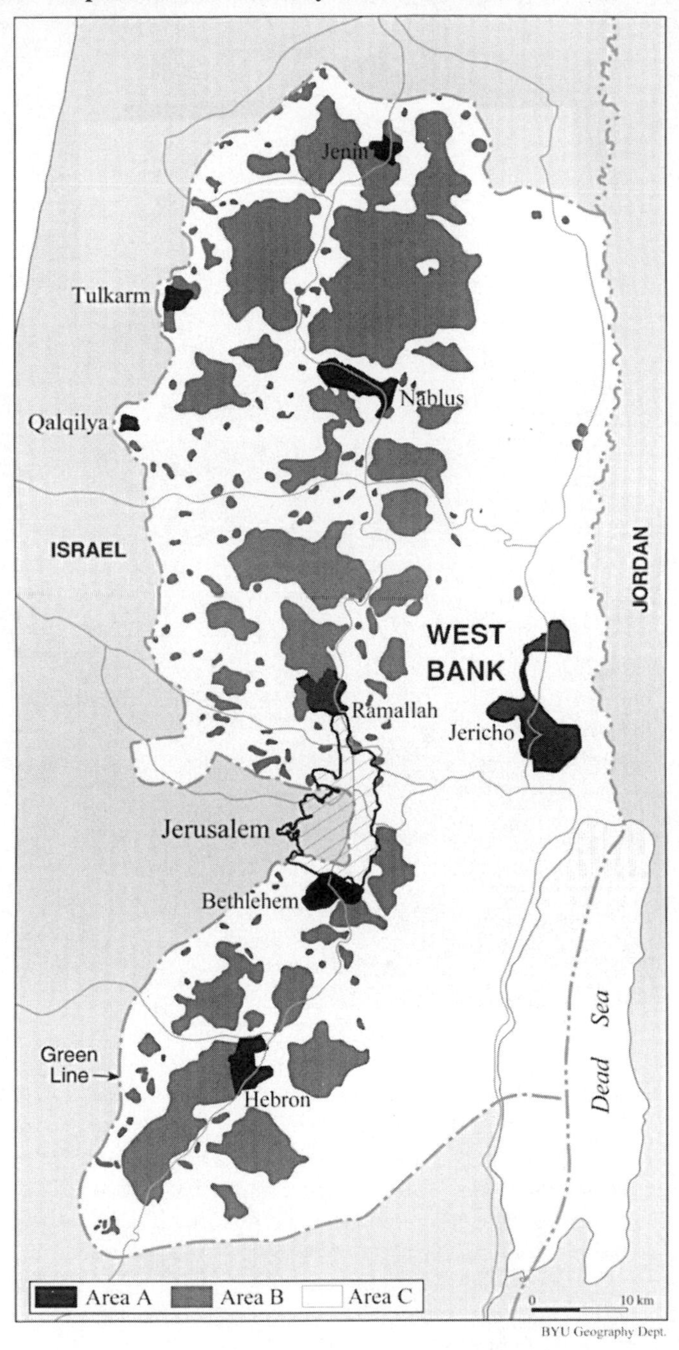

them, provide control of access and drive home the differences in authority (Figure 13.5). Within the Palestinian autonomous zones there are armed Palestinians in a wide array of uniforms. Until recently, such a display was the rare domain of nationalist demonstrations and an invitation to a violent clash with the Israeli army. Today, it represents the implementation of peace negotiations, but, more importantly, it symbolizes the beginning of a shift from occupation to post-occupation. So, too, a change in signage reflects burgeoning Palestinian autonomy, as Israeli-placed bi- or trilingual signs are replaced by those without any Hebrew script. Indeed, though some merchants have yet to do away with the visible vestiges of an Israeli clientele, in Palestinian cities the landscape is becoming increasingly devoid of the superficial displays of occupation.

In some respects, these autonomous areas reflect the withdrawal of Israeli troops in terms of social and economic activities as well. Palestinians, both merchants and customers, are able to engage in daily trade without the interference of Israeli troops or the oversight of Israeli tax authorities.[9] The absence of an Israeli military presence has all but removed the threat of clashes in the street, and the rerouting of Israeli traffic via bypass roads around Palestinian cities has eliminated stone throwing, tire burning, and street barricades, and eased Palestinian insecurity vis-à-vis hostile Israeli settlers. As a result of these factors, Area A zones have become a safe haven of sorts for Palestinians, a bracketed respite from immediate contact with Israel and Israelis.

In the Palestinian villages of Area B, there is a measure of freedom from Israeli

Figure 13.5 Israeli Checkpoint on the Bethlehem Road.

policing, but it is not one that provides real security. Three factors continue to impinge on Palestinians in a threatening way. First, there is the presence, albeit diminished, of Israeli troops in joint patrol with Palestinian police, and the ongoing work of Israeli undercover squads that circulate in the area. Next is the impact of the various Palestinian security forces, which work in at least partial cooperation with Israel, a fact which has greatly embittered the Palestinian population. This comes in conjunction with the continued menace of informers/collaborators and destabilizes Palestinian society, as people feel vulnerable to intimidation and political machination that can offer them up to Israeli or Palestinian forces at any time, for any reason. Finally, the passage of Israeli settlers—often in convoys and always armed—on roads that transect Palestinian villages creates a recurring point of friction and, not infrequently, violence. Thus, even though the Israeli presence has lessened, and Palestinians have assumed control over many bureaucratic issues, enormous dissonance still marks daily life.

Area C can be thought of as the gauntlet that Palestinians run between their islands of measured refuge. Area C is still fully under Israeli administrative control, and subject to an ongoing campaign of land use restriction for Palestinians, accompanied by an expansion of Israeli settlements, road building, and infrastructure development. Indeed, the landscape of the West Bank in areas under Israeli control is being rapidly stamped with Israel's presence. Through a massive infusion of funds, signs, and structures intended to support and highlight Israeli settlement patterns, the parts of the West Bank that are for Israeli use are becoming less and less distinguishable from areas of modern Israel that are of recent construction, save perhaps that the West Bank areas are often better developed.

In some respects, the areas of authority in the West Bank are as distinct and discrete as their nomenclature implies, particularly in legal terms. The presence of outposts is a good indicator of the zonation, although the Israeli army reserves the right to enter Zone A should it be deemed necessary. This has not yet happened, and there are, therefore, visible clues that key one's orientation to a given place, though they are far from infallible. Indeed, outside of Area A, which is demarcated by checkposts, the lines sometimes blur, particularly at the margins or outside of built-up areas. The patchwork pattern of authority in the West Bank has created an almost innumerable set of frontiers, within which the boundaries frequently exist less in any concrete way than they do in the drawers of government bureaucrats.

The inability to pinpoint the status of an exact location has long been problematic for Palestinians in one particular respect. Planning law has been used to severely limit construction of Palestinian housing, with the aim of increasing density within Palestinian villages and preventing their expansion onto open or adjacent lands (Cohen 1994; Coon 1992). Being "inside" or "outside" of the village can be the critical distinction between having a building license approved or rejected. Since much of the construction is done without a legal permit (which is often impossible to get), being inside or outside can be the difference between having one's home overlooked by regulatory authorities, or destroyed. Palestinians have two ways of knowing whether they are inside or outside the Israeli-determined planning areas: they can visit a government office to inspect an aerial photograph (a problematic option in a number of ways), or they can learn through bitter experience what is tolerated, and what is not, in terms of housing construction. This situation has changed somewhat since the peace process began, as permits are nearly always granted by Palestinian authorities for construction in areas under their control, and retroactive approval is also possible (Dudin 1998). House demolitions by Israel continue apace, however, providing an informal delimitation of authority in the West Bank in a way that penetrates to the core of the indigenous population. The area most closely contested between Israelis and Palestinians today is that around (and in) the city of Jerusalem, and the lines there are both distinct and blurred, depending on location and purpose. The issue of zonation and place comes into sharp relief there.

THE INS AND OUTS OF JERUSALEM

For both Israelis and Palestinians, Jerusalem is the penultimate "prize" in their competition and conflict. Inasmuch as open warfare is not currently an option, both sides seek to advance their claims to and hold on the city through other means (Cohen 1993). The status of Jerusalem is a question that has been left to the final stages of negotiation in deference to its complexity and sensitivity. As Israel sees it, the Jerusalem municipality, including the Palestinian villages that were incorporated with the annexation and expansion of the eastern part of the city in 1967, is wholly within Israel. According to the Palestinians, East Jerusalem, at the least, is part of Palestine and should become their national capital. Yet while negotiations are in the future, concrete acts are being taken to direct the outcome of the negotiations. One tactic that Israel has pursued is restricting access to Jerusalem by prohibiting the entry of Palestinians from the West Bank, except according to special permits.

This measure has caused enormous disruption in the lives of Palestinians who live in and around the city, as people are faced with a difficult daily choice. They can either cede access to the city that has served as their economic, political, and spiritual center, or risk fines, imprisonment, and possibly beatings for what is categorized as illegal entry. Jerusalem, however, is permeable in a variety of ways, making it elastic as a place, at least at the margins. While the main roads connect with Area A cities in the north (El Bireh/Ramallah) and the south (Beit Sahour/Beit Jalla/Bethlehem), and are thus manned by roadblocks—as are the primary east–west routes—the roads from adjacent West Bank villages are only sporadically watched. The sprawl of the city and the surrounding communities has created a situation wherein Jerusalem, nominally a distinct municipal entity under Israeli control/sovereignty, is enmeshed with areas that are under the auspices of Areas A, B, and C of the West Bank under the Oslo Accords.

Perhaps the most extreme example of this comes with the Ramallah road, in northern Jerusalem, which leads to the area of the municipal airport and the Atarot industrial area, and to the Palestinian cities in the West Bank beyond them. For a portion of this route, the road itself and the land to the west of the pavement is part of the Jerusalem municipality, whereas the land to the east of the pavement is in a Palestinian town in the West Bank. Israeli law applies to the pavement and the area to the west, including restrictions on Palestinian entry, while on the other side of the street West Bank rule applies (Figure 13.6).

While the above example represents a rare and particularly absurd case, Palestinians are subject to rapid transition from one zone to the next, and often they can see from one zone into the next, or even several others, when they are stationary. The intense nature of Israel's security efforts in Jerusalem has always cast the adjacent villages in sharp relief for Palestinians, thousands of whom are able to gaze upon the city from their homes but are not able to enter it.

The absence of Israeli forces in Palestinian cities that bracket Jerusalem serves to emphasize a message that Israel is eager to deliver, a tangible demonstration of territorial control (Sack 1986). In light of the difficulty of entering Jerusalem (Palestinians face long delays for document inspection at the "border"), conducting business or social affairs in the Palestinian cities, instead of Jerusalem, has gained tremendous practical appeal, and the lack of activity on the streets and in

Figure 13.6 The Israeli Checkpoint at Ram, with Palestinian Housing Beyond.

the shops of East Jerusalem is a reflection of this shift. Yet to get to those cities, Palestinians may have to drive from their village in Area B through Area C toward Jerusalem, pass an Israeli checkpoint on the way into the city, then leave Jerusalem and enter Area A, which comes with a Palestinian checkpoint for emphasis. This will encompass shifting from joint Israeli-Palestinian patrols to the Israeli army, then the Israeli border guards (Mishmar HaGvul), to Israeli police, then to the Palestinian police.

As a result of the mosaic of authority, what appears to the untrained eye as a unified landscape is, to both Palestinian and Israeli, a fragmented yet interlocked series of challenges and transitions. Mobility is marked by obstacles and menace at the practical level, and at the psychic level as well. In part due to the insecurity that Israelis feel, despite their sovereign state and occupying army, Palestinians are effectively under siege, in enemy territory at home, and mundane affairs are carried out in the shadow of a never-distant disruption, ranging from the inconvenient at best, to the fatal in severe cases.

All of this conflicts with what Palestinians, like other nations, hope for and have been led to expect by the discourse of self-determination that has been a primary political theme of this century. The Palestinian national movement has elevated the rhetoric of legitimate rights to include a restoration of the past. Even in the era of political compromise, public oration speaks of the conquest of Jerusalem and the liberation of Palestine through holy war. Meanwhile, the daily experience of Palestinians mocks those promises, and each day they see more of the past, idealized or not, being covered over by the manifestations of growth and development, whether it be practically or politically inspired. Increasingly, the images that correspond with Palestinian constructions of the past exist as exclaves, functioning as refugia containing endangered relics and patterns. The landscape has thus become a set of islands, upon which Palestinians can find partial relief from the nemesis of occupation.

THE MALFUNCTION OF PLACE

Place, particularly when associated with home, is generally thought to convey an element of security. It is only in situations that are judged to be problematic that home does not overlap with this sense of security. Issues leading to such dissonance occur across a variety of scales, and between them as well. According to Agnew and Duncan (1989: 7),

> place, both in the past and in the present . . . serves as a constantly re-energized repository of socially and politically relevant traditions and identity which serves to mediate between the everyday lives of individuals on the one hand, and the national and supra-national institutions which constrain and enable those lives, on the other.

For a people whose land is occupied, and who are in a variety of ways displaced, place may seek to mediate the relationship with the dominating culture, but it may fail for a number of reasons. Two such reasons can be seen in the case of the Palestinians and Israel. The first, and most damaging, is the ubiquity of the Israeli presence, which disrupts the Palestinian landscape and overlays Palestinian places with a constant and invasive message of occupation. Lustick (1993) argues that such a presence can goad a subject people into revolt, and, indeed, the Palestinians have risen up in an interesting and somewhat effective manner against Israeli rule. But this raises a second way in which place has failed the Palestinians in that, at least to date, their future has been incrementally realized in ways that fall far short of promises and expectations that have played a central role in the development of Palestinian identity. Those promises related to place, and served as a sustaining and unifying force in Palestinian society. They helped to shape an understanding and anticipation of place that was far along the spectrum in the direction of a local version of utopia.

And, while the peace process may be far from its ultimate outcome, it is already clear to many Palestinians that the end of occupation, the inception of Palestinian self-rule, and the realization of their national ambitions do not necessarily coincide. For those rooted in areas that will remain in Israeli hands, displacement will be formally (and "permanently") conceded. For those coming from the Gaza Strip and the West Bank, the territorial boundaries of Palestine will have closed in upon them far more closely than they had anticipated in earlier notions of statehood. Many Palestinians will not return, whether by choice or circumstance, thus maintaining a diasporic community that will grapple with complex issues of identity and loyalty, bonded to their homeland but visiting it as tourists.

In a certain respect, the Palestinian people are now involved in a process of state building in which they play both an active role and, at the same time, move about almost as tourists in their own land. The lack of enfranchisement comes at two levels, one obvious—their relative lack of power vis-à-vis Israel—and one more subtle. Prior to the peace process, Palestinians engaged their land in the venues of loss and resistance. The ability to create their own place was doubly circumscribed, because they could be masters of their own domain only at the microlevel, as even the interior of their homes was vulnerable. At any level beyond that, they were limited to creating place in their imagination and in the future, and, to some extent, this entailed recreating the past—not replicating it, but recreating it to address current needs and future uses. As Frank (in press) points out, the very language of engaging Israel in the peace process shattered the concept of the past idyll as the future goal, and nominally shifted Palestinian politics from reactive to proactive. Yet in giving up the place of the past in order to achieve a place in the future, the Palestinians became vulnerable in a way that was new to them but which they share with other nations that are in the process of realizing their independence.

Palestinians are now faced by a democratically elected (Palestinian) government, exercising its (limited) authority in (truncated) Palestinian territory. As designed by the Oslo Accords, they are undertaking an experiment in self-rule during a probationary period that ran its course in May 1999. Current external factors obviously prohibit the free realization of any Palestinian conception of statehood, but, at the same time, the nascent outlines of Palestinian rule can be seen. The early outcome of this new active role is less than satisfactory to many Palestinians, and therein lies a new challenge to Palestinian sense of place. Whereas politicians promised a Palestinian Eden of the past, the early inceptions of Palestinian rule cruelly mimic the textures of Israeli occupation: economic hardship, denial of free speech and due process, harsh repression by security forces, inadequate provision of services, political manipulation of social cleavages, and, in a parallel to earlier times, the consolidation of power in the hands of a small ruling class. It is no surprise, then, that religious forces continue to oppose the peace process, betting on an eschatology that seems more rewarding. For them Palestine is sacred, but its administration must follow Islamic form for it to realize its significance.

The remaining Palestinians are left with the task of reworking their myths, tempering their nostalgia, and forging a viable future, in both practical and symbolic terms. Certainly Israel is complicating the task. A new zone for the West Bank has been approved, Area D, which is being described as a nature preserve (see note 8). It served to bridge Palestinian and Israeli proposals on the scope of a second-phase troop withdrawal at the Wye River Accords of 1998. Under the agreement, Palestinians will take control of the area, which falls in the harsh and nearly unpopulated Judean Wilderness. In that Area D, Palestinians are not allowed to undertake any form of development so as not to harm the local environment. The Palestinians thus control the area but are technically able to do almost nothing with it. It is hard to invest territory with meaning under such circumstances, and it is hard to imagine that it will afford Palestinians much comfort or practical utility if they abide by the terms of the Wye Accords.

THE ROADS AHEAD

The case of the West Bank presents a challenge for cultural geographers: how do we address our understanding of place in a world experiencing frequent political change and cultural turmoil at the local level (along with significant territorial change and discourses of globalization more broadly)? Place, for many people, is increasingly circumscribed or dictated, rather than being created by and for individuals and communities or nations. How do we adjust our notions of history and tradition in a world of rapid, if sometimes passing, transformation?

An easy answer might be to accede to the postmodern, with multiple spatial and temporal scales assuaging some of the vacuum that can be created when our definitions of place do not serve those experiencing displacement. Yet, as geographers, we are (or ought to be) committed to an appreciation of scale that does not lose sight of the very real need that is associated with place, one that is not mitigated by "*isms*." The Palestinian case is a candidate for postmodern interpretation, with its flows of territory and time. Yet the *experience* of Palestinian life (and Palestinians themselves) demands concrete solutions based on values and promises of a pre-postmodern era. This is the case for too many people in the world, in a variety of manifestations, and our research agenda must not lose sight of this fundamental circumstance.

What role and at what scale, then, can place play in Palestinian nationhood? While I have argued that the idealized version of Palestine, in both the past and promised future, has not been and cannot be realized, the mobilization of Palestinian patriotism and the strengthening of national identity are affiliated factors that should not be disparaged. If Palestinian writers and politicians overplayed their hand, or were caught up by image and rhetoric, they were at least in part indulging a legitimate need for national focus while providing an outlet for both grief and rage, all within culturally valued and logical modes of expression. It seems likely that, despite the peace process, indeed because of it, Palestinians will continue to require such outlets, even if they have something to celebrate as well.

One might expect, therefore, a reaggregation of Palestinian places, as the islands of refuge are linked together territorially and, hopefully, as the threat from the hostile areas separating them subsides. Place can then become an amalgam of national and subnational associations, dispersed throughout the patchwork of authority and sovereignty that seems likely to be the Palestinian future. Practically, this may be the best that they can do, at least for now. This is because for Palestinians, place has been exchanged, the mythic and sustaining image has been tendered, and, in return, they are left with the bitter reality of their current predicament, trapped between the past that was not and the promised future that will not be. Al-Barghouti suggests (1998: 61) that the last fifty years have changed Palestinians from "sons of Palestine into the sons of the idea of Palestine," and reveals his sense that his "relationship to place is, in fact, a relationship to time," saying, "I live in islands of time, some of which I already have lost, others I possess for a moment, then lose them, because I am always placeless" (1998: 64). Perhaps through sovereignty some Palestinians will find sustenance in their remaining places, even if they are islands, realizing the sentiment expressed by the Palestinian poet al-Hut (Sulaiman 1984: 120), "O Lost Paradise! for us you were never too small."

NOTES

1. The West Bank is an area created by the Israeli-Jordanian armistice agreement of 1949. From that time through the Six Day War of 1967 when it was captured by Israel, it

was ruled by Jordan, and nominally annexed to it since 1951. I am taking it as an administrative unit, and not problematizing it as a region, though there would be merit in that exercise (Murphy 1991). Though the Gaza Strip is an obvious companion to the West Bank in terms of issues of place and the Palestinian experience, conditions are sufficiently different between the two to merit separate treatment. Moreover, my experience in Gaza is slight compared to my years of research in the West Bank.

2. Anthropologists are flocking to place, alerted more by the prompting of social theorists than by an awareness of the work of geographers. Indeed, with rare exceptions (for instance, Olwig and Hastrup 1997), it seems that anthropology is merrily reinventing the wheel that geography keeps in spin.

3. From *The Ship* by Jabra I. Jabra, trans. Adnan Haydar and Roger Allen. Washington, DC: Three Continents Press, 1985. Copyright 1996 by Lynne Rienner Publishers. Used with permission of the publisher.

4. There were, however, military patrols on the lookout for such "infiltration," and many Palestinians paid with their lives for trying to reach their land, even if only to harvest the orchards that they had planted and then return back across the border. See Morris (1993).

5. This is a poem of Mahmoud Awad Abbas entitled "Pieces Played on the Psalms of Winter" from his work "Melodies with a Sharp Rhythm," pp. 88–89.

6. The PLO was formed while the Gaza Strip was under Egyptian control and the West Bank and East Jerusalem were in Jordanian hands. Palestinians had long suspected Jordan of cooperating with Israel in 1948 to the detriment of the Palestinians, and there is evidence to support that claim (Shlaim 1990). The claim to all of Palestine was sometimes interpreted as including not only the territory of Israel, the West Bank, and Gaza, but all of Jordan as well. Though Jordan is demographically dominated by Palestinians, the imagery of loss and place only rarely relates to land that is east of the Jordan River, that is, in the state of Jordan.

7. The use of the term "Yahud" or Jew, is still predominant among Palestinians in the Occupied Territories, though "Israeli" is more common than in the past.

8. The newest category of land in the melange of the West Bank, Area D, is discussed below. It is slight in area and practical importance. At the same time, it represents an interesting mechanism whereby both sides can make concessions that are symbolically significant yet which both sides also understand are effectively hollow. Palestinians may well ignore the restrictions imposed on Area D, and Israel is unlikely to do anything about any such breach of the accords.

9. Tax evasion is a time-honored custom in Palestine, but the strict enforcement of tax law is sometimes used by Israel in a punitive way. The Palestinian Authority has taxation as well, but for the time being evasion is the norm. Palestinian stores continue to carry Israeli products; in that regard there is no differentiation among zones in the West Bank.

REFERENCES

Aburish, Said K. 1988. *Children of Bethany: The Story of a Palestinian Family.* Bloomington: Indiana University Press.

Adams, Paul C. 1997. "Cyberspace and Virtual Places." *Geographical Review* 87 (2): 155–71.

Agnew, John, and James Duncan, eds. 1989. *The Power of Place: Bringing Together Geographical and Sociological Imaginations*. Boston: Unwin and Hyman.

Ashrawi, Hanan Mikhail. 1978. "The Contemporary Palestinian Poetry of Occupation." *Journal of Palestine Studies* 7 (3): 77–101.

al-Barghouti, Mureed. 1998. "Songs for a Country No Longer Known." *Journal of Palestine Studies* 27 (2): 59–67.

Baxter, Diane. 1991. "Living the Uprising: Palestinian Lives During the Intifada." Department of Anthropology Ph.D. dissertation. Los Angeles: University of California.

Cohen, Saul, and Nurit Kliot. 1992. "Place-names in Israel's Ideological Struggle Over the Administered Territories." *Annals of the Association of American Geographers* 82 (4): 653–80.

Cohen, Shaul. 1993. *The Politics of Planting: Israeli-Palestinian Competition for Control of Land in the Jerusalem Periphery*. Geography Research Paper No. 236. Chicago: University of Chicago Press.

———. 1994. *Village Planning in the West Bank: An Analysis of Israeli Policy and Practice.* Unpublished Report, International Committee of the Red Cross, Jerusalem.

———. 1998. "Terra Sancta: Maps, Politics, and History." *Mercator's World* 3 (5): 16–21.

Committee on International Relations. 1978. *Self-involvement in the Middle East Conflict.* New York: Group for the Advancement of Psychiatry, vol. 10, no. 103.

Coon, Anthony. 1992. *Town Planning Under Military Occupation: An Examination of the Law and Practice of Town Planning in the Occupied West Bank.* Dartmouth: Aldershot.

Dudin, Fuad. 1998. Personal interview with the Director of Planning, Hebron Region, Palestinian Authority.

Falah, Ghazi. 1996. "The 1948 Israeli-Palestinian War and Its Aftermath: The Transformation and De-Signification of Palestine's Cultural Landscape." *Annals of the Association of American Geographers* 86 (2): 256–85.

Frank, David. In press. "The Mutability of Rhetoric: Haydar 'Abd al-Shafi's Madrid Speech and Vision of Palestinian-Israeli Rapprochement." *Quarterly Journal of Speech.*

Gruffudd, Pyrs. 1995. "Remaking Wales: Nation-building and the Geographical Imagination, 1925–50." *Political Geography* 14 (3): 219–39.

Jabra, Jabra. 1985. *The Ship.* Trans. by Adnan Haydar and Roger Allen. Washington, DC: Three Continents Press.

Khalidi, Rashid. 1988. "Palestinian Peasant Resistance before World War I." In *Blaming the Victims: Spurious Scholarship and the Palestinian Question*, ed. Edward W. Said and Christopher Hitchens, 207–233. New York: Verso.

———. 1997. *Palestinian Identity: The Construction of Modern National Consciousness.* New York: Columbia University Press.

Khalidi, Walid, ed. 1992. *All That Remains: The Palestinian Villages Occupied and Depopulated by Israel in 1948.* Washington, DC: Institute for Palestine Studies.

Lustick, Ian S. 1993. *Unsettled States, Disputed Lands: Britain and Ireland, France and Algeria, Israel and the West Bank-Gaza*. Ithaca: Cornell University Press.

Moore, John Norton, ed. 1974. *The Arab-Israeli Conflict Volume III: Documents.* Princeton: Princeton University Press.

Morris, Benny. 1993. *Israel's Border Wars, 1949–1956 : Arab Infiltration, Israeli Retaliation, and the Countdown to the Suez War*. New York: Oxford University Press.

Murphy, Alexander B. 1991. "Regions as Social Constructs: The Gap Between Theory and Practice." *Progress in Human Geography* 15 (1): 22–35.

Olwig, Karen Fog, and Kirsten Hastrup, eds. 1997. *Siting Culture: The Shifting Anthropological Object*. London and New York: Routledge.

O'Sullivan, Patrick. 1992. "On Place and Being." *Geographical Perspectives* 63 (Spring): 1–12.

Parmenter, Barbara. 1994. *Giving Voice to Stones: Place and Identity in Palestinian Literature*. Austin: University of Texas Press.

Relph, Edward. 1980 ed. *Place and Placelessness*. London: Pion.

Sack, Robert D. 1986. *Human Territoriality: Its Theory and History*. Cambridge: Cambridge University Press.

Said, Edward W. 1980. *The Question of Palestine*. New York: Vintage Books.

Shinar, Dov. 1987. *Palestinian Voices: Communication and Nation Building in the West Bank*. Boulder: Lynne Rienner.

Shlaim, Avi. 1990. *The Politics of Partition: King Abdullah, the Zionists and Palestine 1921–1951*. New York: Columbia University Press.

Starrs, Paul F. 1997. "The Sacred, the Regional, and the Digital." *Geographical Review* 87 (2): 193–218.

Sulaiman, Khalid A. 1984. *Palestine and Modern Arab Poetry*. London: Zed Books.

Taylor, Jonathan. 1997. "The Emerging Geographies of Virtual Worlds." *Geographical Review* 87 (2): 172–92.

Turki, Fawaz. 1988. *Soul in Exile: Lives of a Palestinian Revolutionary*. New York: Monthly Review Press.

Withers, Charles W. 1996. "Place, Memory, Monument: Memorializing the Past in Contemporary Highland Scotland." *Ecumene* 3 (3): 324–44.

Conclusion

Contemplating Enduring Themes and Future Trajectories

Alexander B. Murphy and Douglas L. Johnson

In a book on the so-called culture wars that have raged in literary studies circles, Gerald Graff (1993: 52–53) observed that "theory is what erupts when what was once silently agreed to in a community becomes disputed, forcing its members to formulate and defend assumptions that they previously did not even have to be aware of." This observation is of considerable relevance to the practice of cultural geography over the past twenty-five years. The decade and a half immediately following the publication of *Readings in Cultural Geography* (Wagner and Mikesell 1962) saw much work building on the themes outlined in that volume, but relatively little explicit theoretical writing about cultural geography itself. By the late 1970s, however, the situation was changing as theoretical developments in economic and social geography combined with an intensifying extradisciplinary discussion of culture to challenge that which was "silently agreed to." This, in turn, precipitated the theoretical turn in cultural geography of the past two decades.

The intellectual ferment of recent years has brought to the fore new themes and orientations of the sort discussed in the introduction to this book. It has also precipitated an ongoing theoretical dialogue about the nature of cultural geography itself. The very persistence of that dialogue is an indication that a new community of agreement has not emerged in cultural geography. Yet the discussion of the nature and meaning of cultural geography has matured to the point that some thoughtful "stock taking" is increasingly possible. A number of commentators have taken up this challenge, focusing principally on the problems and prospects of reconciling different theoretical positions that have emerged in recent decades (see, for example, some of the essays in Kobayashi and Mackenzie

1989). But it is also useful to consider the changing thematic emphases of studies in different eras and from different traditions—for these too offer insight into where an intellectual project has come from and where it is going.

The essays in this volume have much to say about cultural geography's past and future, for each is born of a conviction that certain traditional themes merit continuing attention, yet each seeks to treat those themes in ways that draw from the substantive and theoretical emphases of recent cultural geographic work. The very choice of subject matter reflects the authors' notions of the types of long-standing themes that deserve enduring attention yet are at some risk of marginalization. At the same time, none of the authors seeks to treat those themes in a nostalgic fashion. Instead, their empirical and analytical orientations are indicative of where they see cultural geography going.

A better appreciation of the larger lessons of these papers comes from situating them in relation to the themes that have arguably been "pushed somewhat to the side in the wake of more recent ideas and orientations" (see the introduction to this volume). These are: "(1) the interplay between the evolution of particular biophysical niches and the activities of the culture groups that inhabit them; (2) the diffusion of cultural traits; (3) the establishment and definition of culture areas; and (4) the distinctive mix of geographical characteristics that gives places their special character in relation to one another." Interest in these themes does not generally coincide with the large-scale thematic orientations around which the three major parts of this book are structured. Instead, these themes cut across efforts to understand the changing nature of space, environment, and place.

Turning to the first theme, recent work in cultural geography has devoted significant attention to the material arrangements and discursive practices that shape human–environment relations. Yet the problematization of environmental categories and constructs has sometimes shifted attention away from exactly what is happening where. Peter Goheen, James Schmid, and John Kirchner offer an important counterpoint to this state of affairs through work that begins with a particular feature of the physical landscape—a waterfront, a set of rivers and a complex of wetlands—and then focuses on changes over time. None of these essays advocates objectifying these features of the biophysical environment, however. Instead, they look at how societies have both materially and perceptually constructed these features—and the concrete environmental implications of those constructions. Such an approach is evident in many other essays in this volume as well—even those more centrally concerned with different themes.

Interest in the diffusion of cultural traits—the second of our identified themes—has long characterized work in cultural geography, yet cultural diffusion studies have not attracted a great deal of attention of late. The chapters by Karl and Elisabeth Butzer and by James Wescoat remind us of the power and importance of such studies, providing fundamental insights into the evolution of places in the process. Yet in exploring their topics, the authors are not content merely to describe the spread of cultural attributes; in keeping with recent interest in the development of ideas about place, they seek to unravel the cultural and social significance of particular diffusions. This latter concern also marks two essays that take up diffusion of a different sort—the diffusion of particular ideological representations across time and space. This is the focus of attention of both David Lowenthal and Charles Good, and their chapters reflect the larger contemporary concern with situating ideas in social and cultural context. Yet in approaching their topics they are unusually sensitive to the ways in which particular practices in particular places shape the evolution and diffusion of the ideological constructs they examine.

Perhaps no realm of cultural geography is more associated with the early development of the field than is the one that focused on the establishment and development of culture areas. In the hands of Michael Conzen, Carville Earle, and Chauncy Harris, however, this third thematic concern takes on a new and vibrant character. In each of their studies, the contemporary interest in the social importance of ideology and belief is fused with more traditional "culture region" interests in a manner that offers provocative new insights. Their studies suggest that a productive fusion can arise out of a concern with the role of beliefs in conflicting efforts to construct culture areas. This fusion can develop when one considers how dominant understandings or beliefs shape our understanding of historical developments (Earle), and it can follow from a juxtaposition of demographic circumstances with ideological representations of ethno-cultural patterns (Harris).

The fourth of our themes, the distinctive mix of geographical characteristics that gives places their special character in relation to one another, harkens back to cultural geography's long-standing effort to make sense of place. This is perhaps the theme that has remained the strongest in the cultural geography literature, yet it is also one that has evolved greatly in recent years. Anne Buttimer, Chad Emmett, and Shaul Cohen are clearly attuned to that evolution, pointing to the importance of such cultural constructions as literature, built form, and sense of territory in the development of place images. Yet each is also interested in connecting such images to the type of knowledge about place that is sometimes thought to be at risk in the face of the current enthrallment with discourse. In the process, the contributors to this volume suggest how the experience of richly textured places finds its way into cultural forms, as well as how discursive-cum-ideological struggles over place are played out in concrete physical settings.

Taken as a whole, the book's contributors point to a set of themes of enduring importance to our ongoing effort to make sense of ourselves and our world. At the same time, they suggest how those themes can be constructively engaged. Yet the implications of this collective undertaking extend beyond the identification of important research themes. The contributions to this volume also highlight the importance of a creative tension between a careful rendering of the details of a research topic and contemplation of where that rendering might lead. An ending that terminates conclusively, summarizing past activities but leaving few openings for future work, may present a certain logical coherence, but probably is not very memorable. Far better is an ending that points to useful insights with applicability to the future, that highlights paths not taken, that reminds how often old, frequently unanswered questions remain relevant and suggest new beginnings.

This volume offers a picture of how a loosely connected group of scholars, sharing diverse backgrounds but common origins, interests, inspirations, and purposes, can develop a set of synergistic scholarly insights over more than three decades of intellectual activity. This body of scholarship is suggestive of at least four overarching lessons with clear relevance for the future growth and development of cultural geography.

First, it is clear that historical geographic scholarship does not simply offer insights into the past; it provides important perspectives on contemporary dilemmas. Whether the issue is waterfront land-use policy, irrigation system organization and development, or the intellectual ethos that gives character to place, past action shapes contemporary context. Understanding the deep structural legacy associated with such matters helps us avoid the repetition of past mistakes and suggests constructive approaches to contemporary problems.

Second, there is no substitute for meticulous empirical research. From the vantage point of the end of the twentieth century, it sometimes seems that we are too busy or too wrapped up in reflection to undertake the plodding work of exploring, counting, digging, mapping, and describing in clear but painstaking detail the changing components of the world around us. Many of the contributions in this volume remind us of the rewards of such work—not because it adds to some encyclopedic body of unexamined "facts," but because our understandings are enriched when theory and observation are inextricably interwoven, each informing the other.

Whether in the archive or in the field, whether using traditional methods or the latest technology, there is no substitute for direct experience of place and problem. Many traditional methods of collecting data may be out of favor, but their disuse is not necessarily a function of their inappropriateness or inapplicability. More often traditional methods that fall out of use do so because new generations of scholars shift their attention to new problems and technologies. The more traditional practices are not disproven; they are simply ignored. Yet the type of careful examination of house types and construction techniques, used by Marvin Mikesell to understand regional patterns in northern Morocco (Mikesell 1961), can provide the foundation of a study of colonial Mexico three decades later that helps resolve the long-simmering debate over the role of independent invention versus diffusion in the evolution of that place. Similarly, the careful examination of settlement patterns can provide a basis for contemplating whether the frontier had the salient role in the development of American culture that it is often claimed to have had. In these examples, traditional topics and approaches can be deployed in ways that provide important insights into contemporary debates.

Third, values constitute the fundamental foundation upon which behavior is constructed. Values shape the historical and contemporary environments and landscapes that inform our lives. Our often contradictory values fundamentally influence humankind's environmental history, contemporary landscape, and potential future. Values support the view that wetlands are neutral commodities that can be altered with impunity in pursuit of progress and development. Values, when reflected in the creations of artists and writers, offer windows into the souls and objectives of both individuals and societies. Values about development and historical use condition our cultural perception of the myth and reality of past riverine navigation and the patterns and practices of potential future river use. Values about self, sex, and society establish basic patterns of individual and group identity and shape the cultural rituals that distinguish one group and place from another. We have far too limited an understanding of the geographic implications of the values that shape contemporary life and livelihood in rural, urban, and industrial worlds.

Finally, events in far-off or remote places that may seem inconsequential to urban-dwelling scholars are immediate and important to those directly engaged. These events are also distant reflections of generic processes and problems that afflict the urbanized and industrialized world. In an age when language skills atrophy, geographic illiteracy abounds, and fieldwork in foreign places declines, developing the kinds of knowledge about the world that can only be gained by direct encounter is a pressing challenge. How are North Americans to grasp what motivates ethnic conflict in distant lands if scholars do not examine the conditions and the consequences of conflict over place? At the same time, how can North Americans make judgments about ethnic conflicts around the world if they do not look within their own society and reflect on what motivates race riots or discrimination based on sexual orientation? The distant and the near are inextricably linked, and no geography of culture can be complete that looks at just one or the other.

In the new millennium, humankind's ongoing encounter with both the physical and the humanized environment will shape the future of our species. Culture will play an extraordinarily important role in determining the nature of that encounter. Our search for understanding must reach across traditions and perspectives so that we may look broadly and deeply at how cultural encounters with the environment have unfolded in the past, and how they will continue to unfold in the future.

REFERENCES

Graff, Gerald. 1993. *Beyond the Culture Wars*. New York: W. W. Norton.

Kobayashi, Audrey, and Suzanne Mackenzie, eds. 1989. *Remaking Human Geography*. Boston: Unwin Hyman.

Mikesell, Marvin W. 1961. *Northern Morocco: A Cultural Geography*. Berkeley, CA.: University of California Press.

Wagner, Philip L., and Marvin W. Mikesell, eds. 1962. *Readings in Cultural Geography*. Chicago: University of Chicago Press.

Epilogue

Each Particular Place: Culture and Geography

Philip L. Wagner

Of a sudden,
A shift in the light
Turns the window we looked through
into a mirror.
And we see not the world
But ourselves.

Who has reflected so deeply on culture and geography than my old friend and collaborator, Marvin Mikesell? He knows more of us and our work in the Berkeley tradition than I could pretend to. Dedicating this essay to him, the unchallenged Defender of the Faith, I cannot match his erudition and insight, but must content myself with a brief confession of my own idiosyncratic, possibly heretical views on the past and current condition of cultural geography and what is to come of it. I begin with what everyone now seems to know: The human use of the earth has acquired new and ominous dimensions; relationships between environments and peoples have entered a startlingly novel, truly global rather than regional phase; and cultural encounters with the environment will in future almost certainly decide the fate of humanity.

PANANTHROPY

The biggest new fact in the world is pananthropy (humanity unified)—the rise of our species worldwide to imperious ecological dominance, and progressive incorporation of its powers into a single global entity.

In geography classes at Berkeley we complacently learned of the greenhouse effect; now it vexes scientists and policy makers. We heard in cultural geography lectures about depletion of plant, animal, and other natural resources, beginning in ancient times; this old theme of Berkeley geographers now urgently intrudes into any serious discussion of humanity's future prospects. Our culture history contemplated the gradual diffusion and consolidation of constellations of knowledge, action, and power, moving out of Stone Age origins in obscure tropical and subtropical localities ("hearths") via the Middle East, China, and centers in Mexico and Peru to form ever growing "culture worlds" and contest for supreme dominance; today that theme, refined and brought up to present, appears again under the guise of "world systems." Carl Sauer's and Jim Parsons' historical geography also paid heed to economic institutions and their expansion; now we watch with foreboding the spread of multinational enterprise. Alongside these characteristic emphases of Berkeley geography one can cite many other prophetic intimations of pananthropy, from Marshall McLuhan's "global village" to the "anthropic principle" espoused by some cosmologists. But perhaps the crucial recognition of developing, and possibly dangerous, human dominance inhered in the forceful rejection of environmental determinism (and also vapid, pedantic regional geography) by Mr. Sauer and his coterie, eloquently documented, for example, in the landmark *Man's Role in Changing the Face of the Earth*. Rather unheeded at publication—and lamentably inspiring little activism on the part of contributors and admirers—that volume still speaks to our time and concerns. The concept of pananthropy presaged in Berkeley tradition, then, presents a comprehensive aggregative perspective on today's global change. It does not, however, explain it. In order to discern the motives and mechanisms that have played the decisive part in that change, it behooves us to look more minutely not only at cultural encounters with the environment, but also at encounters among people themselves, and at their mutual dependence.

INTERDEPENDENCE

I fondly remember Carl Sauer's cluttered office, redolent of Revelation pipe tobacco, stuffed with paperbound foreign books, and overflowing with somewhat infrequently answered correspondence from far and improbable places. I recall seeing him closeted often with unidentified but obviously distinguished visitors, as he amiably but doggedly extracted whatever special exotic, arcane knowledge they might possess, just as he did with graduate students bold enough to enter his den. When I dutifully digested his writings and those of scholars he championed, they led me all over the map of academic disciplines and into polyglot patches of foreign authority. His skilled interaction in discourse and discovery with other people—from humble Mexican village Indians to world-renowned scholars, stumbling students, and unorthodox travelers—furnished much of his stupendous knowledge and fueled his vast imagination. He possessed the peerless virtue of listening carefully. What he learned, he transmitted, through a process of gentle diffusion, in classes and conversations and seminars, not merely to students but also to campus colleagues and a worldwide interlocutor network. And, preciously, he furthermore let his students proceed on their own, fostering independence and encouraging individual initiative. Certainly he wanted to found no "school": he preferred leaders to followers, always. I have the impression that Marvin Mikesell works in a similarly generous way with his students.

This tale is a pretext for making a crucial point: no one can do much entirely alone. Whatever form they may take, abundant human interaction and collaboration underlie any considerable achievement.

That same observation, furthermore, applies to even the humblest of human endeavors. Consider the question of how you could speak, and thereby fully enter humanity, without someone's showing you how. What could you make out of an environment lacking the tools made by somebody else? How long a road can you build if nobody helps?

Human evolution has led us away from prowling the forest alone. Even hunters and gatherers must learn particular skills from others and work together for survival. The perilous plight of contemporary urbanites confronted with natural catastrophes illustrates our constant and urgent requirement for help from our fellows. Cultural geography records no credible cases of Crusoes who lived very long.

Inborn ecological adaptation dictates our working closely together as natural beings and, notably, as intensely social ones. If we wish to comprehend well the regional "sequent occupance" phenomenon that Mikesell once wrote about, we had best pay close heed to the social arrangements that governed each respective stage. And we might even audaciously amend the title of one of Carl Sauer's provocative essays, to read, "The Agency of Collaborating Humans on Earth."

Collaboration does not come easily. It emerges from strenuous discussion, debate and resistance, and painful decision, carried out during purposeful encounters among specific individuals at given times in particular places; and decisions, in turn disseminated, become somehow digested and somewhat deconstructed, and somewhere determine the projects of people working in concert. Such human encounters, repetitive episodes of decision and diffusion lacing through the cellular fabric of places in earth space—and further subordinate but also often rather subversive responding moments of decision and diffusion—must precede any telling encounter with the environment. Whatever the case considered, agricultural origins and dispersal or the daily conduct of business in multinational corporations works the same way.

Only what happens in each particular place can account for our ways in the world. Only deep lessons yet to be learned regarding the regularities of interaction among individuals as focused through encounters in place can inform us fully of our human historical, ecological heritage or award us control of our destiny. The familiar but underexploited concept of place discovers the ground of human encounter upon which we base encounters with the environment.

CLOSE ENCOUNTER

The fieldwork of the master and most of his students, and their students in turn, has tended to thrive on homely, close human encounters, out in the margins and backwoods of Latin America and elsewhere. And archival research that often accompanied it brought the investigator face to face with long-ago squabbles among petty petitioners and minor colonial officialdom. In addition, fieldworkers not only stood present to watch clearing and planting, tillage and harvest under archaic agricultural systems, but listened attentively and respectfully to farmers' accounts of the rules and ideas guiding their efforts. The geographer could witness personal interactions taking place in tribal markets, as Mikesell did, or discuss local cuisine with its cook, or sit in, while community elders resolved disputes over land or formulated festivals. A few non-Berkeley geographers have gotten that close to such interaction, but not very many.

So despite the arrogant disdain directed at their supposed exoticism and remoteness from matters and problems more urban, commercial, and central to life in modern, highly developed countries, at least a substantial portion of the students trained in the Berkeley tradition gained privileged insight, or some sort of rich glimpse, into how people deal with and learn from their fellows in seeking to manage and valorize environments. I doubt that my coreligionists have ever given much thought to this aspect of their experience, and yet I think they would readily perceive their advantage in seeing how things actually work. If only we could know as much about how small encounters in each particular place affect, say, national policies or the business of General Motors as we can know about how the elders (and occasional rebels) of indigenous villages, or even the women semi-sequestered, in fact preside over action! And how much more enlightenment we might gain about great global questions, if we could see directly how specific encounters in particular places, not very different from those in a village, guide observance of custom as well as adoption of innovation, and both the resolution and creation of problems, and how they channel the action, whether individual or collective, that impacts on the environment.

By particular places I do not intend here whole Indian villages, for instance, or substantial areas of similarity such as regions. Understanding human behavior and agency requires a much finer scale of resolution—blotted out all too often by our lazy taste for easy statistical summaries. The Berkeleyans learned most when sitting with the old men, chatting with the women at home, or venturing out into the very field of the farmer, attending the rare ceremonials, or perhaps buying beer for the boys in local cantinas. If they kept eyes and ears open, they could garner at least an impression of the "where-when-how-and-who" and "with-what," and conceivably even the "why" of proceedings of human encounter that managed change and continuity and set standards for acceptable action; and then they could watch resulting decisions played out (as the "culture" of the moment) in encounter with the environment.

The foregoing depiction of fieldwork assuredly strays very far from the vision we had of our enterprise then. Probably most of us strove to act "scientific" more than just human, and so of course we undoubtedly missed a great deal of significance. But still, the immersion in the immediate that belonged to the Berkeley ethos (something rather postmodern, at that) contributed a kind of insight not vouchsafed often to geographical contemporaries east of the Sierra Nevada. We had not merely done field research, we had really "been somewhere," and came away marked by it.

All pride aside, I have to acknowledge that other scholars and writers have done the same sort of thing, equally well or maybe even better, for centuries already. We might find it hard to catch up with the anthropologists, foremost; with careful historians who scrupulously reconstruct the circumstances of decisive moments; with biographers who frame the actions of significant individuals in proper context; or with novelists who cast their tales, in observant detail, in each particular place of encounter. Such other students of interactive human behavior excel in grasping the implications of place and do not hesitate to get particular. From them we can profit in learning much about uses of place.

USES OF PLACE

Geographers, as well as novelists and poets, have exalted individual attachment to place, at scales which vary from the whole home country to the lesser region of childhood experience, the favorite city, the urban neighborhood or native village, and the old family house. They have also not overlooked feelings evoked by great memorial sites and revered supernatural shrines. Individual apperception, association, and sentiment have chiefly attracted attention, but the places considered vary widely in scale. Here I intend to dwell neither on personal emotional states and imaginative individual impressions nor on places conceived at such various magnitudes. Rather, I present place as a locus of meaningful human dialogue, social and physical: place as potential for action.

Place, in this perspective, amounts to exactly where (and when) people meet and interact, among themselves as well as with their physical surroundings. We can describe it at only a single specific spatial scale—the immediate. It likewise implies sociality, whether under the form of face-to-face encounters or behavior under socially imparted norms. The focus falls on the individual only as a social (and hence cultural) being, significant just as participant in some interaction.

The foregoing perspective on place has major implications, not only for understanding of geography but also for our concept of culture. I think, too, that it comes far closer to the ordinary layperson's everyday notion of place than do explorations of subtle sentiment.

Consider, for instance, colloquial usage. More or less universally, people tend to talk about places in terms of what they can do there: not only of what they regard as possible physically, but as well of what they accept as permitted and proper. The lifeworld consists of concrete, discrete sites, with appropriate uses well known to the individual. (Fantasies of contrasting, vague large realms of escape, adventure, and romance often fascinate people too, precisely because they feature release from all the constraints known to attach to any real place.)

Languages tend to lay out a matrix of usable places, well bounded off and distinguished according to function. Take, for example, rooms in American houses: kitchen (from the old root of "to cook"); dining room (notice, the breakfast nook is time specific, like "dining" as well); parlor ("place to talk") or living room (for a social species, talking means living); bedroom (whose bedroom?); and bathroom (certain further uses politely unspecified). One does well not to transgress the various specifications for behavior assigned to each particular room, or place.

Terms applied to other enclosures analogous to rooms carry a similar message: "hall" (beer hall, dance hall, meeting hall, lecture hall) for large congregations, or "-house" (courthouse, whorehouse, jailhouse, poorhouse) denoting somewhat reprehensible functions. Such common usage reveals a well-rooted practical concept of place.

The nearest parallel to the kind of functional place classification laid forth here surely occurs, at a wider focus of scale, in urban geography, where districts distinguished according to function (for example, light or heavy industrial, commercial, transportation oriented, residential in various categories) parcel out cities and towns. The old regional tradition, at a much more general scale, similarly did consult function. And the rarified realm of Central Place likewise used (rather skeletal) marketing functions as a basis for areal classification. All the above schemes of districts, regions, or centers and ranges of function certainly somehow alluded to place, although in a thoroughly attenuated way. The average citizen would probably not speak of their component units as places (apart from those of Central Place Theory), however, but would go a step higher, designating whole cities, states or provinces, and countries as places again—but notably not on the basis of function.

Only the vernacular vision of functional place at the miniature scale exemplified in rooms and halls corresponds to the conception I wish to espouse. And even so, the functional aspect itself can serve as no more than heuristic guideline. In fact, to see clearly the nature of place we have, respecting the scale chosen, to look on all places as sharing a single common character and constitution.

THE CONSTITUTION OF PLACE

Some geographers, using fashionable terminology, favor the phrase "construction of place." Let us instead avoid misleading implications of mere building, and consider a componential specification of what must necessarily enter into creation and use of a place. I propose that one complex configuration of inherent features and properties alone qualifies a site as the scene of enacted behaviors that, so to speak, bring it alive and make it significant. The life that goes on in a place responds to and depends on certain attributes of it, which I shall summarize schematically under ten rubrics: venue, enclosure, internal plan, impedimenta, incumbents, procedure, agenda, program, linkage, and outreach. For each I shall give some examples.

During graduate student days at Berkeley, Mr. Sauer's course on "Conservation of Natural Resources" met, if I remember aright, from two to three on Monday, Wednesday, and Friday afternoons in a certain sunny westerly room in Ag Hall. The hour counted just as much as spatial location in making the venue, for had I appeared at a different time, I might have heard lectures on clay soils or alfalfa, instead of Sauerian retrospect on resource employment that started approximately with anthropoid ancestors and only got about as far, by semester's end, as the close of the Neolithic or slightly beyond. The venue of place subsumes time, in either perspective.

Furthermore, the same enactment of scholarly discourse—in effect the same Sauer course—could well have proceeded anywhere else anytime, for venue can vary without vitiation of that vital substance of encounter that confers reality and meaning on place. Portability of place, under which enough other componential attributes persist to uproot the anchor of actual location, helps to comprehend the character and operation of place in the cyber age in which electronic virtual places play a large role. Virtual place, though, must await further probing.

Place can either construct or appropriate boundaries, but it must have enclosure. Indeed, students used to "go to school" with Mr. Sauer even in dubious dim restaurants or out in a milpa someplace in Mexico, where he held forth almost as he did in the classroom to whomever among the elect had luckily come on his summer field trek. That sort of incident would initially appear to impugn the relevance or validity of the second attribute of place, its enclosure. However, few locals we ran into on those trips ever cared to intrude on the evidently sacrosanct circle of adepts in rapture around the odd but awesome elderly gringo. The group, by its manner, simply enacted its separateness and reserved a place for itself.

That rapt circle of listeners also illustrates another componential feature of place. Students arranged themselves informally, but observing diffident distance, around their mentor, concentrically. Back in the classroom, he took his position, facing the row-seated students over a lab table, back against the chalkboard where his wisdom congealed in white scrawls. The layout of classrooms, like that of courtrooms or churches, declares who shall preside. The internal plan of a place thus makes explicit an order of precedence, and furthermore frequently implements a whole processional performance, a sequence of small spatio-social "rites of passage," ordaining the stations or posts through which people must pass during the course of total encounter. Minute spatial progressions resulting embody a sequence of statuses thereby assumed by incumbents. This expressive and directive spatial ordering even becomes evident to a careful observer during brief conversational clottings occurring in passage along city streets.

Our favorite impedimenta, in the places constituted as Sauerian lectures, consisted of his spectacles, which he perilously twirled in his hand all the while he talked and inadvertently would once in a while shoot off into a corner. The unwieldy Justus Perthes maps he and his colleagues carried to class and hung up also helped consecrate the place as one for geography, as did other component attributes.

PEOPLE IN PLACE

At the very core of this hasty componential prescription for place comes the incumbent company admitted to and involved in the place and its interactive performances, as in the case of designated professors and qualifying students. The effective enclosure of any sort of place precludes random entry: people come in on tolerance or by invitation, if not chartered as legitimate participants. The walls and doors of the world screen out and filter potential entrants and actors. Only such exclusivism, apparently, can ensure the orderly—and frequently secretive—conduct of business. Attendance at seminars always, at Berkeley, hinged on some qualification as graduate student or welcome visiting scholar. This rule did not hold, though, for the very different venues of Mr. Sauer's office or the weekly Geography Tea, both much more open.

Thus each particular place, whether only instantaneously constituted or enduring, belongs in a way exclusively to meticulously selected incumbents. Human beings conduct transactions in privileged precincts, protected from unwanted incursions and interference by outsiders. This selective, exclusive componential feature of place institutes and sustains, of necessity, social disparities, domination, and power.

The concept of place, componentially interpreted and applied, helps visualize how spatial and behavioral discrimination between, literally, insiders and outsiders impinges on options and initiatives open to individuals, and thus on their social potential and environmental influence. The place concept should illuminate many of the painful, pressing issues and problems of our day and inspire promising strategies for attacking them. Its relevance to society's dilemmas becomes clear when we reflect on what transpires within place so convened and furnished and arranged—when we consider the rest of its componential attributes: procedure, agenda, program, linkage, and outreach, remembering always that they refer now to interactions among a definite, previously selected company of insiders.

The five further aspects of place just listed might do as an outline for studying culture. For I think we can grasp that slippery foundational notion most firmly and surely and deconstruct it most fruitfully if we boldly assume culture to mean nothing more than the interactively crafted standards for correct and effective behavior proper in specific context to each particular place. If driven to the risky task of defining culture, I should prefer to disregard its putative character as common heritage, for it exists as something alive and contested and changeable; and I surely should not presume to assign any human individual simply to one given, totalistic culture, for we all sense too well that we ourselves behave, and perhaps even think, dissimilarly in differing encounters and in many diverse places. If one can define culture at all, it might consist in the highly contingent, unstable specifications of the behavior permissible, proper, and productive in each particular place, and in each position within it. Constantly innovated, propounded, explicated, negotiated, debated, resisted, interpreted, and revised in ongoing encounter, such models of legitimate behavior, generated and evolving in a given place, intend both application to goings-on within its precincts, and extension to behavior elsewhere in the possible diffusional domain consisting of the other places somehow linked to that venue.

The Berkeley graduate department acted as a sort of a club when I was there. I hope it still does, for whichever faculty greats inspired it, the intergenerationally ongoing company of aspirants embodied the soul of its glory. We had our scrupulous inmate etiquette (as in prisons!); our accustomed significant seating pattern around the seminar table; a muted but imperative pecking order; our little cooperative cigarette supply at a penny a smoke (which, once established, would in those days periodically earn enough itself to buy a new pack). Our ethos prescribed a feigned nonchalance about future job prospects and a discriminating choice of pretentious and weighty topics for polite conversation, just as in a fabled English gentlemen's club.

All that cozy clubbishness depended on a greater purpose, though. One might well compare it with the atmosphere prevailing in the fighter pilots' mess, or perhaps among certain of the regular clergy of Catholic orders. All our corporateness and conviviality would have meant mere frivolity had we not considered ourselves as having a lofty mission, a mission of outreach.

THE OUTREACH OF CULTURE

All internal procedures, agendas, programs established in a place make for nothing else than self-indulgent fun unless they ultimately help to reach beyond to other places, and only when interactions under them can aim to influence some actions and conditions elsewhere. What good does the sermon in church do if nobody takes home the pious message?

Even a relaxed and friendly social gathering inevitably has lasting repercussions and results. When I slipped on a throw rug at one of Mr. Sauer's magisterial parties and lay grinning guiltily before him on the floor, I knew I risked regrettable demotion.

But we had serious business other than parties. In geography at Berkeley, the heavy industry consisted of creating cultural origins and dispersals. We had not only to absorb and ponder facts and concepts laid before us by the faculty. We felt ourselves called to explore and invent on our own, and had timorously to proclaim our findings and constructions in the seminars, after having had them well chewed over beforehand by peers. We even slipped our presumptuous profundities into agendas for conversation at parties, so that the latter would almost sound like amateurish academic conclaves.

Some students experienced Berkeley as more or less timeless, an intellectual Nirvana. It took pitiful pushing by partners, as well as sensed faculty pressure to produce against the chastening constraint of constant frank critique and disputation, to propel many of us forward and outward, out of our unworldly, conceited complacency. But we finally emerged from that comfortable chrysalis as professional creatures able to fly on our own and display brilliant colors. The academic reproductive cycle spewed us forth to colonize new intellectual habitats and engender newly begotten cultural geographers.

This account presents a fair picture of what sorts of things go on in a place, and what they imply for the concept of culture. I deliberately emphasize the revelation that we students felt as though we had found a home, in which we kept eagerly busy obeying the program ordained and practicing correct procedures, but neglected to think much about outreach. I, for myself, had only a misty idea of a teaching career and how one might enter and pursue it, and I suspect that a similar ignorance existed among some of my peers. It satisfied, nay, very much gratified us to perfect our in-house rituality and to toy or toil with practices and projects already native to that habitat—and somehow for their own sake, as if none might call for application. That we may indeed confess as something of a failing of the Berkeley way. Alas, for all our grandiose talk, we did not unravel the concept of culture, for instance (which presumably "superorganic" act we would have estimated as anathema). So a prominent critic's assertion that we had bypassed the "inner workings of culture" hit home, despite his mistakenly reading an accurate descriptive admission of this neglect for a prescriptive admonition.

I dwell at length upon our careless period of bliss because it nicely exemplifies something important concerning both culture and place. Places can become bastions of comfortable routine almost as an end in itself; they then give rise to bureaucracy or something akin. As long as they continue to command effective access to channels of diffusion, without intense contestation, such centers of cultural influence can, despite their sometime contribution to a beneficial degree of social stability, induce such sluggishness and soporific inattentiveness into a society that inventive, prudent adaptation to insistent exogenous change becomes unfeasible.

Did the Berkeley tradition founder on those rocks? We continued in adherence to our creed and custom and our calling as we saw it, as if we held them immutable. We faithfully followed familiar programs and agendas, and in doing so we documented ably, in our articles and books and dissertations, accreted and substantial institutions such as marked by language or religion, or inveterate technical practices. We explored and mapped and described those very ingrown, laggard social and productive systems which of late fall into shambles. And surely we performed an important and permanent service in doing so. But what then?

WHITHER HENCE

Allow me a bit of harmless mischief, please: let me give a title to this section that most younger readers may not quite fully understand or feel easy with; if you like, you may rename it "Where do we go from here?" But hang on to this little lesson: language, like everything else, can change pretty fast, and not even professors can stop it. As Jim Blaut argues so tellingly, largely anonymous interactions going on in each particular place, anywhere, enter into the cultural flux and possess the power to diffuse new rules for proper behavior that ooze, sooner or later, into even the procedures of staid and stodgy bastions, for instance of formal writing.

I could rewrite my crude conception of culture to read, "culture means how to behave in some particular place, here and now, in some selected company." The formula for it keeps changing, and in fact our major reason for getting together in given, constituted places and discussing and arguing and issuing innovative pronouncements has to do with getting other people to accept instructions from us on how to behave.

In this very fashion, our Berkeley folk succeeded for a long while at impressing and intimidating other academics with their exotic erudition and literary energy (as postmodern pundits do now), to the point that departments all over rushed to capture and collect a token specimen cultural geographer, as one might sweep up and pin a beautiful, eccentric butterfly. I do not think that many such collectors understood what they had caught or what it might be good for, but also I regret to say that the scorn so prevalent among the Berkeleyans for anything resembling theory (or what postmodernly might figure as grand narrative) may have prevented them from knowing quite what they had done or could do. When I once unguardedly told Mr. Sauer that I considered him a true philosopher, he only snorted in contempt. The reflections on place and its constitution reported herein should enliven, if not enlighten, discussion of what made Berkeley powerful, and provide a perspective on geography as a whole.

The "tale of culture living, growing, moving through communication," resuscitated in my *Showing Off: The Geltung Hypothesis*, needs to incorporate more fully the effects of agency and structure simultaneously enacted in each particular place, whereupon it might reveal how most geographers, of whatever persuasion, have tended toward the same single goal.

I regard neither landscape nor space as the keystone concept of geography, and not culture either, as such, but just place. With place as my vantage point, I survey both the world and the discipline, describing them as cohesive, consistent, comprehensible realities. In the real world and in academic bastions, cultural encounters with the environment are now potent factors in planetary physical events and bioevolution, and those encounters always focus in each particular place with its own constitution and linkages.

Academic conditions have mightily changed, and so have societies. With procedures, programs, and agendas radically rewritten, with new standards for admission to our hallowed enclosures, with a challenging and more generous mission of outreach—in response to new social imperatives—contemporary cultural geography, confronted with the threat of obsolescence, has had to reinvent itself. Berkeley no longer presides or prescribes, or even counts very much. We can see, though, even in articles included in this volume, signs of coalescence among all geographic traditions, and of a sharpened focus on place.

We must change the world.
We know where to go;
Whom we must join with;
How to behave;
What really matters;
What we must learn about;
Whom we must reach.
But how to begin?
Place is the place.

Index

About the Contributors

Anne Buttimer is Professor of Geography and Head of the Department of Geography, University College Dublin. Her doctoral research on the conceptual foundations for social geography was inspired in large part by a summer course on cultural geography offered by Marvin W. Mikesell. She is the author of *Geography and the Human Spirit* (Johns Hopkins University Press, 1993) and many other books and articles on subjects ranging from social space and urban planning to the history of ideas and environmental policy.

Elisabeth K. Butzer is Research Fellow in the Institute of Latin American Studies at the University of Texas, Austin. Her archival research in Spain focused on medieval Muslim communities; in Mexico she has worked in manuscript repositories on colonial land grants as well as civil conflicts between indigenous people and Spaniards. A book on a Tlaxcalan community of northern New Spain (1686–1820) is currently in press.

Karl W. Butzer is Dickson Centennial Professor of Liberal Arts at the University of Texas, Austin, and a Fellow of the National Academy of Sciences. His research on environmental history and cultural ecology is based on fieldwork in Africa, Spain, and Mexico. He is presently working on indigenous maps and the indigenous imprint on church architecture in Colonial Mexico, as examples of transculturation.

Shaul E. Cohen is Assistant Professor of Geography at the University of Oregon. His research focuses on the interface of politics and environment, particularly in relation to forest issues. He has also written extensively about Jerusalem and the West Bank and is the author of *The Politics of Planting: Israeli-Palestinian Competition for Control of Land in the Jerusalem Periphery* (University of Chicago Press, 1993).

Michael P. Conzen is Professor of Geography at the University of Chicago. His chief interests are American historical and urban geography, landscape history, and the development of commercial cartography. He is the editor of *The Making of the American Landscape* (Unwin Hyman, 1990) and coeditor of *A Scholar's Guide to Geographical Writing on the American and Canadian Past* (University of Chicago Press, 1993).

Carville Earle is the Carl O. Sauer Professor of Geography at Louisiana State University and a former editor of the *Annals of the Association of American Geographers*. He is the author and editor of various articles and books on the historical geography of the United States, most recently *Geographical Inquiry and American Historical Problems* (Stanford University Press, 1992). His published works also include commentaries on geographical ideas, most notably *Concepts in Human Geography* (Rowman & Littlefield, 1996).

Chad F. Emmett is Associate Professor of Geography at Brigham Young University. His research focuses on the political and cultural geography of the Middle East. He is the author of *Beyond the Basilica: Christians and Muslims in Nazareth* (University of Chicago Press, 1995).

Peter G. Goheen is Professor of Geography at Queen's University, Kingston, Ontario, Canada. He studied under Marvin Mikesell and Brian Berry at the University of Chicago. His interest at present is how public space in the modern North American city became a significant public resource and he resulting contest among contending parties to control its disposition.

Charles M. Good is Professor of Geography at Virginia Tech, Blacksburg, Virginia. His main interests include the geography of health and disease; the cultural and historical geography of Africa; and the role of culture, health, and place in the transition of new immigrant communities in North America. He is the author of *Ethnomedical Systems in Africa* (Guilford, 1987).

Viola Haarmann is an independent editor and writer in Worcester, Massachusetts, with a doctoral degree in geography from the University of Hamburg, Germany, where she held a research appointment. She has published in both English and German and currently specializes in editing manuscripts by authors whose native language is not English. She has recently contributed to the *Columbia Gazetteer of the World* (Columbia University Press, 1998).

Chauncy D. Harris is the Samuel N. Harper Distinguished Service Professor Emeritus of Geography at the University of Chicago. His principal interests lie in urban, economic, and cultural geography of Russia and the former Soviet Union.

Douglas L. Johnson is Professor of Geography at Clark University, Worcester, Massachusetts. His major interests are cultural ecology, pastoral nomadism, desertification, and environmental change, with special reference to the Middle East and North Africa. He is the coauthor of *Land Degradation: Creation and Destruction* (Blackwell, 1995).

John A. Kirchner is Professor of Geography and Transportation at California State University, Los Angeles. His recent research focuses on two aspects of developing nations: transportation and tropical agriculture. His interests in cultural–historical geography are expressed through applied research in both the United States and Latin America, where his current transportation research includes investigation of the Ecuadorian and Central American railroad systems.

David Lowenthal is Professor Emeritus at University College London and visiting professor of heritage studies, St. Mary's University College, Strawberry Hill, England. His foci have been the history of geography, landscape attitudes, the Caribbean, the role of the past in national identities, and the environmental movement. His new biography, *George Perkins Marsh: Prophet of Conservation*, will be published by the University of Washington Press in April 2000.

Alexander B. Murphy is Professor of Geography, Department Head, and Rippey Chair in Liberal Arts and Sciences at the University of Oregon. His primary research interests are political and cultural geography, with a regional emphasis on Europe. He is the author of *The Regional Dynamics of Language Differentiation in Belgium* (University of Chicago Geography Research Series, 1988) and a coauthor of *Human Geography: Culture, Society, and Space* (John Wiley & Sons, 1999).

James A. Schmid taught environmental science, biogeography, and cultural geography at Barnard College and Columbia University in the early 1970s. His primary research interests are urban vegetation, environmental impact assessment, and wetlands. For the past two decades he has headed his own consulting firm in Media, Pennsylvania.

Philip L. Wagner, Professor Emeritus of Geography at Simon Fraser University, received his Ph.D. at the University of California, Berkeley, specializing in cultural geography and Latin America. His books include *Nicoya: A Cultural Geography*; *The Human Use of the Earth*; *Environments and Peoples*; and *Showing Off: The Geltung Hypothesis*.

James L. Wescoat Jr. is Associate Professor of Geography at the University of Colorado, Boulder. He works on water resource issues in South Asia and the American West. He conducted a 10-year program of research on Mughal gardens and urban water systems, and he has recently chaired a National Research Council committee review on "Downstream Adaptive Management of Glen Canyon Dam and the Grand Canyon Ecosystem."